# NEUROMETHODS ■ 38

# Patch-Clamp Analysis

## Second Edition

# NEUROMETHODS

## Series Editor: *Wolfgang Walz*

# NEUROMETHODS ∎ 38

# Patch-Clamp Analysis

## ADVANCED TECHNIQUES

Second Edition

*Edited by*

# Wolfgang Walz

*Department of Physiology, University of Saskatchewan,
Saskatoon, Canada*

HUMANA PRESS ✳ TOTOWA, NEW JERSEY

© 2007 Humana Press Inc., a division of Spring Science+Business Media, LLC
999 Riverview Drive, Suite 208
Totowa, New Jersey 07512

www.humanapress.com

For additional copies, pricing for bulk purchases, and/or information about other Humana titles, contact Humana at the above address or at any of the following numbers: Tel: 973-256-1699; Fax: 973-256-8341; E-mail: humana@humanapr.com or visit our website at http://humanapress.com

**Photocopy Authorization Policy:**

Library of Congress Control Number: 2006939959

10  9  8  7  6  5  4  3  2  1

ISBN: 978-1-58829-705-1          ISBN 978-1-59745-492-6  (eBook)
ISSN: 0893-2336

# Preface

Neher and Sakmann were the first to monitor the opening and closing of single ion channels in the membranes of cells by conductance measurements. In 1976 they used fire-polished micropipettes with a tip diameter of 3 to 5 μm to record currents from a small patch of the membrane of skeletal muscles, thereby decreasing background membrane noise. To reduce the dominant source of background noise (the leakage shunt under the pipette rim between membrane and glass), the muscle membrane had to be treated with an enzyme. Despite these early limitations, a new technique was born—the patch-clamp.

The final breakthrough came in 1981 when the same investigators, in collaboration with Hamill, Marty, and Sigworth, developed the gigaohm seal. Not only did this improve the quality of the recordings, it was now possible to gently pull the pipette with an attached patch of membrane of the cell and to study its trapped ion channels in isolation. Another offshoot of the gigaohm seal technique was the whole-cell patch-clamp technique, in which the attached patch of membrane is ruptured without breaking the seal. This technique is really a sophisticated voltage-clamp technique and it allows for the altering of cytoplasmic constituents if the investigator wishes.

This is the third edition of this best-selling neuroscience book by Humana Press. The rationale for its design was to represent any patch-clamp method that has been in more than 10 to 15 publications over the last three years. As well, newly emerging techniques, with future potential, such as uncaging experiments with lasers and high throughput techniques, have also been represented.

Thus, the reader will find the latest developments in the traditional patch techniques like whole cell and single channel as well as perforated patch, fast drug application, loose patch, and macro-patch techniques. The fields of internal pipette perfusion techniques and patch techniques combined with molecular biology represent major innovations. Three technical developments are brand new: (1) the combination of patch clamp and optical physiology has seen the introduction of two-photon lasers and uncaging experiments; (2) it is now possible to patch in animals in vivo; and (3) in pharmacological testing, high throughput techniques are making their appearance with both automated glass pipettes and planar patch electrodes. Thus, the arrival of the planar patch electrodes has, for the first time, enabled patch clamping without glass pipettes.

It is obvious that patch clamping is a technique that is here to stay. We will probably see future developments in addition to the glass pipette. As well, the glass pipette will be used more and more as a tool to make discrete changes to the *milieu interieur* of cells.

*Wolfgang Walz*

# Contents

# Contributors

JAN C. BEHRENDS • *Department of Physiology, University of Freiburg, Freiburg, Germany*

ARUNI BHATNAGAR • *Department of Medicine, University of Louisville, Louisville, KY*

THOMAS H. BROWN • *Department of Psychology, Yale University, New Haven, CT*

GUIYING CUI • *School of Biology, Georgia Institute of Technology, Atlanta, GA*

UWE CZUBAYKO • *Flyion GmbH, Tubingen, Germany*

JENS EILERS • *Carl-Ludwig-Institut fur Physiologie, Universitat Leipzig, Germany*

NIELS FERTIG • *Nanion Technologies GmbH, Munchen, Germany*

MATTHEW D. FULLER • *Program in Molecular and Systems Pharmacology, Emory University, Atlanta, GA*

HIDEMASA FURUE • *Department of Integrative Physiology, Kyushu University, Fukuoka, Japan*

MICHAEL FEJTL • *Flyion GmbH, Tubingen, Germany*

STEFAN HALLERMANN • *Institut fur Klinische Neurobiologie, Bayerische Julius-Maximilians Universitat, Wurzburg, Germany*

MANFRED HECKMANN • *Institut fur Klinische Neurobiologie, Bayerische Julius-Maximilians Universitat, Wurzburg, Germany*

ALEXANDER HÜMMER • *Flyion GmbH, Tubingen, Germany*

TOSHIHIKO KATAFUCHI • *Department of Integrative Physiology, Kyushu University, Fukuoka, Japan*

TOBIAS KRAUTER • *Flyion GmbH, Tubingen, Germany*

JOSÉ R. LEMOS • *Department of Physiology, University of Massachusetts, Worcester, MA*

RICHARD A. LEVIS • *Department of Physiology, Rush Medical College, Chicago, IL*

ALBRECHT LEPPLE-WIENHUES • *Flyion GmbH, Tubingen, Germany*

HÉCTOR G. MARRERO • *Department of Physiology, University of Massachusetts, Worcester, MA*

NAEL A. MCCARTY • *School of Biology, Georgia Institute of Technology, Atlanta, GA*

JAMES R. MOYER, JR. • *Department of Psychology, University of Wisconsin, Milwaukee, WI*

MICHELLE L. OLSEN • Department of Neurobiology, University of Alabama, Birmingham, AL

JAMES L. RAE • Department of Physiology, Mayo Clinic College of Medicine, Rochester, MN

CONSTANTINE SARANTOPOULOS • Department of Anesthesiology, Medical College of Wisconsin, Milwaukee, WI

DMITRY V. SARKISOV • Department of Molecular Biology, Princeton University, Princeton, NJ

HARTMUT SCHMIDT • Carl-Ludwig-Institut fur Physiologie, Universitat Leipzig, Germany

GERALD SEIFERT • Department of Neurosurgery, University of Bonn, Bonn, Germany

CHRISTIAN STEINHÄUSER • Department of Neurosurgery, University of Bonn, Bonn, Germany

HARALD SONTHEIMER • Department of Neurobiology, University of Alabama, Birmingham, AL

CHRISTOPHER H. THOMPSON • School of Biology, Georgia Institute of Technology, Atlanta, GA

SRINIVAS M. TIPPARAJU • Department of Medicine, University of Louisville, Louisville, KY

WOLFGANG WALZ • Department of Physiology, University of Saskatchewan, Saskatoon, Canada

SAMUEL S.-H. WANG • Department of Molecular Biology, Princeton University, Princeton, NJ

DAVID J.A. WYLLIE • Centre for Neuroscience Research, University of Edinburgh, Edinburgh, UK

MEGUMU YOSHIMURA • Department of Integrative Physiology, Kyushu University, Fukuoka, Japan

ZHI-REN ZHANG • School of Biology, Georgia Institute of Technology, Atlanta, GA

# 1

# Technology of Patch-Clamp Electrodes

## Richard A. Levis and James L. Rae

## 1. Introduction

The extracellular patch voltage clamp technique has facilitated studying the currents through single ionic channels from a wide variety of cells. In its early form (Neher and Sakmann, 1976), the resolution of this technique was limited by the relatively low [~50 megohm (M$\Omega$)] resistances, which isolated the interior of the pipette from the bath. The high resolution that can presently be achieved with the patch-clamp technique originated with the discovery (Neher, 1981) that very high resistance seals [tens or even hundreds of gigaohm (G$\Omega$)] can form between the cell membrane and the tip of a clean pipette when gentle suction is applied to the pipette interior. Although the precise mechanisms involved in this membrane-to-glass seal are still not fully understood, the importance of the gigaohm seal is obvious. The high resistance of the seal ensures that almost all of the current from the membrane patch flows into the pipette and to the input of the current-sensitive head-stage preamplifier. It also allows the small patch of membrane to be voltage-clamped rapidly and accurately via the pipette, and the mechanical stability of the seal is vital to the whole-cell voltage clamp technique. Of equal importance, the high resistance of the seal greatly reduces the noise it contributes to single-channel measurements. Although the seal can often represent only a small fraction of total patch-clamp noise (particularly as the bandwidth of recording increases), its importance should never be minimized. Without such high-resistance seals, most of the steady progress to reduce background noise levels would not have been possible.

From: *Neuromethods, Vol. 38: Patch-Clamp Analysis: Advanced Techniques, Second Edition*
Edited by: W. Walz @ Humana Press Inc., Totowa, NJ

Of course the patch pipette is not simply a tool in the formation of gigaohm seals. The pipette serves as a fluid bridge that connects the current-sensitive head-stage amplifier input to the surface or interior of the cell. The insulating properties (both resistive and more importantly, capacitive) of the glass that forms the wall of the pipette are also crucial to the ability to measure current originating in the patch and to the background noise levels that can be achieved.

For any patch-clamp measurement, several steps are required to construct a proper glass electrode. First, a glass that has optimal properties is selected. The required properties differ substantially for single-channel recordings and whole-cell current recordings. For single-channel measurements, low noise is the most important electrical parameter, whereas for whole-cell measurements, dynamic performance is more important than the contribution of the electrode to the background noise. This is simply because the background noise in a whole-cell recording is dominated by the noise from the electrode resistance (actually the access resistance) in series with the capacitance of the entire cell. The dynamic bandwidth of a whole-cell recording also depends on the same factors. Therefore, the goal in constructing an electrode for whole-cell recording is simply to make it as blunt and as low in resistance as is compatible with sealing it to the cell. In single-channel recordings, the pipette is a major contributor to the background noise and so requires many subtle considerations to produce an electrode optimal for recording single-channel currents.

As a second step in pipette construction, the electrode glass stock is pulled into a pipette with a tip of optimal geometry. This geometry differs for whole-cell and single-channel recordings. In a third step, the outside wall of the pipette is coated with a hydrophobic elastomer possessing good electrical properties. This procedure is essential for low-noise, single-channel recordings but can be done much less carefully for whole-cell recordings. Fourth, the tip is fire-polished to round it and clean its surface of any thin film of elastomer coating. This step can also be used to adjust the final tip diameter. Fire-polishing promotes seal formation but often is not required. After all these procedures, the electrode can be filled and used.

Several general properties of glasses must be considered when trying to construct optimal electrodes for patch clamping (Table 1). Thermal properties determine the ease with which desired tip

Table 1
General properties of types of glass

| Glass | LF | $Log_{10}$ vol. R. | Diel. const. | Softening temp. (C°) | Description |
|---|---|---|---|---|---|
| 7940 | 0.0038 | 11.8 | 3.8 | 1580 | Quartz (fused silica) |
| 1724 | 0.0066 | 13.8 | 6.6 | 926 | Aluminosilicate |
| 7070 | 0.25 | 11.2 | 4.1 | — | Low loss borosilicate |
| 8161 | 0.50 | 12.0 | 8.3 | 604 | High lead |
| Sylgard | 0.58 | 13.0 | 2.9 | — | No. 184 coating compound |
| 7059 | 0.584 | 13.1 | 5.8 | 844 | Barium-borosilicate |
| 7760 | 0.79 | 9.4 | 4.5 | 780 | Borosilicate |
| EG-6 | 0.80 | 9.6 | 7.0 | 625 | High lead |
| 0120 | 0.80 | 10.1 | 6.7 | 630 | High lead |
| EG-16 | 0.90 | 11.3 | 9.6 | 580 | High lead |
| 7040 | 1.00 | 9.6 | 4.8 | 700 | Kovar seal borosilicate |
| KG-12 | 1.00 | 9.9 | 6.7 | 632 | High lead |
| 1723 | 1.00 | 13.5 | 6.3 | 910 | Aluminosilicate |
| 0010 | 1.07 | 8.9 | 6.7 | 625 | High lead |
| S-8250 | 1.08 | 10.0 | 4.9 | 720 | — |
| 7052 | 1.30 | 9.2 | 4.9 | 710 | Kovar seal borosilicate |
| EN-1 | 1.30 | 9.0 | 5.1 | 716 | Kovar seal borosilicate |
| 7720 | 1.30 | 8.8 | 4.7 | 755 | Tungsten seal borosilicate |
| 7056 | 1.50 | 10.2 | 5.7 | 720 | Kovar seal borosilicate |
| 3320 | 1.50 | 8.6 | 4.9 | 780 | Tungsten seal borosilicate |
| 7050 | 1.60 | 8.8 | 4.9 | 705 | Series seal borocsilicate |
| S-8330 | 1.70 | 8.0 | 4.6 | 820 | — |
| KG-33 | 2.20 | 7.9 | 4.6 | 827 | Kimax borosilicate |
| 7740 | 2.60 | 8.1 | 5.1 | 820 | Pyrex borosilicate |
| 1720 | 2.70 | 11.4 | 7.2 | 915 | Aluminosilicate |
| N-51A | 3.70 | 7.2 | 5.9 | 785 | Borosilicate |
| R-6 | 5.10 | 6.6 | 7.3 | 700 | Soda lime |
| 0080 | 6.50 | 6.4 | 7.2 | 695 | Soda lime |

shapes can be produced and they determine how easily the tips can be heat-polished. Optical properties often result in a distinct visual end point so that tips can be fire-polished the same way each time. Electrical properties are important determinants of the noise the glass produces in a recording situation and determine the size and number of components in the capacity transient following a change of potential across the pipette wall. Glasses are complex substances composed of many compounds, and most of their properties are determined to a first order by the composition of the glass used.

Glass composition may also influence how easily a glass seals to membranes and whether or not the final electrode will contain compounds leached from the glass into the pipette-filling solution, which can activate, inhibit, or block channel currents.

## 2. General Properties of Pipette Glass

Before proceeding to the details of electrode fabrication, it is useful to consider in more detail glass properties that are important for patch-clamp pipette construction. We begin with thermal properties. It is important that glasses soften at a temperature that is easily and reliably achieved. This formerly was a stringent constraint since glasses like aluminosilicates, which melt at a temperature in excess of 900°C, would shorten the lifetime of a puller heating filament so much that their use was unattractive. Quartz, which melts above 1,600°C, could not even be pulled in commercially available pullers and so was not used at all. Today, at least one puller exists that will do these jobs easily (P-2000, Sutter Instruments, Novato, CA) and so virtually any kind of glass can be used routinely. It is generally true that the lower the melting temperature of the glass, the more easily it can be fire-polished. Low melting temperature glasses such as those with high lead content can be pulled to have tip diameters in excess of 100 μm and still be fire-polished to a small enough tip diameter that the pipette can be sealed to a 7- to 10-μm diameter cell. With such glasses, one has greater control over the final shape of the tip than is possible with higher melting temperature borosilicate glasses. Quartz pipettes cannot be fire-polished with a usual fire-polishing apparatus, although with care they can be fire-polished in a temperature-controlled flame.

Electrical properties are most important for providing low noise, as well as low amplitude, simple time course capacity transients. As will be discussed later, it is not possible to achieve low background noise without an elastomer coating the outside of the pipette. In general, glasses with the lowest dissipation factors have minimal dielectric loss and produce the lowest noise. There is a wide variety of glasses to choose from that will produce acceptable single-channel recordings, although quartz is clearly the best material to date. Good electrical glasses are also necessary for whole-cell recordings, not because of noise properties but because they result in the simplest and most voltage- and time-stable capacity transients.

Major chemical constituents in glass are important since they determine the overall properties of the glass and because they are potential candidates to leach from the glass into the pipette-filling solution where they can interact with the channels being studied. No glass can be deemed to be chemically inert since even tiny amounts of materials leached in the vicinity of the channels may produce sufficient local concentrations to interact with channels and other cellular processes. Again, quartz would be expected to have fewer chemical impurities than other glasses, but every kind of glass should be suspected of having an effect on the channels being measured.

# 3. Whole Cell Pipette Properties: Practical Aspects

## 3.1. Choice of Glass

Modern computerized pipette pullers are capable of pulling glass with almost any thermal properties (with the exception of quartz) into the proper blunt-tipped geometry that is ideal for whole-cell recording. Therefore, almost any glass can be used to form whole-cell pipettes. Nevertheless, we feel that some types of glass usually should be avoided, while others have some particularly useful properties for this application.

Soda lime glasses like Kimble R-6 and Corning 080 generally should not be used because of their high dielectric loss. When a voltage step is applied across a patch pipette fabricated from one of these glasses, there will be a large slow component in the resulting capacity transient (Rae and Levis, 1992a). For a 2-mm depth of immersion with a moderate coating of Sylgard No. 184 (Dow Corning, Midland, MI) to within ~200 µm of the tip, we have found following a 200-mV voltage stop that is a slow component for a soda lime pipette can be as large as 50 picoampere (pA) 1 ms after the beginning of the step. The slow tail of capacity current can still be as much as 10 pA 10 ms after the step and may require as much as 200 ms to decay to below 1 pA. The time course of this slow tail is not exponential, but more closely approaches a logarithmic function of time. In addition, we have observed that for soda line pipettes the magnitude of the slow component of capacity current is not always constant during a series of pulses that occur at rates faster than about 1 to 2 per second. Instead, the magnitude of this component is sometimes observed to decrease with successive pulses. Because of these characteristics, these capacitive currents can possibly be mistaken for whole-cell currents. Heavy Sylgard coating can reduce the amplitude of the slow component of capacity current for soda lime glasses, but it is generally better (and certainly more convenient) simply to use glasses with lower loss factors (see Rae and Levis, 1992a, for further discussion).

High-lead glasses such as 8161, EG-6, EG-16, 0010, 0120, and KG-12 possess much lower dissipation factors than soda lime glasses and are particularly useful due to their low melting point. This property allows the construction of initially very large-tipped pipettes that can be subsequently fire-polished to blunt bullet-shaped tips offering the lowest possible access resistance. This, of course, minimizes series resistance. In addition, pipettes of this shape also draw in the largest surface area patch of membrane when suction is applied. This is useful in perforated patch recordings since the larger area of membrane available for partitioning by amphotericin or nystatin results in the maximum incorporation of perforation channels and thus the lowest access resistance. KG-12 (Friedrich and Dimmock, Millville, NJ) is a good choice for glasses of this class since it seals well, has good electrical properties, and is readily available.

Pipettes for whole-cell recording can be thin walled in comparison to those for single-channel recording. In whole-cell measurements, other sources of noise far outweigh the contribution from the pipette per se (see below). In terms of total background noise, the major consideration in pipette fabrication is simply achieving the lowest possible resistance. Glass with an outer diameter/inner diameter (OD/ID) ratio of 1.2 to 1.4 has lower resistance for a given outside tip diameter than does thicker walled glass, and is therefore useful for whole-cell recording. Some precautions are necessary, however, since if the walls become too thin, the pipette will more easily penetrate the cell during the attempt to form a seal.

Other glasses that have been successfully used by many laboratories for whole-cell recording include Pyrex (Corning No. 7740), Kimble's Kimax, and Corning No. 7056. Schott No. 8250 or No. 8330 are also good choices and readily available from Garner Glass (Claremont, CA). Although we usually prefer the high-lead glasses described above, these other glasses have produced perfectly acceptable results.

### 3.2. Pulling Whole-Cell Electrodes

This work can be done on any commercially available electrode puller. Here one simply strives for as blunt a taper and as large a tip diameter as is compatible with sealing of the electrode to the cell.

### 3.3. Elastomer Coating Whole-Cell Electrodes

Elastomer coating of electrodes reduces electrode noise in single-channel recordings. In whole-cell recordings, the noise associated with electrode glass is usually insignificant in comparison to other noise sources and so elastomer coating is not required for noise reduction. Elastomer coating also reduces electrode capacitance. Commercial patch-clamp amplifiers have the ability to compensate about 10 picofarad (pF) of electrode capacitance. For pipettes made from glasses with high dielectric constants (e.g., soda lime and high-lead glasses) immersed deeply into a tissue bathing solution, the electrode (and holder) capacitance may exceed the compensation range of the electronics. Elastomer coating helps to keep the total electrode capacitance within the compensation range. For whole-cell recordings, it is not usually necessary to paint the elastomer close to the tip. Coating that extends from the top of the shank to 1 mm from the tip is sufficient for whole-cell recordings. Many investigators do not use elastomer coating for whole-cell recordings.

### 3.4. Fire-Polishing Whole-Cell Electrodes

Finally, to promote gigaohm seals and to reduce the possibility of tip penetration into the cell during seal formation, electrode tips should be fire-polished. In some cells, fire-polishing has proven unnecessary, but we have found that sealing is generally promoted by fire-polishing the electrode tip, particularly for cells where seal formation is difficult. Whole-cell and single-channel electrodes are fire-polished with the same basic apparatus. Fire-polishing can be done either using an upright or an inverted microscope. In fact, many investigators have chosen to coat their pipettes and fire-polish them using an inverted microscope with approximately a

40× long working distance objective. Another very useful approach is to utilize a standard upright microscope equipped with an objective for metallurgical microscopes. Several microscope companies (e.g., Nikon, Olympus, Zeiss) make extra-long working distance high magnification metallurgical objectives. Most noteworthy are the 100× objectives that have 3- to 3.5-mm working distances and numerical apertures of 0.8. With these objectives and 15× eyepieces and with the electrode mounted on a slide held in the mechanical stage of the microscope, it is possible to move the electrode tip into the optical field and visualize directly the electrode tip at 1500× magnification. At such high magnifications, it is possible to fire-polish the tip to a very distinct optical end point under direct visualization. This approach ensures very repeatable results from one electrode to the next. The only drawback is that the objectives are quite expensive.

The fire-polishing itself is accomplished by connecting to a micromanipulator a rod of inert material to which has been fastened a short loop of platinum iridium wire. The ends of this wire must be soldered to two other pieces of wire that can be connected to a voltage or current source to allow current to be passed through the platinum wire. The platinum loop is generally bent into a very fine hairpin so that it can be brought to within a few micrometers of the electrode tip under direct observation. Because of early reports that platinum can be sputtered from the wire onto the electrode tip and prevent sealing, the platinum wire is generally coated with a glass like Pyrex (Corning No. 7740) or Corning No. 7056 to prevent such sputtering. This is done by overheating the platinum wire and pushing against it a piece of electrode glass that has been pulled into an electrode tip. At high temperatures, the glass melts and flows over the platinum wire and ends up thoroughly coating it and forming a distinct bead of glass. If the elastomer has been coated too near the tip, fire-polishing causes the tip to droop downward at the juncture where the coating ends. If one desires to paint elastomer extremely close to the tip, it may be necessary to do the majority of the fire-polishing before coating and then fire-polish lightly again afterward. As a general rule, fire-polishing with the electrode tip close to the heating wire at low temperature produces a tip whose inner walls are parallel and relatively close together. With a hotter heating element and the tip farther away, the tip tends to round more and end up quite blunt.

# 4. Patch Electrode Fabrication for Single-Channel Recording

## 4.1. Choice of the Glass

There is a limited number of glasses available for single-channel patch clamping. Perhaps the most important feature to consider is the amount of noise in the recording that is due to the pipette itself. This subject is sufficiently important that we include an entire section dealing with noise sources in pipettes so that readers can use these principles to make optimal pipettes for their own recording situation. There is no longer any question, however, that quartz is the best glass if noise performance is important. Quartz itself is quite expensive and requires an expensive laser-based puller, and so probably is not the glass for routine studies. Therefore, we consider other glasses here as well. Garner Glass has been particularly helpful in the development of specialty glasses for patch clamping, although the company no longer provides any of the high-lead glasses that we find so useful.

The glass tubing selected for the fabrication of single-channel patch electrodes should have walls of substantial thickness. Wall thickness results in decreased electrical noise and increased bluntness at the tip, which prevents penetrating the cell during seal formation. Glass tubing with an OD/ID of 2.0 to 3.0 is easily obtainable and is expected to yield the lowest background noise levels. Generally, the outside diameter chosen is 1.5 to 1.7 mm. For single-channel recordings, only the glasses with the best electrical properties should be used if optimal noise performance is desired. Corning glasses Nos. 8161 and 7760 were particularly good in this regard, but Corning no longer makes them. Corning No. 7052 was also quite acceptable but also is not available. Sadly, most of the options for particularly low-noise glasses are running out, and so quartz is expected to become increasingly more attractive, even given its cost. Readily available glasses like Corning No. 7740 or Kimble's Kimax are not particularly quiet glasses. High-lead glasses like Kimble's KG-12 have a better signal-to-noise ratio than the Pyrex-type glasses but are substantially worse than the best glasses mentioned above. Schott No. 8250 is readily available and has thermal and noise properties that are about as good as those of Corning No. 7052. Low-noise recordings have also been obtained with Schott

No. 8330, a glass with electrical properties about midway between those of Corning No. 7740 and Schott No. 8250.

For single-channel recordings, one should not choose a glass with a filament fused to the interior wall. Such a filament promotes filling but presumably serves as an internal conduit, which allows fluid to creep up the inside of the pipette even if the pipette has been vacuumed with a suction line. This fluid near the internal filament presumably serves as a noise source, much like the external fluid film that forms in the absence of elastomer coating.

In our experience, it is usually unnecessary to clean electrode glasses prior to pulling. On occasion, however, normally quiet pipette glasses are found to be noisy in use, and then it is imperative to clean the glass for the best noise performance. Sonicating the glass in 100% ethanol or methanol in an ultrasonic cleaner is effective for this purpose. Following any cleaning procedure, it is a good idea to place the glass in an oven at around 200°C for 10 to 30 minutes to achieve complete drying. Heat treatment of this sort has also proven necessary if low-noise recordings are required in environments where the humidity is exceptionally high.

## 4.2. Pulling Single-Channel Electrodes

Single-channel pipettes made from glasses other than quartz can be pulled on any commercially available patch electrode puller. Here the tips can be less blunt and higher in resistance. In fact, they can be pulled with a single-stage pull so that the tips are almost perfectly conical. They look similar to tips that might be used for impaling cells, although their resistance is only 10 to $50\,M\Omega$ or so. The electrode resistance in series with the patch capacitance is a potential noise source (see $R_e$-$C_p$ Noise, below). However, as will be seen, this source of noise may actually be minimized by using high-resistance pipettes insofar as such high resistance correlates with a small patch area. In addition, sharper tip tapering often leads to higher resistance seals to the membrane. Thus for the best noise performance for single-channel recording, it is better not to use the blunt electrode tips that are good for whole cell situations.

## 4.3. Coating Single-Channel Pipettes with Elastomers

For the lowest noise recordings, electrodes must be coated with a hydrophobic elastomer to within $100\,\mu m$ or less of their tip. The closer it can be painted to the tip, the better. This coating prevents bathing solution from forming a thin fluid film along the outer surface of the electrode. This thin film of bathing solution would be a substantial noise source. A commonly used compound is Sylgard No. 184. Sylgard also has exceptional electrical properties (see Table 1), and so it improves the electrical properties of most glasses when a thick coat covers the glass surface. Sylgard, meticulously mixed, can be stored at −20°C in small, capped centrifuge tubes. The thorough mixing is required to prevent pockets of the compound that are not adequately exposed to polymerizer. This unpolymerized elastomer can flow to the electrode tip (even against gravity) and render the tips difficult to seal.

At freezer temperatures, the mixed Sylgard can be stored for several weeks. A tube of this freezer-stored Sylgard, when brought to room temperature for use in painting electrodes, lasts for several hours before it begins to polymerize. Care must be taken not to open the tube until the contents have reached room temperature, to prevent water condensation. Condensed water can degrade the electrical properties of the elastomer and increase noise. The Sylgard is applied to the electrode tip with a small utensil such as a piece of capillary tubing pulled to a reasonably fine tip in a flame. Sylgard is applied using dissecting microscopes at magnifications of 10× to 30×. It is useful, but not required, to modify the dissecting microscope to work in a dark field. This can be done inexpensively with a fiber optic ring illuminator connected to a fiber optic light source. The ring illuminator is placed under the stage of the microscope. Three to four inches above the ring light, dark-field illumination is achieved and the walls of the electrode glass show up as bright lines of light against a dark background. Both the Sylgard coat and the tip of the electrode are easily seen with this dark-field illumination. The Sylgard must be directed away from the tip by gravity at all times during the painting procedure or the Sylgard may flow over the tip to make fire-polishing or sealing impossible. The Sylgard can be cured by holding the tip for 5 to 10 seconds in the hot air stream emanating from a standard heat gun like those used in electronics to heat shrink tubing. Again, the Sylgard must be gravitationally directed away from the tip during this curing process.

While Sylgard is the most commonly used elastomer, there are a number of other elastomers available that are as good as Sylgard in most respects and better in others. RTV615A from General

Electric (GE) has properties nearly identical to Sylgard and can be used in exactly the same way as Sylgard. Dow Corning Medical Silastic MDX-4 has dielectric properties slightly better than Sylgard No. 184 but polymerizes more rapidly at freezer temperatures. To date, it has not offered any obvious improvement in noise on a day-to-day basis, but several of the lowest noise measurements done with quartz electrodes utilized this elastomer. It is considerably more expensive than Sylgard No. 184.

Dow Corning No. R-6101 is another excellent elastomer that costs more to buy but probably not to use than Sylgard (Rae and Levis, 2000). It has the useful property of not polymerizing appreciably at room temperature and so can be used for up to 12 months without freezing. Its noise properties are as good as (but should be a little better than) those of Sylgard No. 184, and it does result in low noise when used with quartz or some other very good electrode glass. It is particularly useful because it does not require mixing and it can be partially cured in an oven to produce any desired viscosity. The viscosity remains quite stable, increasing only slightly over a year's time. Heating at 95°C for 48 hours yields about as high a viscosity as one might desire. Lower temperature or shorter heating times give intermediate viscosities. Because R-6101 shows little tendency to run into the tip of the electrode, it is possible to coat the electrode and then cure the R-6101 in a hot air stream with the tip pointing down rather than up. This procedure results in a bulbous mass of the elastomer that is much thicker along the shank of the pipette than is possible by heat curing with the tip up. This approach is also helped by painting a thin ring of the elastomer as close to the tip as possible, curing it, and then painting the electrode shank with a much thicker layer of elastomer. When this thicker layer is cured tip down, there is very little tendency for the R-6101 to get into the tip, and so thicker layers are possible near the tip than with most other coating schemes. We have now switched to using this elastomer for all of our studies because of its obvious advantages.

### 4.4. Fire-Polishing Single-Channel Pipettes

The same principles apply here as in the fire-polishing of whole-cell electrodes. The same apparatus is used for both. In general, patch electrodes are fire-polished with the tip close to the heating filament, with the goal of thickening the glass near the tip in addition to rounding it. For high-resistance seals, it may be useful to fire-polish so that the internal walls of the tip become parallel for several micrometers. This mode of fire-polishing increases the tip resistance a few megohms but often results in lower noise due to higher resistance seals (see $R_e$-$C_p$ Noise, below).

## 4.5. Fabrication Methods Specific to Quartz

Quartz softens at about 1600°C, and so no platinum or nichrome wire–based heat source will melt it. Both of these materials disintegrate long before 1600°C is reached. Quartz can be pulled in a flame but the tip geometry is unreliable with such fabrication techniques. The new laser-based P-2000 electrode puller from Sutter Instruments generates enough heat to pull quartz fairly easily. It begins to have trouble when the glass OD exceeds 1.5 mm. It has no difficulty pulling quartz tubing with an OD/ID = 3, as long as the OD does not exceed 1.5 mm. Since the major reason to use quartz patch pipettes is for the reduction of single-channel background noise currents, it is best to use quartz with as thick a wall as possible. A 1.5-mm OD with a 0.5-mm ID produces about the smallest bore that is practical. Even at 0.5-mm ID, there is some difficulty with the internal silver–silver chloride (Ag-AgCl) electrode since it must be made of such flimsy silver wire that it is often damaged (bent) or denuded of silver chloride as the electrode is placed into the small bore. An ID of 0.6 to 0.75 mm makes the pipettes much easier to use.

Quartz cannot easily be fire-polished with any presently available commercial apparatus. Those apparatuses that fire-polish other glasses, including aluminosilicate, do not generate enough heat to fire-polish quartz. It is possible to fire-polish it in a carefully controlled Bunsen burner, but that approach is sufficiently unreliable that it is best to try to pull tips whose geometry is good enough to allow sealing without fire-polishing. That places an additional constraint on the puller since most other glass pullers need only to produce electrode tips that are approximately correct since the final tip geometry can be customized while fire-polishing. With quartz, the tips must be good enough for use immediately after pulling.

Because of the noise produced by a thin film bathing solution creeping up the outer surface of an electrode, quartz must be elastomer coated like any other glass. This bathing solution film is such a large noise source that if an elastomer coating is not used to reduce it, there is absolutely no reason to use quartz electrodes for patch clamping. It does not perform appreciably better than poor glasses if this noise source is not eliminated or minimized. Because quartz must be elastomer coated, it must also be subjected to the heat-polisher. While the polisher cannot smooth or round the quartz tip as it does with other glasses, it can burn off any residual elastomer and so should be used with quartz electrodes just before filling.

## 4.6. Low-Noise Recording

Low-noise recording requires meticulous attention to detail. Even with an electrode optimally pulled, coated, and fire-polished, there are still many ways in which excess noise can creep in. It is important that the electrodes be filled only to just above the shank. Fluid in the back of the electrode can cause internal noise generating films and allow fluid into the holder. It is important for low-noise recordings that a suction line with a syringe needle that is the correct size to fit into the bore of the pipette be maintained near the experimental setup. This suction line can be used to vacuum fluid from the pipette and ensure none gets into the holder or coats the majority of the back of the electrode. Alternatively, silicone fluid or mineral oil can be used to fill the electrode for a short distance in back of its filling solution. These "oils" are somewhat messy and not really required if a proper suction line for vacuuming the electrode interior is used. The internal electrode should be adjusted in length until its tip just comfortably is immersed in the filling solution. In general, the shorter the length of the internal electrode (and of the pipette), the lower the noise will be. Therefore, it is best to use the shortest possible holder and electrode that is practical.

During experiments where low noise is required, it is best to test the noise at intermediate stages. Most modern patch-clamp amplifiers have a root mean square (rms) noise meter that can be checked to determine the noise levels at any time. This meter should be checked immediately after inserting the electrode into the holder and placing the electrode tip over the bath but before actually immersing the tip in the bath. Poorly filled electrodes, fluid in the holder, a generally dirty holder, and pickup from the environment show up as elevated noise. What the actual level of the noise will be depends on the noise of the patch clamp, the kind of holder and electrode glass that are used, and how well one has shielded against pickup of electrical interference. Specific examples appear in Levis and Rae (1993). But as a general rule, total noise in this situation should not be more than approximately 10% to 20% above that of the open circuit head-stage. If you see excess noise, you can remove the electrode, dry the internal electrode, and then test the noise with only the head-stage and holder placed above the bath. If this is elevated above what is normal for your setup, either your holder is dirty or you are experiencing pickup from the environment. Environmental pickup can often be seen as noise spikes at discrete frequencies, whereas a dirty holder contributes noise across a broad range of frequencies. You can try to dry the holder by blowing dry, clean air through it, but it is possible that you will have to clean the holder before the noise will be reduced. This can be done by disassembling it, sonicating it in ethanol, and drying it for several hours in an oven at 60° to 70°C. Because of the time involved in cleaning the holder, it is wise to have two or more holders available when attempting very low-noise recordings.

The noise of the electronics, holder, electrode glass, and elastomer can be determined by making a thin pad of Sylgard and placing it in the bottom of the chamber. You then seal your electrode to it, much as you would seal to a cell. No suction, however, is required to make the seal. You simply push the tip against the Sylgard and a seal forms. The seal should be 200 G$\Omega$ or more if you have done it correctly. Under these circumstances, the seal noise is essentially negligible and you are able to quantify the remaining composite noise sources. This noise depends on how deep the bathing solution is; the deeper the bathing solution, the greater the noise. For most purposes, the bath depth need not be more than 1 to 3 mm. This simple procedure will let you know what is routinely possible with your setup and give you a baseline for comparing the noise you actually get in experiments. A good seal to a cell will often produce noise that is about the same as the noise you get when the cell is sealed to Sylgard.

Note, however, that as soon as the electrode tip is placed in the bath, the noise will be enormous since you are now measuring at best the thermal noise of a 1- to 10-M$\Omega$ resistance tip. The readings on the noise meter will not be meaningful until you have obtained a gigaohm seal. If the seal resistance is <20 G$\Omega$, the majority of the noise will be due to the seal, and really low noise recordings cannot be achieved.

## 5. Electrodes for Single-Cell Electroporation

Electrodes for single-cell electroporation, a relatively new patch-clamp technique (Rae and Levis, 2002a,b) are very simple to fabricate by comparison to other patch-clamp electrodes. Since no seal is formed, the background noise is dominated by the thermal noise of the tip resistance. Therefore, electrodes do not have to be elastomer coated and do not need to be made of high-quality glass. They require a glass with an internal glass filament fused to the wall to facilitate filling the tip while using minimum volumes of solution. Internal filament electrode glass is supplied by every company that sells electrode glass. The pipettes should be pulled in a single stage with tips rather like those used for intracellular recording. Resistances in the range of 30 to 50 M$\Omega$ for 150-mm NaCl–filled electrodes are ideal. This ensures both a small tip area and power dissipation at the tip well below that which produces membrane-tip adherence.

## 6. Noise Properties of Patch Pipettes

### 6.1. Noise Contribution of the Pipette

The earliest patch pipettes were fabricated from "soft" soda lime glasses. Such glasses were easy to pull and heat-polish to any desired tip geometry, primarily because they soften at relatively low temperatures. Unfortunately, such pipettes introduced relatively large amounts of noise into patch-clamp measurements. It was soon found that "hard" borosilicate glasses produced less noise, but, owing to their softening at higher temperatures, were somewhat more difficult to pull and heat-polish. Probably as a result of these early findings, it has sometimes been assumed that "hard" high melting temperature glasses necessarily had better electrical properties than "soft" low melting temperature glasses. However, there is no obligatory relationship between the thermal and electrical properties of glass. For example, several low melting temperature high-lead glasses (e.g., 8161, EG-6) have been shown to produce less noise than a variety of high melting temperature borosilicate and aluminosilicate glasses (e.g., 7740, 1720). The reason for these findings becomes clear when the electrical properties of the glasses are considered.

The electrical properties of glass that are important to its noise performance are its dielectric constant and its dissipation factor; the bulk resistivity of a glass might also be important, but is usually sufficiently high to be ignored. The dielectric constant of a substance is the ratio of its permitivity to the permitivity of a vacuum. Thus for pipettes of equivalent geometry and depth of immersion, the higher the dielectric constant of the glass, the higher the pipette capacitance. The dielectric constants for glasses commonly used for patch pipette fabrication range from 3.8 for quartz to more than 9 for some high-lead glasses. The dielectric constant of borosilicates is typically 4.5 to 6, while that of soda lime glasses is near 7. The pipette capacitance generates noise by several mechanisms that are described below. The dissipation factor is a measure of the lossiness of a dielectric material. Ideal capacitors display no dielectric loss and do not generate thermal noise. However, all real dielectrics are lossy and do produce thermal noise; we refer to this as dielectric noise. Glasses with the lowest dissipation factors are the least lossy and generate the least dielectric noise. Quartz is among the least lossy of all practical dielectrics; its dissipation factor, which is in the range of $10^{-5}$ to $10^{-4}$, is far lower than that of other glasses used for patch pipettes. Several high-lead glasses have dissipation factors of ~$10^{-3}$. The dissipation factor of borosilicates that have been successfully used to fabricate patch pipettes varies from about 0.002 to 0.005. Soda lime glasses have the highest dissipation factor (~0.01), which is the principal reason for their high noise.

The best glasses for patch pipette fabrication are those with the best electrical properties, that is, low dissipation factor and low dielectric constant. However, understanding pipette noise requires more than simply understanding the electrical properties of glass. A variety of other factors also influence the noise performance of the patch pipette, for example, pipette geometry, depth of immersion, and the type and extent of elastomer coating. Here we summarize our present understanding of all major pipette noise sources; more detailed discussions can be found elsewhere (Levis and Rae, 1992, 1993, 1998; Rae and Levis, 1992a,b).

Attaching the electrode holder to the head-stage input slightly increases the noise above its minimum level associated with an open-circuit input. The mechanisms involved in generating this noise are discussed elsewhere (Levis and Rae, 1993, 1998). Here we only note that the contribution of the holder by itself to total patch-clamp noise should be very small. Holder noise is minimized by constructing the holder from low-loss dielectric materials, minimizing its size, and always keeping it clean. Shielded holders produce more noise than unshielded holders.

Simply adding the pipette to the holder (attached to the head-stage input) slightly increases the capacitance at the amplifier input. After the pipette has been immersed into the bath and a gigaohm seal has been formed, the capacitance at the head-stage input is further increased. As will be seen, the capacitance of the immersed portion of the pipette is a consideration in several sources of noise. Here, however, we begin by noting that all of this capacitance, at a minimum, produces noise because it is in series with the input voltage noise, $e_n$, of the head-stage amplifier. The current noise produced has a power spectral density (PSD, Amp$^2$/Hz), which rises as $f^2$ at frequencies above roughly 1 kHz. Of course this noise is correlated with noise arising from $e_n$ in series with other capacitance (amplifier input capacitance, stray capacitance, capacitance of the electrode holder). The total amount of capacitance associated with an immersed pipette can vary from a fraction of a pF up to 5 pF or more. Low capacitance is associated with heavy elastomer coating and shallow depths of immersion. The amount of noise arising from this mechanism increases as the capacitance associated with the pipette increases. However, regardless of the value of the pipette capacitance, the noise it contributes in conjunction with $e_n$ is small in comparison with other pipette noise sources described below. For low-noise patch-clamp measurements, it is imperative that the pipette capacitance be minimized. The reason for this will become clearer as other noise sources associated with this capacitance are described.

In addition to the mechanism just described and to noise arising from the membrane to glass seal, which will be discussed separately, the pipette contributes noise by at least four mechanisms. Each mechanism is described below, followed by a summary of pipette noise sources. Our emphasis is on the minimization of each noise, rather than simply its description.

### 6.2. Thin Film Noise

Thin films of solution are capable of creeping up the outer surface of the pipette from the bath (Fig. 1A). The noise associated with such films has previously been shown to be very significant (Hamill et al., 1981). Such a film has a relatively high distributed resistance, and the thermal voltage noise of this resistance is in series with the distributed capacitance of the pipette wall. It is expected that the PSD of this noise will rise at low to moderate frequencies and then level out at frequencies in the range of several kilohertz to several tens of kilohertz. We have estimated with uncoated pipettes made from several types of glass that the noise associated with such a film of solution is usually in the range of 100 to 300 pA rms in a bandwidth of 5 kHz. Evidence for such films has been found in

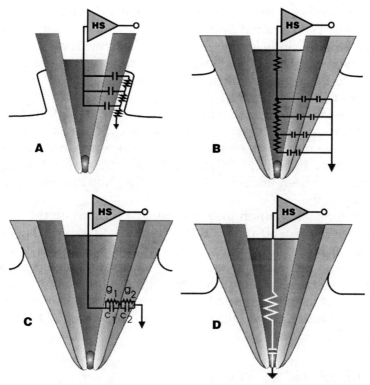

Fig. 1. Simplified circuit representations of the major noise mechanisms of the patch pipette. (**A**) Thin solution film on the exterior surface of an uncoated patch pipette; noise arises from the thermal voltage noise of the distributed resistance of this film in series with the capacitance of the pipette wall. In B, C, and D, the pipette is shown coated with a suitable elastomer. (**B**) Distributed RC noise arising from the thermal voltage noise of the distributed resistance of the pipette filling solution in series with the distributed capacitance of the immersed portion of the pipette wall and its elastomer coating. (**C**) Dielectric noise of the series combination of the pipette ($g_1$, $C_1$, where $g_1 = \omega C_1 D_1$) and the elastomer coating ($g_2$, $C_2$, where $g_2 = \omega C_2 D_2$). In the region immersed in the bath, the glass wall of the pipette and its elastomer coating are represented by ideal lumped capacitances $C_1$ and $C_2$, respectively, in parallel with loss conductances $g_1 = 2\pi f C_1 D_1$ and $g_2 = 2\pi f C_2 D_2$. The thermal noise (dielectric noise) of the coated pipette is then $4kT$ multiplied by the real part of the admittance of the series combination of dielectrics. (**D**) $R_e$-$C_p$ noise arising from the thermal voltage noise of the entire (lumped) resistance, $R_e$, of the patch pipette in series with the patch capacitance, $C_p$. See text for further details.

pipettes fabricated from all glasses we have tested when elastomer coating has been omitted. However, pipettes pulled from GE quartz produce significantly less noise without elastomer coating than any other type of glass. Apparently the surface of this glass is less subject to the formation of such thin films.

Coating the pipette with Sylgard 184 or other suitable elastomers can essentially eliminate the formation of external films of solution and eliminate the otherwise large amounts of noise they produce. These elastomers have a hydrophobic surface, which prevents the formation of such films. Sylgard 184 is so effective in this regard that we have been unable to detect any thin film noise in properly coated pipettes.

Thin films of solution may also be able to form on the interior surface of the pipette and inside of the holder. To avoid the formation of such films, it is possible after filling the pipette with the desired amount of ionic solution to layer a few millimeters of paraffin oil or silicone fluid on top of the filling solution. However, we have found that this is usually unnecessary (and it can get messy) if excess solution is carefully suctioned from the back of the pipette as described above.

### 6.3. Distributed RC Noise

Noise also arises from the thermal voltage noise of the resistance of the pipette-filling solution in series with the capacitance of the immersed portion of the pipette (Fig. 1B). Most of the resistance of the pipette resides at or near its tip. However, significant resistance is distributed along the shank distal to the tip. This resistance and its thermal voltage noise are in series with the capacitance of the pipette wall distributed along the portion that is immersed in the bath. We refer to the noise that results as distributed RC noise. In the frequency range of greatest interest to patch clamping (DC to 100 kHz or more), the PSD of this noise is expected to rise as $f^2$. Most of our theoretical predictions of the noise arising from this mechanism (e.g., Levis and Rae, 1992) have relied on idealizations of the pipette geometry. More complicated real-world geometrics and factors such as nonuniform thinning of the pipette wall, which often occurs during pulling, are expected to make such predictions rather imprecise (see also Levis and Rae, 1998). Because of this, we chose to study distributed RC noise directly. These experiments used quartz pipettes pulled from OD/ID = 2.0 tubing that were coated with Sylgard No. 184 only to the point where the electrode entered the bath (i.e., most or all of the immersed portion of the pipette was uncoated); immersion depth was ~1.8 mm, and the pipettes were sealed to Sylgard (seal resistance >200 GΩ). Our strategy was to vary the ionic strength of the internal filling solution. Changing the ionic strength of the filling solution changes the pipette resistance, but it has no effect on the pipette capacitance. Because of this, it is expected that for pipettes of equivalent geometry and with the same depth of immersion into the bath, the PSD of distributed RC noise will vary as $1/M$, where M is the ionic concentration of the filling solution. The rms noise in any particular bandwidth is expected to vary as $1/M^{1/2}$. In our study of this noise, we used NaCl solutions with concentrations from 1.5 mM to 1.5 M to fill the pipette. As expected, the noise increased as the ionic strength of the filling solution decreased. When the noise component attributable to distributed RC noise was parsed from total noise (and it was the dominant noise source for ionic strength of 15 mM or less), the predicted behavior was reasonably well confirmed. Also, as expected the PSD of this noise component increased approximately as $f^2$ as frequency increased.

On the basis of these experiments, we concluded that for uncoated quartz pipettes that were pulled from OD/ID = 2 tubing and immersed to a depth of about 1.8 mm and filled with 150 mM NaCl (i.e., the ionic strength typical of most experiments), the PSD of distributed RC noise was approximated by $2.5 \times 10^{-38} f^2$ amp$^2$/Hz. The rms noise contribution in a bandwidth $B$ is then $(8 \times 10^{-39} c_3 B^3)^{1/2}$ amps rms, where $c_3$ is a coefficient that depends on the type of filter used ($c_3 \approx 1.9$ for an eight-pole Bessel filter). This equation predicts a noise component of ~44 pA rms for a 5-kHz bandwidth (−3 dB, eight-pole Bessel filter), or about 123 femtoampere (fA) rms in a 10 kHz bandwidth.

It must be remembered, however, that these results were for relatively thick-walled pipettes fabricated from quartz, which has a low dielectric constant of 3.8. It must also be remembered that the pipettes were not coated with Sylgard (or other suitable elastomer) in the region immersed in the bath. The capacitance of the wall of the pipette is expected to vary directly with the dielectric constant of the glass (for pipettes of the same geometry) and vary inversely roughly in proportion to the log of the OD/ID ratio. The PSD of distributed RC noise should vary in proportion to the pipette capacitance ($C_e$) squared; rms noise in a given bandwidth will therefore vary linearly with $C_e$. Thus for an uncoated pipette fabricated from OD/ID = 1.4 tubing from a glass with a dielectric constant of 7.6 (twice that of quartz), the numbers given above would be expected to increase by a factor of about 4. On the other hand, coating the immersed portion of a pipette with a suitable elastomer will thicken its walls and therefore reduce $C_e$. Thus very heavy coating of the pipette with an elastomer such as Sylgard No. 184 can dramatically reduce distributed RC noise, and, with such a coating, the amount of this noise will become almost independent of the type of glass used. In the experiments described above, we measured $C_e$ to be in the range of 1.4 to 1.8 pF. We have found that using the tip-dip elastomer coating method (Levis and Rae, 1993) to build up a heavy coat of Sylgard all the way to the tip of the pipette, we can obtain values of $C_e$ as low as ~0.35 pF for a comparable depth of immersion. This should reduce distributed RC noise to less than 10 fA rms in a 5-kHz bandwidth. Of course, shallow depths of immersion can also reduce distributed RC noise.

From the preceding discussion, it should be clear that the reduction of distributed RC noise is one of the major benefits of coating the immersed portion of the pipette with a low dielectric constant elastomer such a Sylgard No. 184. This noise component can also be minimized by using thick-walled tubing of glasses with low dielectric constants and by shallow depths of immersion of the pipette into the bath. Distributed RC noise is also expected to depend on pipette geometry, and should be minimized by shapes that reduce the distributed resistance distal to the pipette tip.

### 6.4. Dielectric Noise

Dielectric noise (Fig. 1C) also arises from the capacitance of the pipette wall over the region that is immersed in the bathing solution. For pipettes fabricated from glasses other than quartz, dielectric noise is likely to be the dominant source of noise arising from the pipette. For a single dielectric with a capacitance $C_d$ and a dissipation factor $D$, the PSD of dielectric noise is given by

$$S_d^2 = 4kTDC_d(2\pi f) \text{ Amp}^2/\text{Hz} \tag{1}$$

and the rms noise in a bandwidth $B$ is given by

$$I_d = (4kTDC_dc_2\pi B^2)^{1/2} \text{ Amp rms} \tag{2}$$

where k is Boltzman's constant and $T$ is absolute temperature (°K). $c_2$ is a coefficient that depends of the type of filter used; for an eight-pole Bessel filter with B as the −3 dB bandwidth, $c_2 \approx 1.3$. It is important to note that the PSD of dielectric noise rises linearly with increasing frequency and that the rms value of this noise is proportional to filter bandwidth. This is quite unlike the other noise sources discussed, and is very useful in experimentally parsing dielectric noise from other types of noise generated by the pipette.

For an uncoated pipette, these equations can be applied simply by noting that $C_d$ is the capacitance of the immersed portion of the pipette (denoted by $C_e$ above), and that $D$ is the dissipation factor of the glass. It is instructive to consider two uncoated pipettes with the same geometry both pulled from OD/ID = 1.4 tubing and both immersed to a depth of about 2 mm. One pipette is fabricated from quartz ($D = 0.0001$, dielectric constant = 3.8), and the other pipette is fabricated from a borosilicate with factor $D = 0.005$ and a dielectric constant of 5.0. The capacitance ($C_d$ or $C_e$) of the quartz pipette should be about 1.5 pF, while that of the borosilicate pipette will be about 2 pF due to its higher dielectric constant. Using these numbers, it can be estimated that the uncoated quartz pipette will produce about 16 fA rms dielectric noise in a 5-kHz bandwidth (−3 dB, eight-pole Bessel filter), while the borosilicate pipette would produce 128 fA rms dielectric noise in the same bandwidth. The superiority of quartz is clear in this case.

Of course the importance of coating the pipette with a suitable elastomer has already been demonstrated regardless of the type of glass used. Therefore, it is necessary to consider the dielectric noise in this more complicated situation. We have presented a more detailed analysis of the dielectric noise in this case elsewhere (Levis and Rae, 1993, 1998). Here we summarize our most important conclusions. When the pipette is coated with an elastomer, it is necessary to derive equations that describe the dielectric noise of the series combination of two different dielectrics with capacitances $C_1$ and $C_2$ and dissipation factors $D_1$ and $D_2$ (see Fig. 1C and its legend). For $D_1$, $D_2 \gg 1$, the dielectric noise PSD of the elastomer coated pipette is well approximated by

$$4kT(2\pi f)\frac{D_1 C_1 C_2^2 + D_2 C_2 C_1^2}{(C_1 + C_2)^2} \text{ Amp}^2/\text{Hz,} \qquad (3)$$

and the rms noise in a bandwidth $B$ is approximated by

$$\left\{4kTc_2\pi B^2\frac{D_1C_1C_2^2+D_2C_2C_1^2}{(C_1+C_2)^2}\right\}^{1/2} \text{ Amps rms.} \tag{4}$$

In these equations $C_1$ and $D_1$ are the capacitance and dissipation factor of the glass wall of the pipette and $C_2$ and $D_2$ are the capacitance and dissipation factor of the elastomer coating. The capacitance $C_1$ depends on the depth of immersion, the thickness of the pipette wall, and the dielectric constant of the glass. For a 2-mm depth of immersion, $C_1$ can vary from as little as about 1 pF for very thick walled quartz pipettes, to more than 6 pF for thin-walled pipettes made from glasses with high dielectric constants (e.g., soda lime and high-lead glasses). Of course the capacitance $C_2$ of the elastomer coating also depends on the depth of immersion, the dielectric constant of the elastomer, and the thickness of the elastomer coating. Heavy elastomer coating leads to the smallest values of $C_2$. However, it is important to realize that the thickness of the elastomer coating is not uniform. In particular, it is hard to achieve very thick elastomer coatings near the tip of the pipette. The dip method of elastomer coating (Levis and Rae, 1993) has proved to be useful in building up relatively heavy coats of elastomer all the way to the tip of the electrode, but even with this method the thickness of the coat is still not uniform. Because of this, it is difficult to predict the value of $C_2$. However, we have measured the value of $C_2$ (see Levis and Rae, 1993) to be as little as 0.4 to 0.5 pF for a 2-mm immersion depth when heavy coatings of Sylgard were applied with the dip method. With lighter coating, the value of $C_2$ can easily be much higher (2 pF or more).

The dissipation factor $D_1$ of the glasses used to fabricate patch pipettes has already been discussed; reported values range from as little as $10^{-5}$ to $10^{-4}$ for quartz to as much as 0.01 for soda lime glasses. The dissipation factor $D_1$ of the elastomer is also very important. Sylgard No. 184 has a dissipation factor of about 0.002, which is lower than that of most glasses with the notable exception of quartz. Because of this, coating pipettes fabricated from glasses other than quartz will significantly reduce their dielectric noise and the relative reduction will be greatest for the poorest (most lossy) glasses. However, the dissipation factor of Sylgard No. 184 is a factor of 20 or more higher than that of quartz, and predictions based on equations 3 and 4 indicate that for all realistic values of $C_2$ coating, a quartz pipette with Sylgard actually increases its dielectric noise relative to that which would have been produced by the pipette alone. This is true despite the reduction in overall capacitance produced by the Sylgard coating. Thus for a quartz pipette with $C_1 = 1.5 \, pF$ and $D_1 = 0.0001$ and a Sylgard coating with $C_2 = 0.5 \, pF$ and $D_2 = 0.002$, equation 4 predicts $30 \, fA$ rms of dielectric noise in a 5-kHz bandwidth (−3 dB, eight-pole Bessel filter), while as described above, the same pipette without the Sylgard coating would have produced only about $17 \, fA$ rms dielectric noise in this bandwidth. Estimates of the dielectric noise of quartz pipettes coated with Sylgard and several similar elastomers and sealed to Sylgard have produced values that are in good agreement with the predictions of equations 3 and 4.

It is apparent from the above discussion that coating a quartz patch pipette with Sylgard No. 184 is not desirable in terms of dielectric noise. Nevertheless, coating with Sylgard or some other suitable elastomer is necessary to eliminate thin film noise and to minimize distributed RC noise. If fact, very heavily Sylgard-coated quartz pipettes display the least noise of all pipettes, so the small increment in dielectric noise resulting from such coating is more than offset by the benefits in terms of reduction of other types of noise. It is also important to realize that even though Sylgard-coating a quartz pipette increases its dielectric noise, the final dielectric noise of such a pipette still remains significantly below that of Sylgard-coated pipettes fabricated from any other type of glass we have tested. If elastomers with dissipation factors significantly less than that of Sylgard No. 184 can be found that are otherwise suitable for coating pipettes, they could be effective in lowering the dielectric noise of quartz pipettes. It can be appreciated from examination of equations 3 and 4 that the dissipation factor of such an elastomer need not be less than that of quartz to lower total dielectric noise of a heavily coated pipette. Such elastomers (if found) should also be very beneficial for other types of glass. Dow Corning R-6101 and Q1-4939 both are reported by the manufacturer to have dissipation factors of 0.00025. However, our preliminary measurements of pipettes coated with R-6101 have failed to demonstrate any significant noise advantage over pipettes coated with similar thicknesses of Sylgard No. 184. Although we are unable to account for this finding, it certainly seems possible that the true dissipation factor of this elastomer exceeded the value in the manufacturer's data sheet.

Due to the volume of material presented regarding dielectric noise, it is probably worthwhile to summarize our conclusions. For thick-walled quartz pipettes with heavy Sylgard coating to the tip and seal to Sylgard at an immersion depth of ~2 mm, our estimates of dielectric noise have generally been in the range of 20–35 fA rms in a 5-kHz bandwidth. On occasion, with actual excised patches and heavily Sylgard-coated patch pipettes, our estimates of dielectric noise have been less than 15 fA rms in a 5-kHz bandwidth when the electrode tip has been withdrawn close to the surface of the bath. For other types of glasses, dielectric noise is significantly higher. Our previous measurements of the noise arising from light to moderately Sylgard-coated pipettes made from more than 20 different glasses (Rae and Levis, 1984, 1992a), indicated that in a 5-kHz bandwidth and with a ~2-mm depth of immersion, dielectric noise varied from about 100 to 200 fA rms. The lowest noise was associated with glasses with the smallest loss factor (i.e., dissipation factor multiplied by dielectric constant), while the highest noise arose from the very lossy soda lime glasses. Recently we have measured a few pipettes made from Corning Nos. 7052 and 7760 (tubing OD/ID = 1.4) that were heavily coated with Sylgard No. 184 to the tip by the dip method described above. These measurements indicated that dielectric noise could be as low as ~70 fA rms in a 5-kHz bandwidth for these glasses (with heavy Sylgard coating) at a 2-mm immersion depth. This is somewhat more noise than would be predicted from equations 3 and 4, but less than we have estimated previously.

### 6.5. $R_e$-$C_p$ Noise

The last pipette noise mechanism that we will consider is the noise that is expected to arise from the thermal voltage noise of the entire lumped pipette resistance, $R_e$ in series with the capacitance of the patch membrane, $C_p$; we refer to this noise source as $R_e$-$C_p$ noise (Fig. 1D). This noise is expected to have a PSD that increases as $f^2$ up to frequencies of about $1/2\pi R_e C_p$ (which is usually several hundred kilohertz); at frequencies below this the PSD is expected to be

$$S_{ep}^2 = 4\pi^2 e_e^2 C_p^2 f^2 \text{ Amp}^2/\text{Hz} \tag{5}$$

where $e_e^2 = 4kTR_e$, that is, the thermal voltage noise PSD for the pipette. The rms noise attributable to this mechanism in a bandwidth $B$ is then given by

$$I_{ep} = \{1.33\pi^2 c_3 e_e^2 C_p^2 B^3\}^{1/2} \text{ amps rms} \qquad (6)$$

where $c_3$ is a coefficient that again depends on the type of filter used to establish the bandwidth; for an eight-pole Bessel filter with a $-3\,\text{dB}$ bandwidth of B Hz, $c_3 \approx 1.9$.

For single-channel measurements, patch capacitance typically ranges from approximately 0.01 to 0.25 pF for pipette resistances in the range of 1 to 10 MΩ. As expected, higher values of patch capacitance are associated with lower resistance pipettes. Because of the inverse relationship between $R_e$ and $C_p$, equations 5 and 6 predict that the smallest amount of $R_e$-$C_p$ noise arises from the smallest patches, even though such patches are obtained with higher resistance pipettes. For example, with $R_e = 10\,\text{M}\Omega$ and $C_p = 0.01\,\text{pF}$, equation 6 predicts a noise contribution of only about 6 fA rms in a 5-kHz bandwidth ($-3\,\text{dB}$, eight-pole Bessel filter). On the other hand, with $R_e = 2\,\text{M}\Omega$ and $C_p = 0.25\,\text{pF}$, the predicted noise is more than 60 fA rms in the same bandwidth. This latter amount of noise can exceed the total of all other pipette noise sources for quartz pipettes, and remains significant even for pipettes fabricated from other glasses. Obviously, however, this noise source itself does not depend on the type of glass, but rather on the geometry of the pipette (and, to some extent, on luck).

### 6.6. Seal Noise

The noise associated with the membrane-glass seal is less easily predicted. It is expected that the PSD of this noise for zero applied voltage will be given by $4kT\,\text{Re}\{Y_{sh}\}$, where $\text{Re}\{Y_{sh}\}$ is the real part of the seal admittance. The minimum value of $\text{Re}\{Y_{sh}\}$ is $1/R_{sh}$, where $R_{sh}$ is the DC seal resistance, and this leads to a minimum estimate of the seal noise in a bandwidth $B$ of $(4kTB/R_{sh})^{1/2}$. This is just the thermal current noise of the seal resistance and can be very small for high-resistance seals. For example, for a 200-GΩ seal and a bandwidth of 5 kHz, $(4kTB/R_{sh})^{1/2} = 20\,\text{fA}$ rms. Our measurements from several patches with seal resistances in the range of 40 to 100 GΩ have shown that the noise attributable to the seal is often indistinguishable from the predicted thermal current noise of $R_{sh}$

(Rae and Levis, 1992b). Nevertheless, it is certainly possible that seal noise may sometimes exceed this minimum prediction. As anyone who has spent much time trying to achieve low noise with the patch-clamp technique knows, there is a great deal of variability in the noise achieved even when all of the precautions we have described have been followed and when very high seal resistances have been obtained. It is certainly tempting to blame some of this variability on the noise associated with the seal.

## 6.7. Summary of Pipette Noise Sources

It is important to realize that the noise sources described above (with the exception of noise arising from various capacitances in series with the amplifier's input voltage noise $e_n$) are all uncorrelated. Uncorrelated noise sources add in an rms fashion. For example, if four uncorrelated noise sources have rms values denoted by $E_1$, $E_2$, $E_3$, and $E_4$, then the total rms noise resulting from the summation of these sources is given by $(E_1^2 + E_2^2 + E_3^2 + E_4^2)^{1/2}$. Because of this, the largest individual source of noise tends to dominate the total noise.

Of the noise sources described above, only thin film noise can be completely eliminated (or in any case reduced to negligible levels). Distributed RC noise, dielectric noise, and $R_e$-$C_p$ noise can never be eliminated, but they can be minimized. In many cases, precautions taken to reduce one noise source are also beneficial in reducing other sources of noise. Thus with any type of glass, thick-walled pipettes, all else being equal, have less capacitance and therefore display less distributed RC noise and dielectric noise. Similarly, shallow depths of immersion also reduce pipette capacitance and simultaneously reduce distributed RC and dielectric noise. Coating pipettes with a heavy layer of a low-loss elastomer such as Sylgard No. 184 also reduce the pipette's capacitance and reduce distributed RC noise for all types of glass. For all types of glass other than quartz, a heavy coat of Sylgard No. 184 extending as close to the tip as possible also significantly reduces dielectric noise. In the case of quartz pipettes, elastomers with dissipation factors comparable to that of Sylgard No. 184 actually somewhat increase dielectric noise. However, within the range of realistic thicknesses of the coating, even for quartz, heavy coatings generally lead to the least dielectric noise; this is compatible with the requirements for minimizing distributed RC noise in quartz pipettes.

It is also important to recall that even though Sylgard coating somewhat increases the dielectric noise of quartz pipettes, the final dielectric noise of a heavily Sylgard-coated quartz pipette is still much less than that of pipettes made from any other type of glass. The major distinction in terms of noise between pipettes fabricated from quartz and other types of glasses is, in fact, the much lower dielectric noise of quartz. $R_e$-$C_p$ noise has often been ignored in the past, and in many situations it is sufficiently small to still be ignored. However, when all other sources of noise have successfully been reduced to the lowest limits presently achievable, it can become significant, and even dominant at very wide bandwidths (Levis and Rae, 1993). $R_e$-$C_p$ noise is minimized by forming the smallest patch areas that are consistent with the goals of the experiment being undertaken. Although the data are widely scattered, patch area (or patch capacitance) decreases as pipette resistance increases. The net result is that it is predicted that higher resistance patch pipettes with small tips will tend to produce the least amount of $R_e$-$C_p$ noise. The geometry of such electrodes is not necessarily the best selection for minimizing distributed RC noise, but this can be overcome by heavy elastomer coating. Although we have not systematically studied the relationship between pipette resistance (tip diameter) and noise, it is our experience that the lowest noise patches are usually obtained from small-tipped, high-resistance pipettes.

It is difficult to assign values to what can be expected as typical or "best-case" noise from pipettes fabricated from different glasses in different situations. Nevertheless, some rough estimates can be provided. For low-loss borosilicate, aluminosilicate, or high-lead glasses with moderate Sylgard coating extending to within ~100 µm of the tip, it is reasonable to expect that in a 5-kHz bandwidth total pipette noise (excluding seal noise) as low as 100 to 120 fA rms can be achieved with a ~2 mm depth of immersion. With very heavy Sylgard coating all the way to the tip, this value should fall to somewhat less than 100 fA rms in this bandwidth. With quartz pipettes that are heavily Sylgard-coated to the tip, we have been routinely able to keep total pipette noise to ~40 fA rms in a 5-kHz bandwidth for a 2-mm immersion depth. With the quartz pipette raised to the surface of the bath with an excised membrane patch, we have occasionally achieved a total noise of the pipette plus seal as low as 30 to 35 fA rms in a 5-kHz bandwidth; subtracting the thermal current noise of the seal (as judged from its measured resistance) yields an estimate of 10 to 20 fA rms for total pipette noise in these cases.

## 6.8. Noise Sources for Whole-Cell Voltage Clamping

While all of the pipette noise mechanisms described above with the exception of $R_e$-$C_p$ noise are present in whole-cell voltage clamping, their relative importance is very much less than is the case for patch voltage clamp measurements. Of course, this is not because these pipette noise sources have become less in the whole-cell situation, but rather because other noise source have become much higher. In the first place, most whole-cell voltage clamp measurements are made with a patch-clamp head-stage amplifier configured with a 500-MΩ feedback resistor. In a 5-kHz bandwidth, this resistor alone will produce 400 fA rms noise, which is more than even soda lime pipettes produce, provided they are reasonably Sylgard-coated. Under most situations, however, the dominant source of noise in a whole-cell voltage clamp is the thermal current noise of the pipette resistance $R_e$ in series with the cell membrane capacitance $C_m$.

As just noted, the whole-cell voltage clamp lacks $R_e$-$C_p$ noise. The reason for this is simply that the patch membrane has been disrupted, or shorted out, as is the case for perforated patch measurements. However, in the whole-cell situation, the entire cell membrane is in series with the pipette resistance and with the thermal voltage noise of this resistance. The noise produced by this has precisely the same mechanism that underlies $R_e$-$C_p$ noise, but, since $C_m \gg C_p$, it is of far greater magnitude. It might also be recalled that the time constant $R_eC_p$ is typically a microsecond or less and so can usually be neglected. However, the time constant $R_eC_m$ is much larger and its effects cannot be ignored, either in terms of noise or dynamic performance.

Of course, the electrode resistance $R_e$ is the series resistance in the whole-cell variant of the patch voltage clamp, and many of its effects are well known and need no further comment here. But it seems that some of its effects can never be emphasized often enough. One of these is the filtering effect that uncompensated series resistance has on the measured current. In the absence of series resistance compensation, this filtering effect (equivalent to a simple RC low-pass filter) limits the actual bandwidth of current measurement to $1/2\pi R_e C_m$. For example, with $R_e = 10$ MΩ and $C_m = 50$ pF, this is ~320 Hz, and it should be remembered that $R_e$

after patch disruption or perforation is usually higher than the pipette resistance that was measured in the bath. With series resistance compensation, this bandwidth limit is increased. We define $\alpha$ as the fraction of the series resistance compensated $(0 < \alpha < 1)$, and $\beta = 1 - \alpha$. With series resistance compensation, the uppermost usable bandwidth is extended to $1/2\pi\beta R_e C_m$. So in the previous example, 90% series resistance compensation $(\beta = 0.1)$ extends the actual bandwidth limit to about 3.2 kHz. It also greatly increases the noise at this bandwidth. The PSD, $S_{em}^2$, of noise arising from the thermal voltage noise of $R_e$ in series with $C_m$ is given by

$$S_{em}^2 = \frac{4\pi^2 e_e^2 C_m^2 f^2}{1 + 4\pi^2\beta^2 R_e^2 C_m^2 f^2} \tag{7}$$

where $e_e^2 = 4kTR_e$ is the thermal voltage noise PSD of $R_e$. Note that this expression takes into account the effects of series resistance compensation. For 100% series resistance compensation $(\alpha = 1, \beta = 0)$, equation 7 reduces to $4\pi^2 e_e^2 C_m^2 f^2$, which has exactly the same form as equation 5.

From equation 7 it can be seen that the PSD of the noise arising from $R_e$ and $C_m$ rises with increasing frequency as $f^2$ until it reaches $f = 1/2\pi\beta R_e C_m$. Thereafter, this noise plateaus to a value of $4kT/\beta^2 R_e$, which, of course, is many times larger than the thermal current noise of the feedback resistor. This plateau level of the PSD is maintained until a frequency where it is rolled off by an external filter (or the inherent bandwidth limit of the electronics). As an example of the magnitude of the noise introduced by this mechanism, consider a rather favorable example for whole-cell voltage clamping with $R_e = 5\,M\Omega$ and $C_m = 30\,pF$. Without series resistance compensation, the "corner frequency" at which the noise PSD plateaus (and the limit of actual bandwidth of current measurement) is at about 1060 Hz. For a −3-dB bandwidth (eight-pole Bessel filter) of current measurement that is only 500 Hz, the noise arising from $R_e$ and $C_m$ would already be nearly 0.5 pA rms, which is more than a very bad electrode would produce in a bandwidth of 5 kHz. By a bandwidth of 1 kHz, the noise would have increased to about 1.3 pA rms. Increasing the bandwidth of current measurement much beyond 1 kHz without series resistance compensation is not justified since the measured current will still be effectively filtered at 1.06 kHz (−3 dB bandwidth of the 1-pole low-pass filter arising from $R_e$ and

$C_m$). This does not mean, however, that setting the external filter to a bandwidth higher than 1 kHz will not add more noise. Increasing the bandwidth of the external filter to 5 kHz increases the noise to more than 3 pA rms, but it provides very little signal information that was not contained when the data were filtered at 1 kHz. Series (pipette) resistance compensation can extend the usable bandwidth, but it significantly increases the noise at external filter bandwidths higher than $1/2\pi R_e C_m$. Thus with 90% series resistance compensation, the maximum usable bandwidth of current measurement is extended to 10.6 kHz. In this case with an external filter (eight-pole Bessel) with a −3 dB bandwidth of 5 kHz, the noise is increased to almost 15 pA rms. For a 10-kHz bandwidth the noise increases to about 40 pA rms. In noises of this magnitude, the pipette noise mechanisms previously discussed become quite insignificant. Therefore, it can be concluded that many of the characteristics of the pipette that were important to patch clamping are not important to a whole-cell voltage-clamp situation.

The noise arising from $R_e$ and $C_m$ in whole-cell voltage clamping can only be minimized by minimizing $R_e$ and/or $C_m$. Of course, minimizing $C_m$ means selecting small cells and this is often not possible. In addition, it should also be noted that if we are studying a particular type of channel in a population of cells of various sizes but the channel density is the same in all cases, there is no clear advantage, in terms of signal-to-noise ratio, in selecting smaller cells. For a constant value of $R_e$, it is simple to show that at a given bandwidth (below $1/2\pi\beta R_e C_m$) the rms noise decreases linearly as $C_m$ decreases, but since the number of channels is also proportional to $C_m$, the signal also decreases linearly with decreasing $C_m$: the signal-to-noise ratio is constant. In this case, the signal-to-noise ratio only depends on $R_e$, and it improves as $1/R_e^{\frac{1}{2}}$. So the most practical way to minimize this source of noise is to use the lowest resistance pipettes that are capable of sealing to the cells and make every effort to minimize the increase in access resistance that often occurs when the patch is disrupted.

Finally, it is worth emphasizing that another important way of minimizing this noise is to not make the mistake of using a bandwidth of the external filter that is not justified by the situation. Increasing the external bandwidth significantly beyond $1/2\pi\beta R_e C_m$ adds essentially no information about the signal, but it does add additional noise.

## 7. Recent Developments

Since the last edition of this book was published, there have been some new developments that warrant at least brief mention here. Specifically, we consider the ongoing efforts in several laboratories to replace present-day patch pipettes with what can generally be called planar electrode techniques. Briefly, these techniques involve a small hole (or holes, a few micrometers down to less than 1 μm in diameter) in a sheet of a low-loss dielectric material (with a typical thickness of a few hundred micrometers). Typically the hole expands from a small opening at the upper surface to a much wider diameter at the lower surface, forming a conical or pyramidal well. This sheet separates two compartments. The uppermost compartment contains a bathing solution and a cell or cells from which measurements are to be made. The lower compartment contains only solution in the small well just mentioned. This very small amount of solution is attached to a patch-clamp amplifier input, while the solution in the upper compartment is grounded. This arrangement can greatly reduce the capacitance associated with the traditional pipette and its holder. It also allows for the possibility of arrays of holes/wells for simultaneous whole-cell recording from numerous cells, with isolation occurring between each of these by the absence of solution except in the wells. More exotic arrangements than those just described are also being considered, but will not be discussed here.

Thus there are two possible benefits of this approach: (1) it can allow for multiple whole-cell measurements to be made simultaneously (ultimately with a high degree of automation), and (2) it can potentially reduce the capacitance—and hence the noise—associated with single-channel patch-clamp measurements. To take full advantage of the latter possibility, new lower-capacitance low-noise head-stage amplifiers must be developed, and it must become possible to form very high resistance seals.

In general the formation of high resistance seals has been the greatest problem encountered in the development of such planar electrode (or "sheet/hole") techniques. However, significant progress has been made over the past year. Klemic et al. (2001) have formed arrays of planar electrodes in a partition formed by micromolding Sylgard into the desired shape; Sylgard has proved successful in forming gigaohm seals. Fertig et al. (2001) have successfully formed tiny (~1-μm diameter) lipid bilayers in "chip electrodes" made from glass or quartz with etched apertures. Finally, Farre et al. (2001) have prepared chips for whole-cell recordings with standard optical lithography and etching of silicon wafers covered by silicon nitride ($Si_3N_4$). More progress in such techniques seems likely in the future.

## References

Farre, C., Olofsson, J., Pihl, J., Persson, M., and Orwar, O. (2001) Whole-cell patch clamp recordings performed on a chip. *Biophys. J.* **80**, 338a.

Fertig, N., Meyer, C., Blick, R. H., and Behrends, J. C. (2001) A microstructured chip electrode for low noise single channel recording. *Biophys. J.* **80**, 337a.

Hamill, O. P., Marty, A., Neher, E., Sakman, B., and Sigworth, F. J. (1981) Improved patch-clamp techniques for high-resolution current recording from cells and cell-free membrane patches. *Pflugers Arch.* **391**, 85–100.

Klemic, K. G., Baker, W. T., Klemic, J. F., Reed, M. A., and Sigworth, F. J. (2001) Patch clamping with a planar electrode array. *Biophys. J.* **80**, 337a.

Levis, R. A. and Rae, J. L. (1992) Constructing a patch clamp setup. *Methods Enzymol.* **207**, 18–66.

Levis, R. A. and Rae, J. L. (1993) The use of quartz patch pipettes for low noise single channel recording. *Biophys J.* **65**, 1666–1677.

Levis, R. A. and Rae, J. L. (1998) Low noise patch clamp techniques. *Methods Enzymol.* **293**, 218–266.

Neher, E. (1981) Unit conductance studies in biological membranes, in *Techniques in Cellular Physiology* (Baker, P. F., ed.), Elsevier, North Holland, Amsterdam.

Neher, E. and Sakmann, B. (1976) Single channel currents recorded from membrane of denervated frog muscle fibers. *Nature* **260**, 799–802.

Rae, J. L. and Levis, R. A. (1984) Patch voltage clamp of lens epithelial cells: theory and practice. *Mol. Physiol.* **6**, 115–162.

Rae, J. L. and Levis, R. A. (1992a) Glass technology for patch electrodes. *Methods Enzymol.* **207**, 66–92.

Rae, J. L. and Levis, R. A. (1992b) A method for exceptionally low noise single channel recordings. *Pflugers Arch.* **42**, 618–620.

Rae, J. L. and Levis, R. A. (2000) An electrode coating elastomer to replace Sylgard 184. *Axobits* **29**, 6–7.

Rae, J. L. and Levis, R. A. (2002a) Single cell electroporation. *Pflugers Arch.* **443**, 664–670.

Rae, J. L. and Levis, R. A. (2002b) Electroporation of single cells. *Axobits* **34**, 7–11.

# 2

# Whole-Cell Patch-Clamp Recordings

## Harald Sontheimer and Michelle L. Olsen

## 1. Introduction

The patch-clamp recording technique measures ionic currents under a voltage clamp and was designed to study small patches of membrane in which near-perfect control of the transmembrane voltage can be readily achieved. Today, this technique is most frequently used to examine currents across entire cells. This application defies many of the original design requirements, such as small size and near-perfect voltage control. Nevertheless, whole-cell recordings are routinely used to characterize current flow through ionic channels, neurotransmitter receptors, and electrogenic transporters in cell types of virtually any origin. Since its introduction in 1981 (Hamill et al., 1981), patch-clamp recordings have essentially replaced sharp electrode recordings, particularly in the study of cultured cells and more recently in brain slice recordings.

Whole-cell patch-clamp recordings and their applications have been the topic of numerous excellent reviews and book chapters (see Recommended Readings at the end of the chapter). This chapter summarizes some basic concepts and describes "hands-on" procedures and protocols, so as to complement previous accounts of the patch-clamp technique.

## 2. Principles (Why Voltage-Clamp?)

Electrophysiologists are especially interested in the activity of membrane proteins that provide conductive pathways through biological membranes: ion channels, transmitter receptors, and electrogenic ion carriers. Channel activity, whether through voltage-dependent or ligand-gated ion channels, results in changes of membrane conductance, which can be most conveniently evalu-

From: *Neuromethods, Vol. 38: Patch-Clamp Analysis: Advanced Techniques, Second Edition*
Edited by: W. Walz @ Humana Press Inc., Totowa, NJ

ated by recording membrane currents at a constant membrane voltage. Under such "voltage-clamped" conditions, current is directly proportional to the conductance of interest.

A two-electrode voltage-clamp design was first introduced in the seminal studies of Hodgkin and Huxley (Hodgkin et al., 1952) for the study of ionic conductances of the squid giant axon. In this application, one of the electrodes serves as voltage sensor and the second functions as a current source, with both interconnected through a feedback amplifier. Any change in voltage detected at the voltage electrode results in current injection of the proper polarity and magnitude to maintain the voltage signal at a constant level. The resulting current flow through the current electrode can be assumed to flow exclusively across the cell membrane and as such is proportional to the membrane conductance (mediated by plasma membrane ion channels). The major disadvantage of this technique, however, is its requirement for double impalement of the cell, which restricts its application to rather large cells (>20 µm) and prevents study of cells embedded in tissue.

In an attempt to solve this problem, single-electrode switching amplifiers were developed that allowed the use of one electrode to serve double duty as voltage and current electrode. For short periods of time, the amplifier connects its voltage-sensing input to the electrode, takes a reading, and subsequently connects the current source output to the same electrode to deliver current to the cell. This approach, however, is limited in its time resolution by the switching frequency between the two modes, which must be set based on the cell's RC time constant (the product of cell input resistance $R$ and capacitance $C$). Both single-electrode switch clamp and double-electrode voltage clamp allow direct measurement of the cell's voltage and avoid the introduction of unknown or unstable voltage drops across the series resistance of the current passing electrode.

The whole-cell patch-clamp technique similarly uses only one electrode. However, in contrast with above techniques, it uses the electrode continuously for voltage recording and passage of current. Consequently, the recording arrangement contains an unknown and potentially varying series resistance in the form of the electrode and its access to the cell. For the technique to deliver satisfactory results, it is essential that this series resistance be small relative to the resistance of the cell. Numerous measures are taken to satisfy these requirements (see below), including the use of blunt, low-resistance electrodes, small cells with high impedance, and electronic compensation for the series resistance error. When effectively utilized, the whole-cell technique can yield current recordings of equal or superior quality to those obtained with double- or single-electrode voltage-clamp recordings.

## 3. Procedure and Techniques

### 3.1. Pipettes

In contrast with sharp electrode recordings that utilize pipettes with resistances of >50 MΩ, comparatively blunt low-resistance (1–5 MΩ) recording pipettes are used for whole-cell patch-clamp recordings. This is done for two reasons: (1) series resistance should ideally be two orders of magnitude below the cell's resistance, and (2) blunt electrodes (1–2 μm) are required to achieve and maintain mechanically stable electrode-membrane seals.

As described in Chapter 1, electrodes can be manufactured from a variety of glass types. Although it has been frequently reported that glass selection has a significant influence on the quality of seal or the frequency at which good seals are obtained, little scientific evidence supports this notion. To achieve the shape ideal for sealing membrane patches, electrodes are pulled from capillary glass pipettes in a two-stage or multistage process using commercially available pullers, such as those of Narashige (Tokyo, Japan), Sutter (Novato, CA), and others. An additional step of fire-polishing the pipette tip may be used to improve the likelihood of seal formation. As an added benefit, the tips of pipettes with very low resistance (steep, tapering tips) can be fire-polished to reduce the tip diameter to appropriate levels. Once an appropriate set of variables, such as cell preparation, glass type, electrode resistance, and shape, is identified, the success rate for stable electrode-membrane seals should be between 50% and 90%.

### 3.2. Electronic Components of a Setup

The electronic components of a patch-clamp setup are comparatively few: a patch-clamp amplifier, a digital-analog/analog-digital converter, and a computer. An oscilloscope, external stimulator, or external signal filters, and recording devices are options. High-quality patch-clamp amplifiers are available from a number of manufacturers. At present, many laboratories utilize either a Heka EPC 9/10 (Southboro, MA), or an Axopatch 200B, or MultiClamp 700B amplifier (Molecular Devices, Sunnyvale, CA). Most manufacturers design their amplifiers to offer a broad range of functions (whole-cell recording, single-channel recording, current clamp, voltage clamp). However, some amplifiers are particularly tailored for specific functions, such as recording from lipid bilayers. Caution must be exercised, especially by the novice electrophysiologist, in using computer-controlled amplifiers such as the Axopatch 700B. Many features including pipette offset, series resistance, compensation, and whole-cell capacitance, which on traditional amplifiers must be adjusted manually, are automatically determined by the amplifier, communicated to the software, and stored. Previous experience on traditional amplifiers provides the users of automated amplifiers with the experience to determine the reliability and validity of these values.

The single most important electronic component of a patch-clamp amplifier is the current-to-voltage converter, which is contained in the head stage (Fig. 1). Its characteristics are described in detail elsewhere (Sigworth, 1983). Until recently, most head stages used resistive technology. Here current flow through the electrode ($I_p$) across a resistor of high impedance ($R$) causes a voltage drop that is proportional to the measured pipette current ($I_p$). An operational amplifier (OpAmp) is used to automatically adjust the voltage source ($V_s$) to maintain a constant pipette potential ($V_p$) at the desired reference potential ($V_{ref}$). As the response of the OpAmp is fast, it can be assumed that for all practical purposes $V_p = V_{ref}$. When

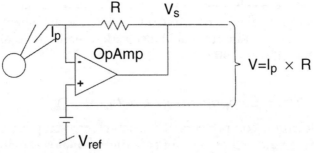

Fig. 1. Scheme of current-to-voltage converter (for details see Sigworth, 1983). $I_p$, pipette current; $R$, feedback resistance; OpAmp, operational amplifier; $V_{ref}$, reference potential; $V_s$, voltage source.

current flows across the membrane through ion channels, $V_p$ is instantaneously displaced from $V_{ref}$. The OpAmp alters $V_s$ to generate an $I_p$ that will exactly oppose the displacement of $V_p$ from $V_{ref}$. Thus, the current measured during a patch-clamp experiment (membrane current flow) is equal and opposite to $I_p$.

In their whole-cell mode, resistive patch-clamp amplifiers use a 500 MΩ feedback resistor, which enables measuring currents of up to 20 nA. For some amplifiers a low-gain 50 MΩ head stage is available that can pass currents of up to 200 nA; however, its use sacrifices the use of capacitance and series resistance compensation. More recently, amplifiers such as the Axopatch 200B or 3900A Integrating Patch Clamp (Dagan, Minneapolis, MN) have head stages that operate in two modes: traditional resistive feedback for whole-cell recording, and capacitor feedback or integrating mode for single-channel recording. Because capacitors generate less noise than resistors (an ideal capacitor generates no thermal noise), the instrument or circuit noise associated with the resistor is eliminated when recording in integrating mode, allowing for extremely low noise during single-channel recording.

Although some amplifiers, such as the Axopatch 1D, have built-in stimulators, most electrophysiologists prefer the use of an external stimulator that offers greater versatility. Low-cost microcomputers serve as both digital stimulator and on-line recorder. While the use of a microcomputer is not essential, it has become the de facto norm, as the convenience of storing and analyzing digital data far outweighs the cost associated with such systems. Moreover, most systems are now highly integrated, such that the amplifiers communicate via telegraphs many important analog settings such as gain, capacitance, series resistance, and filter frequency settings. In addition they provide for easy delivery of stimulus protocols via their digital/analog (D/A) converters. Data can be collected and digitized on-line with sampling rates of up to 5 µs and can be readily stored on the hard disk of a computer. All necessary components can be purchased at a price well below that of an external stimulator alone.

Additional equipment is recommended for specific data collection needs, and it is not uncommon to find an oscilloscope or even a VCR recorder attached to a recording setup. The latter, however, has become unpopular with the introduction of relatively inexpensive large hard drives and convenient inexpensive means to back up data digitally on CDs and DVDs. Even in the presence of a computer-based data acquisition system, oscilloscopes are a convenient means for monitoring data collected by a microcomputer, and are convenient for debugging environmental electrical noise from a recording setup. If nothing else it conveys the "cockpit look" to a patch-clamp setup and provides enough counterweight to the equipment rack to prevent it from tipping over.

### 3.3. Recording Configuration

The whole-cell patch-clamp recording setup closely resembles that used for sharp electrode intracellular recordings. An electrically grounded microscope on an isolation table serves as the foundation of the recording setup. A recording chamber is mounted to the stage of an inverted microscope for isolated cell recording (Fig. 2A) or an upright microscope with a water immersion lens for slice recording (Fig. 2B). Various types of recording chambers are commercially available, all of which serve well for most purposes. If perfusion is desired, several options are available, including constant perfusion via peristaltic pump, gravity-fed perfusion systems, and systems that allow for an ultrafast fluid exchange. Electrodes are typically placed under visual control (400×) onto a cell by use of a high-quality, low-drift micromanipulator.

Numerous hydraulic, piezoelectric, motorized, and mechanical designs are commercially available, each offering unique benefits. Motorized manipulators, such as Sutter model MP-225 (Novato, CA), combine precise movement with large travel, and are very versatile and easy to use (Fig. 2B). These types of manipulators usually have different settings, which allow for both fine and coarse movement in the x, y, and z planes. Piezoelectric manipulators, such as the Burleigh PCS 5000 (McHenry, IL), allow only 150 μm of piezoelectric travel, and as such are useful only in combination with a coarse positioning manipulator. Piezoelectric systems provide excellent stability and are the instruments of choice for excising patches for single-channel recordings. Stable, low-cost mechanical manipulators (such as Soma Instruments series 421, Irvine, CA) can be assembled from single-axis translation stages. The arrangement shown in Fig. 2A includes three 421 stages of which the x and y axes are controlled manually by micrometer screws, whereas the z axis for electrode placement on the cell uses an 860 DC motor controlled by a hand-held battery-operated manipulator (model 861). Their modular design, combined with

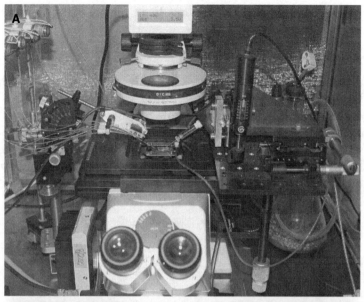

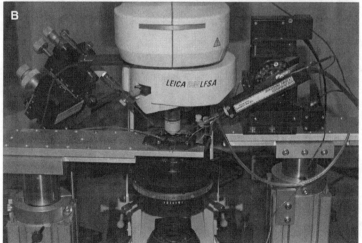

Fig. 2. Whole-cell cultured patch clamp setup (**A**) and slice recording setup (**B**). Photomicrographs of a typical recording setup based on a Zeiss Axiovert inverted microscope with a movable stage. Patch-clamp head stage with electrode holder mounted on a swivel clamp and attached to a three-axis manipulator constructed from three series 420 microtransla- tion stages (Newport Corporation, Irvine, CA). (**B**) Leica DMLFSA (model type) upright microscope with a fixed stage and water immersion lens for slice recording. Patch-clamp head stage mounted to a motorized manipu- lator (Sutter model MP-225).

micrometer screws and DC motors (e.g., model 860A) make them extremely versatile (Fig. 2A).

### 3.4. Experimental Procedure

During electrode placement, electrode resistance is monitored continuously by applying a small voltage pulse (1–5 mV, 2–10 ms) to the electrode (Fig. 3A). Once contact is made with the cell, electrode resistance spontaneously increases by 10% to 50%. Application of gentle suction to the electrode by mouth or a small syringe quickly results in the formation of a gigaseal (Fig. 3B). At this point, seal quality can be improved by applying a negative holding poten-

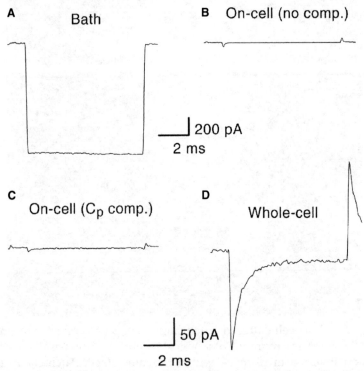

Fig. 3. Oscilloscope traces before and during establishment of whole-cell recording. (**A**) Electrode in bath ($V = 0$ mV). (**B**) On cell after formation of gigaseal ($V = 0$ mV). (**C**) As in **B**, after $C_p$ compensation (comp.). (**D**) After rupturing patch, whole-cell configuration but prior to cell capacitance and series resistance compensation ($V = -80$ mV).

tial to the pipette. In this cell-attached configuration, pipette capacitance transients ($C_p$) are reduced using the fast compensation adjustment at the amplifier (Fig. 3C). Compensation of pipette capacitance is essential for subsequent series resistance compensation. Should compensation be incomplete, coating of future electrodes with Sylgard (Dow Corning, Midland, MI) or lowering the bath perfusion level is recommended to reduce the residual transients and improve $C_p$ compensation.

Following pipette capacitance cancellation, a brief pulse of suction will rupture the membrane patch under the electrode, providing low-resistance access to the cell. This also results in a large-capacity transient arising from the added membrane capacitance (Fig. 3D). Immediately after rupturing the membrane, a reading of the cell's potential should be obtained (at $I = 0$), as this access potential is as close to the actual resting potential reading that can be obtained. Within minutes of establishing a whole-cell configuration, the pipette contents will equilibrate with the cell's cytoplasm and will impose an artificial ionic potential across the membrane. Next, by adjusting the capacitance and series resistance ($R_s$) compensation and gradually increasing the percent of compensation, effective $R_s$ compensation should be possible under most circumstances. Ideally, access resistance should be <10 M$\Omega$ prior to activating $R_s$ compensation. Under these conditions 80% compensation results in a <2 M$\Omega$ residual uncompensated series resistance. Series resistance and capacitance compensation result in a change of the step waveform applied without actually changing access resistance or the cell capacitance per se (see below). The procedure for establishing whole-cell configuration and all necessary compensations are nicely illustrated in the manuals from most amplifiers.

Due to the intracellular perfusion of cytoplasm with pipette solution, it is advisable to wait several minutes prior to obtaining recordings to assure that this dialysis has reached a steady state. Diffusion rates depend on the molecular weight and the charge of the diffusing particle, such that longer diffusion times are expected for relatively large, uncharged molecules as compared to small ions. For substances of molecular weight 23,000 to 156,000, diffusion rates have been determined in adrenal chromatin cells, and these rates suggest that complete dialysis of these small cells occurs on the order of tens to hundreds of seconds (Pusch and Neher, 1988). If dialysis of the cell is incompatible with the experimental design, for example, when ionic currents are studied, which are under control of second messengers, the perforated-patch method should be used instead.

## 4. Data Evaluation and Analysis

### 4.1. Data Filtering/Conditioning, Acquisition, and Storage

#### 4.1.1. Filtering

Data are hardly ever acquired and stored without further modifications. The analog output of the amplifier is typically amplified to make effective use of the dynamic range of the acquisition device. In the case of an A/D converter, this typically translates to an amplification range of −10 to 10 V. At the same time, signals are filtered, often using the built-in signal filter. Filtering of data is both essential and inevitable. Due to the RC components of the cell membrane–series resistance combination, the cell and electrode are essentially a single-stage RC filter. This is important to bear in mind, as it can significantly affect the true time-resolution of a recording. Assuming that $R_s \ll$ membrane resistance ($R_m$), uncompensated $R_s$ will filter any current flow recorded with a −3 dB cutoff frequency described by $F = 1/(2\pi * R_s * C_m)$. Assuming, for example, a cell capacitance of 20 pF and a series resistance $R_s$ of 10 MΩ (values typical of small cell recordings), currents across the membrane will be filtered with an effective $F$ of ~800 Hz.

Irrespective of the intrinsic filter properties of the analyzed cell, data filtering is essential to reduce signal components that are outside the bandwidth of interest. Filtering condenses the time domain of the signal to the domain of interest. The fastest signals recorded under whole-cell conditions are on the order of 200 to 500 μs (2–5 kHz). Note that this is of the same order of magnitude as the membrane-$R_s$ filter time-constant above. To eliminate high-frequency noise, a low-pass filter is used. An ideal filter has a steep roll-off, and does not greatly distort signals. A four- or eight-pole Bessel filter has excellent characteristics for filtering whole-cell currents. The filter characteristics of a four- and eight-pole Bessel filter are demonstrated by comparing the onset response of a square pulse before and after filtering at various cutoff frequencies (Fig. 4). In these examples, an eight-pole Bessel filter clearly provides excellent signal filtering with the least distortions. However, at cutoff frequencies above 2 kHz, the four- and eight-pole filters do not differ significantly in the onset or settling time.

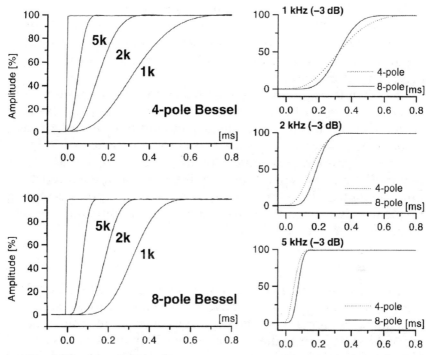

Fig. 4. Frequency response of a four- and eight-pole Bessel filter to square pulse. A 2-ms square pulse was applied to a frequency device model 902 eight-pole Bessel filter as compared to the built-in four-pole Bessel filter of the Axopatch 1D amplifier. The graphs illustrate differences in response characteristics at 3 commonly used 3-dB cutoff frequencies (1, 2, and 5 kHz).

## 4.1.2. Sampling Rate and Dynamic Range of Signal

When using an A/D converter to digitize signals, it is important to select the appropriate filter and sampling rates to accurately represent the analog signal of interest. It is inevitable that A/D conversion will reduce the infinite dynamic range of the analog signal to a well-defined step-like range of the digital signal. Commonly used 12-bit converters (for example, the Axon Digidata 1200, Molecular Devices, Sunnyvale, CA) divide the amplitude range into 4,096 discrete steps, which at a −10 to 10 V signal range will yield steps of 4.88 mV. In theory, signals can be sampled at the highest possible rate supported by the A/D converter. However, practical limitations exist. Most affordable A/D boards sample at

100 to 330 KHz on a single channel, thus allowing sampling in 3- to
10-µs intervals. Depending on the duration of the signal, sustained
sampling at 10 µs generates very large amounts of data, requiring
significant disk space. A vast majority of these data do not contain
necessary information.

The Nyquist sampling theorem states that the minimum sampling rate (Nyquist frequency) required to accurately represent an
analog waveform is twice the signal bandwidth. As a consequence,
if the analog filter is set at a −3 dB cutoff frequency of 3 kHz, a
minimum sampling frequency of 6 kHz or 167-µs intervals is
required. While these minimum requirements allow the reconstruction of data with little error under most circumstances, a sampling
frequency five times the −3 dB frequency is commonly recommended for actual recordings.

### 4.1.3. Aliasing

The Nyquist sampling theorem only applies when sampling
data digitally at frequencies between 0 and the Nyquist frequency.
If frequencies higher than the Nyquist frequency are sampled, they
are "folded back" into the low-frequency domain, a process called
aliasing. Essentially these high-frequency signals affect and distort
those signals within the appropriate frequency domain. Aliasing
can be prevented if signals above the Nyquist frequency are cut off
by a low-pass filter. A proper matching of filter frequency and
sampling rate is thus important to accurately reproduce analog
waveforms. In practical terms this requires that the cutoff frequency
of a low-pass filter be set to no higher than half the sampling
frequency. In the example above, the low-pass filter was set at
3 kHz.

### 4.1.4. Data Storage

On-line digitization has the advantage that data can be directly
stored in computer memory or on a hard drive. This mode of data
storage is preferred, as it gives convenient and fast access to the
data for future evaluation. Using a 12-bit A/D converter, each data
sample uses 2 bytes of information. Thus a 2,048 sample trace
requires about 4 kbytes of memory or disk space. A continuous
sampling of neuronal discharge at a frequency of 50 kHz generates
100 kbytes of data every second. A 5-minute recording would thus
require 300 ∗ 100 kbytes or 30 Mbytes of disk space. It thus becomes
apparent that practical limitations exist for on-line digitization of

data. For prolonged recordings at high frequencies, a VCR recorder (such as the Neurocorder digitizing unit, Pasadena, CA) may be used as an interim storage for data. Segments can subsequently be played back to the A/D converter for data analysis. However, this approach is less popular now that disk space has become so inexpensive. Recordable CD-R disks are a practical and inexpensive way to archive data.

## 4.2. Leak/Subtraction

Currents across a cell membrane consist of two components: ionic current flowing through ion channels of interest, and capacitive current that charges the membrane. Capacitive current contains useful information pertaining to the cell size, as an approximation of cell size (or more precisely, membrane area of the recorded cell) can be derived from the capacitive current. However, in the study of ionic currents, capacitive currents are of relatively little interest. As ideally capacitive currents are linear and not voltage dependent, they can be subtracted from the signal of interest through a process called leak subtraction. This subtraction can be done either on-line or off-line.

Two protocols for leak subtraction are typically used: (1) If the nature of the experiments permits, currents are recorded sequentially in the absence and presence of specific ion channel blockers [e.g., tetraethylammonium (TEA), tetrodotoxin (TTX), and 4-amino pyridine (4-AP)]. As specific blockers eliminate ionic current but should not alter the capacitive or leakage current, subtraction of the two traces or set of traces should result in the removal of capacitive and leakage currents. (2) If the first protocol cannot be used, $P/N$ leak subtraction as first proposed by Bezanilla and Armstrong (1977) can be done. In this subtraction scheme, each test voltage step is preceded by a series of $N$ (typically 4) leak voltage steps of $1/N$ (1/4 or −1/4, dependent on polarity) amplitude of the test pulse activated from a potential at which no voltage-activated currents are activated. In a $P/4$ protocol, these $4*1/4$ amplitude traces are summed together (Fig. 5B) and are subtracted from the actual current trace of interest (Fig. 5A) and isolate the ionic current of interest (Fig. 5C).

The example demonstrated in Fig. 5 actually used a $P/-4$ protocol, in which four hyperpolarizing pulses of −1/4 amplitude were summed and added to the current trace of interest. It is important to obtain the leakage current at potentials at which no voltage-activated currents occur. Most often this can be achieved by stepping

to potentials negative of the resting potential (as illustrated).
However, some cells express inwardly rectifying (or anomalous
rectifying) currents that are active at the resting potential and negative thereof. Under these circumstances leak currents must be
recorded at potentials at which the *I-V* curve is linear and no
voltage-dependent currents are activated. Note that subtraction of
capacitive and leakage currents is purely cosmetic. It does not actually improve the signal recorded. However, it may reveal small
current components that would otherwise be difficult to identify.

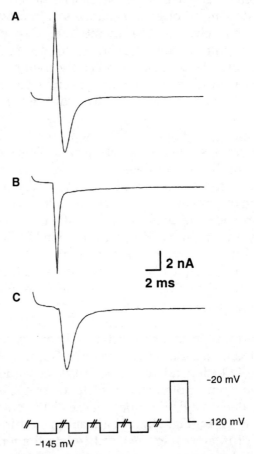

Fig. 5. Capacitive and leakage current subtraction using *P/–4* method.
(**A**) Whole-cell current recorded in response to voltage step from –120 mV
to –20 mV. (**B**) Summed response of 4*–1/4 amplitude voltage steps as
indicated at bottom of **C**. (**C**) Added response of **A** and **B** eliminating
capacitive current.

If capacitive currents are not of interest, it is recommended that $P/4$ leak subtraction be performed on-line by the data acquisition program since it significantly reduces the amount of data stored.

## 4.3. Determination of Cell Capacitance

Biological membranes are lipid bilayers in which membrane proteins (e.g., ion channels and transporters) are contained. The specific capacitance of biological membranes seems to be fairly constant. It is relatively independent of cell type, and a value of $1 \mu F/cm^2$ is typical. The capacitance cancellation circuits of patch-clamp amplifiers are normally calibrated in picofarads (pF), and allow direct determination of the cells' capacitance by adjusting and minimizing the capacity transients in response to a voltage step as discussed previously. The dial reading provides at least a rough determination of membrane capacitance on the basis of which estimates can be made as to the membrane area, as $1 pF$ capacitance represents $100 \mu m^2$ of membrane.

Capacitance can also be derived from the capacity transient at any time during the recording. In response to a voltage step, capacitance is proportional to the integral of the charging transient; thus, it can be derived by determining the area under the transient of a current's trace. Neher and Marty (1982) have developed a very sensitive approach to measuring changes in membrane capacitance using a phase-lock amplifier, which measures currents in and out of phase with a sinusoidal voltage change. This approach can resolve capacitance changes of 10 fempto seimens (fS) and has been used to resolve the fusion of synaptic vesicles by a step increase in capacitance.

## 4.4. Dissecting Current Components

Whole-cell recordings integrate the response of a large number of potentially heterogeneous ion channels. Separation of these ionic current components is a critical step during or following current recordings. Four methods of current isolation are commonly used to isolate voltage-activated ionic currents:

### 4.4.1. Kinetic

Certain types of ion channels activate and inactivate much faster than others. For example, $Na^+$ currents typically activate within 200 to $300 \mu s$ and inactivate completely within 2 to $5 ms$. In contrast, $K^+$

currents may take several milliseconds to activate, and often inactivate slowly, if at all. Simple current isolation can thus be accomplished by studying whole-cell currents at different time points following stimulation, for example, determining $Na^+$ current amplitudes at 300 to 500 µs, and determining $K^+$ current amplitudes after tens or hundreds of milliseconds.

### 4.4.2. Current Subtraction via Stimulus Protocols

Voltage-dependence of the steady-state activation and inactivation of currents often allows selective activation of subpopulations of ion channels. For example, low- or high-threshold $Ca^{2+}$ currents can be activated separately by voltage steps originating from different holding potentials. Similarly, depolarizing voltage steps applied from very negative holding potentials (e.g., −110 mV) can activate both transient A-type ($K_A$) and delayed-rectifying ($K_d$) $K^+$ currents (Fig. 6A). Voltage steps applied from a more positive holding potential (e.g., −50 mV) completely inactivate all $K_A$ channels while not affecting $K_d$ activity (Fig. 6B), such that subtraction of currents recorded with these two protocols effectively isolates $K_A$ currents (Fig. 6C).

### 4.4.3. Isolation Solutions (Ion Dependence)

It is common practice to specifically design the composition of ionic solutions to favor movements of desired ions. As mentioned previously, dialysis of cytoplasm with patch pipette contents occurs rapidly after whole-cell configuration is achieved, thus allowing for manipulation of internal ionic concentrations. This access can be used to block most $K^+$ channel activity by replacing pipette KCl with impermeant $Cs^+$ or N-methyl-D-glucamine (NMDG), thus allowing isolation of $Na^+$ currents. In a similar fashion, acetate, gluconate, or isothionate can each be substituted for $Cl^-$ ions, and tetramethyl-ammonium chloride (TMA-Cl) can be substituted for $Na^+$ ions. It has even been reported that the contribution of one ion channel population can be determined by replacement of all but the desired ion with glucose or sucrose.

### 4.4.4. Current Isolation via Pharmacology

Numerous natural toxins and synthetic pharmacological agents can be used to reduce or eliminate specific voltage-activated ion channel activity. Table 1 lists some of the more commonly used

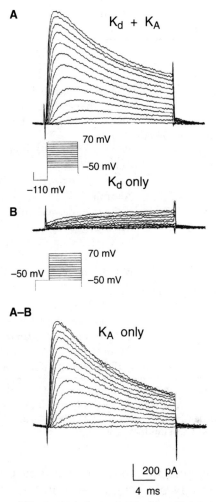

Fig. 6. Isolation of $K_A$ current by subtraction. Current recordings from a spinal cord astrocyte expressing both transient ($K_A$) and delayed rectifier ($K_d$) like K$^+$ currents. (**A**) Currents activated from holding potential of −110 mV. (**B**) Same cell and same voltage step protocol, but steps originated from holding potential of −50 mV. Subtraction of **A–B** yielded $K_A$ currents in isolation.

agents, many of which are effective against one particular type of ion channel, such as tetrodotoxin and dendrotoxin, which target Na$^+$ and K$^+$ channels, respectively. The use of these compounds, either alone or in combination, allows the isolation of specific current(s).

Table 1
Commonly used ion channel blockers

| Channel type | Compound/reagent |
|---|---|
| *K⁺ channels* | |
| Delayed rectifier ($K_d$) | TEA, $Ba^{2+}$, capsaicin, 4-AP, margatoxin, 9-AC, charybdotoxin |
| Inward rectifier ($K_{ir}$) Gaboon viper venom, LY-97241 | TEA, $Cs^+$, $Rb^+$, $Na^+$, $Ba^{2+}$, tertiapin, |
| Fast inactivating ($K_A$) | TEA, 4-AP, dendrotoxin, quinidine |
| K(Ca) | |
| BK | Charybdotoxin, iberiotoxin, paxilline, TEA |
| SK | Apamin |
| IK | Clotrimazole, cetiedil |
| K(ATP) | TEA, $Cs^+$, $Ba^{2+}$ |
| *Na⁺ channels* | TTX, STX, CNQX, agatoxin, scorpion toxin |
| *Ca²⁺ channels* | |
| L-type | Nifedipine, verapamil, BayK8644, |
| $Cd^{2+}$, $La^{3+}$, nimodipine, ω-agatoxin IIIA | |
| T-type | $Ni^{2+}$, $La^{3+}$, mibefradil |
| N-type | ω-conotoxin, $La^{3+}$ |
| P-type | FTX funnel spider toxin, ω-agatoxin IVA |
| *Cl⁻/anion channels* | Chlorotoxin, NPPB, DIDS, niflumic acid SITS, 9-AC, mibefradil, $Cd^{2+}$, $Zn^{2+}$ |

TEA, tetraethylammonium; 4-AP, 4-aminopyridine; 9-AC, 9-aminocamptothecin; TTX, tetrodotoxin; STX, saxitoxin; CNQX, 6-cyano-7-nitroquinoxaline-2,3-dione; NPPB, 5-nitro-2-(3-phenylpropylamino) benzoic acid; DIDS, 4,4'-diisothiocyanatostillbene-2, 2'-disulfonic acid; SITS, 4-acetamido-4'-isothiocyanatostillbene-2,2'-disulfonic acid.

Experimentally, currents are best identified pharmacologically by recording a family of current traces in both the absence and presence of the drug. This is illustrated in Fig. 7, in which the block of spinal cord astrocyte $K^+$ currents by 4-AP is shown. In the control and treated current traces (Fig. 7A,B), the inward sodium current is unaltered. By subtracting the 4-AP–treated current traces from those of control, one can isolate the current component that is 4-AP–sensitive. As discussed previously, a side benefit to such current subtraction is the elimination of capacitive and leakage currents.

Neurotransmitter-activated currents are perhaps easier to iden-tify and isolate than their voltage-dependent counterparts, as these currents are induced in a time-dependent manner based on the application of exogenous ligands. If need be, ligand-gated current

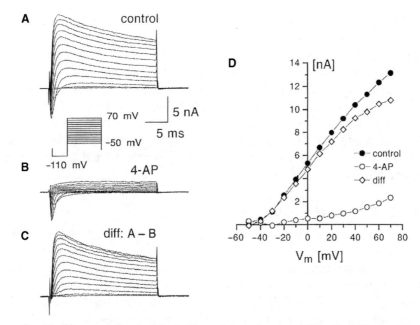

Fig. 7. Pharmacological isolation of 4-amino pyridine (4-AP)-sensitive K+ current. (**A**) Family of current traces recorded from a spinal cord astrocyte. (**B**) Recording in the same cell using the same stimulus protocol 2 minutes after application of 2 mM 4-AP. (**C**) 4-AP–sensitive current isolated by subtraction of **A–B**. (**D**) Current amplitudes determined 8 ms after onset of voltage steps plotted as a function of applied potential for current traces in A–C.

can be isolated from background noise by subtracting recorded currents in the absence and presence of a given ligand.

### 4.5. I-V Curves

Current-voltage (*I-V*) relationships are perhaps the most effective way to summarize the behavior of voltage- and ligand-activated ion channels. A number of important and useful parameters that cannot be easily accessed from the raw data can be readily derived from these plots, including reversal potential, ionic dependence/selectivity, voltage-dependence (rectification), activation threshold, slope and cord conductance, as well as overall quality of voltage-clamp. The *I-V* curves can be determined in various ways, examples of which are discussed below.

The factors that determine current flow through an open channel are conductance and driving force. While conductance is proportional to the number of open channels, driving force is defined as the difference between actual voltage and the equilibrium potential for the ion(s) permeating the channel, also known as reversal potential ($V_{rev}$). Thus current can be described as $I = G (V_m - V_{rev})$.

A plot of $I$ versus $V_m$ is commonly used to derive $G$ (slope) or $V_{rev}$ (x-intercept). The current evoked at a given $V_m$ can be measured using a variety of protocols, as discussed in the following subsections.

### 4.5.1. Peak and Steady-State I-V Curves

Peak currents are measured as the largest current activated by the applied voltage (Fig. 7C). Thus, if a current has a transient peak current amplitude, such as the $K_A$ current in Fig. 7A, analysis software can easily determine maximal current and plot these values against voltage (Fig. 7C). If the current to be studied does not have this defined peak, steady-state values may be used, typically as recorded at the end of a voltage step. This type of analysis was applied to the 4-AP–treated current traces of Fig. 7B, with the resulting amplitudes similarly plotted in Fig. 7C.

### 4.5.2. Continuous Quasi–Steady-State I-V Curve

A convenient way to establish $I$-$V$ curves is by alteration of the membrane potential in a continuous way through a voltage ramp. As this ramp can be applied very slowly, it permits the acquisition of a quasi–steady-state $I$-$V$ relationship. This procedure has proven very useful in determining $I$-$V$ relationships for transmitter responses. Note, however, that this approach assumes that currents do not inactivate during the length of the voltage ramp. An example of a voltage ramp used to determine the reversal potential of γ-aminobutyric acid (GABA)-induced currents is illustrated in Fig. 8. Here, a 200-mV, 400-ms voltage ramp (as indicated in the inset to Fig. 8A) was applied twice—once prior to application of GABA (control) and once during transmitter application (GABA). By subtracting the two responses, one obtains the transmitter-induced current in isolation. As the time axis represents a constant change in the voltage applied, it can be instantaneously replotted as an $I$-$V$ relationship of the transmitter-induced current (Fig. 8B). Using this approach, $I$-$V$ curves of transmitter responses can be easily created

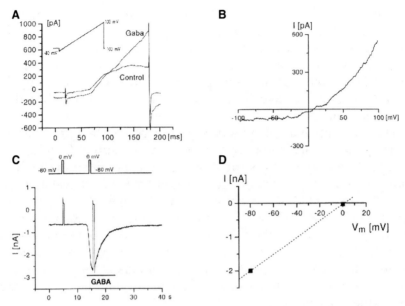

Fig. 8. Stimulus protocols to define reversal potential of γ-aminobu-tyric acid (GABA)-induced currents. (**A**) A voltage ramp (inset) was applied prior to (Control) and at the peak of GABA-induced currents. The difference of these two current ramps represent the GABA-induced current in isolation. (**B**) This current was plotted as a function of applied potential to yield the *I-V* curve of GABA-induced currents. (**C**) A single 50-ms, 80-mV voltage step was applied once prior to and once in the peak of the GABA response. (**D**) The differences of current amplitudes were plotted as a function of applied potential to yield two points of an *I-V* curve for GABA-induced currents. The dotted line extrapolates the current reversal potential.

throughout an experiment. As mentioned above, alteration of ionic composition shifts the reversal potential and thus allows deter-mination of ion specificity and relative permeabilities.

For those situations where currents inactivate rapidly, some indi-cation as to the reversal potential can be obtained using a single-step protocol from which the reversal potential can be extrapolated. An example of this approach is illustrated in Fig. 8C. Here, a larger concentration of GABA was bath applied, which resulted in char-acteristic receptor desensitization. The cell was maintained at −80 mV and a single 80-mV step (50 ms) was applied prior to and at the peak of the transmitter response. Current levels at the two potentials, −80 mV and 0 mV, were then subtracted and plotted as a function of the applied potential (Fig. 8D). The line through the two data points clearly does not reflect the true *I-V* relationship of the response; however, it facilitates determining with fair accuracy the reversal potential of the response. To obtain a more complete *I-V* relationship, the two voltage steps above can be substituted by trains of voltage steps.

The *I-V* plots may also be used to determine the quality of the voltage clamp achieved. If the reversal potential (equilibrium potential) is known, as is the case when isolation solutions limit ionic movements to only one ion, or when it is clearly established, as in the case of GABA$_A$ receptors, currents are mediated predominantly by one ion. Under the imposed ionic conditions, currents must reverse close to the theoretical equilibrium potential for the permeable ion. In poorly voltage-clamped cells, reversal potential is either not achieved or achieved at potentials more positive than the equilibrium potential (see Fig. 11D, below, and next subsection for further discussion).

### 4.5.3. Conductance-Voltage Curves

If the reversal potential is known, a conductance voltage (*G-V*) curve can be readily calculated by dividing current at each potential by the driving force (*V-V$_{rev}$*). These curves are typically sigmoidal and can be fitted by a multistage Boltzmann function. Provided that current through a single channel is linear, conductance is proportional to the number of open channels. Thus, the *G-V* curve resembles an activation curve.

### 4.5.4. Steady-State Inactivation Curves

The voltage dependence of current inactivation allows one to determine the fraction of channels available for activation as a function of voltage. Currents are activated by a voltage step to potentials at which the largest conductance is achieved. This voltage step is preceded by variable prepulse potentials at which the membrane is maintained for 200 to 1,000 ms (Fig. 9A). Current amplitudes at each potential are normalized to the largest current recorded and plotted as a function of prepulse potential (Fig. 9B). These curves can be fitted to a multistage Boltzmann function. In the experiment illustrated, a potential of ~−80 mV yielded about 50% of Na$_+$ channels available for activation (dashed line).

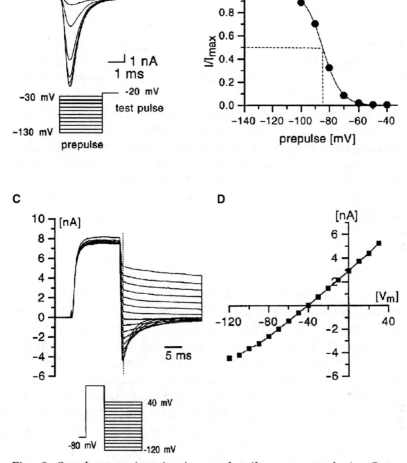

Fig. 9. Steady-state inactivation and tail-current analysis. Current recordings from two different spinal cord astrocytes. (**A**) To study steady-state current inactivation, inward Na$^+$ currents were activated by stepping the membrane to −20 mV for 8 ms. Voltage step was preceded by varying prepulse potential ranging from −130 mV to −30 mV. (**B**) Peak current amplitudes in A were normalized to the largest current amplitude and plotted as a function of prepulse potential. The data were fitted to a two-stage Boltzmann equation (solid line) to yield steady-state inactivation curve. (**C**) Tail current analysis. Outward currents were activated by a 15-ms voltage step from −80 mV to 80 mV. This step was followed by a second step to varying test potentials ranging from −30 mV to −120 mV, resulting in "tail" currents. (**D**) Current amplitude of tail currents was measured 500 µs after stepping potential to second step potential (dotted line) and plotted as a function of applied potential to yield tail current *I-V* curve.

## 4.5.5. Deactivation I-V Curves

As voltage-dependent currents are often also time-dependent, and may activate and/or inactivate in a time-dependent manner, *I-V* curves as described above cannot distinguish between time and voltage dependence. An elegant way to determine conductance independent of its time dependence is by analyzing the deactivation process (Fig. 9C,D). Upon termination of the voltage step, deactivation, which is the reversal of activation, results in tail currents (Fig. 9C, dotted line). Immediately after terminating the voltage step, for a brief period of time current continues to flow through open channels and only subsequently terminates with time-dependent channel closure (relaxation of tail currents). Thus, current amplitudes measured at the peak of these tails resemble time-independent current amplitudes, and these typically yield linear *I-V* curves (Fig. 9D).

## 4.5.6. Fitting of Time Constants

Current activation and inactivation kinetics have been well described by mathematical models. If the model is known, the data can be fitted to the model and will allow the derivation of important kinetic properties, such as time constants for activation ($\tau_m$) and inactivation ($\tau_h$), respectively (Hodgkin and Huxley, 1952). Often the models used are an oversimplification of the true biology. This is particularly true for fitting of whole-cell data for two reasons: (1) whole-cell current may be mediated by the combined activation of numerous channels types; and (2) even if one can be reasonably sure that currents are mediated by a single-channel population, the biophysics of this channel type, for example, the number of open and closed states, may be unknown.

Fitting routines are an integral part of numerous data acquisition or data analysis packages. Most commonly these use either a simplex or a Levenberg-Marquard algorithm to minimize the least squared error. Both algorithms are capable of producing excellent and fast fitting curves to small data sets. Examples of Levenberg-Marquard fits are demonstrated for two examples in Fig. 10. These were obtained using the script interpreter Origin (Orgin Lab, Northhampton, MA) by fitting to user-defined functions. The examples illustrated fitted multiple parameters simultaneously. Thus, in Fig. 10A, transient $K^+$ current activation and inactivation was fitted to a model of the form $f(t) = A_O + A_1 * (1 - \exp(-(t - t_O)/\tau_m))^4 *$

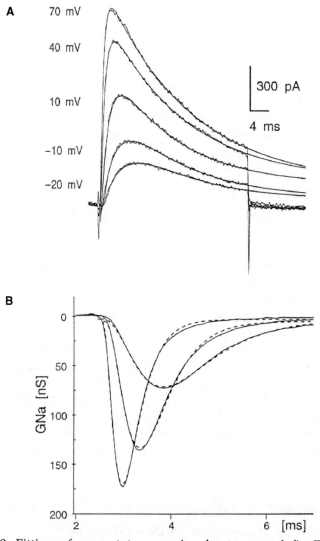

A

70 mV

40 mV

10 mV

-10 mV

-20 mV

300 pA

4 ms

B

GNa [nS]

0

50

100

150

200

2          4          6   [ms]

Fig. 10. Fitting of current traces using least squared fit. Transient outward $K^+$ (**A**) and inward $Na^+$ (**B**) currents were fitted to established kinetic models using a Levenberg-Marquard algorithm to minimize the least squared error. Fitted curves were superimposed on data. The models used were (**A**) current according to Connor and Stevens (1971) (see equation in text) and (**B**) the Hodgkin-Huxley (1952) equation to described $Na^+$ current kinetics (see equation in text). GNa, sodium conductance; [n5], nano seimens.

$(\exp(-(t - t_O)/\tau_h))$ as used by Connor and Stevens (1971) to describe kinetics of A currents. In Fig. 10B, $Na^+$ current activation and inactivation was fitted to the Hodgkin-Huxley equation: $f(t) = A_O + A_1*(1 - \exp(-(t - t_O)/\tau_m))^4*(\exp(-(t - t_O)/\tau_h))$ ($A_O$ and $A_1$ = amplitude factors; $t_O$ = time offset factors; $\tau_m$ and $\tau_h$ = activation and inactivation curves). It is important to keep in mind that data fits are not sufficient to formulate a model, but rather assume that the model is known and used to derive variables such as $\tau_m$ and $\tau_h$ contained in the model.

### 4.5.7. Data Presentation

Digitized data can be easily exported in American Standard Code for Information Interchange (ASCII) format and read into numerous powerful spreadsheet programs (Excel, Lotus, Quattro, SigmaPlot, Origin, Plotlt). There are now programs available that enable access to data without prior conversion. These programs allow electrophysiology data files to be imported for plotting, data analysis, and graphing. The authors' laboratory currently uses Origin, which is based on a scientific scripting language, LabTalk, for which a script interpreter is part of the program. This allows simple programming of frequently used commands or sequences of commands into macros, which can be assigned to visual buttons on the screen. This allows the computer-literate (nonprogrammer) to design custom data analysis and graphing schemes. As most of the graphing programs are Windows-based, merging of graphs into word processors or other drawing programs is as simple as cutting and pasting.

## 5. Limitations, Pitfalls, and Errors

### 5.1. Series Resistance and its Consequences

As mentioned previously, the major limitation of the whole-cell patch-clamp recording technique lies in its design as a continuous single-electrode voltage clamp. The continuous use of one electrode for current passage as well as voltage sensor makes true membrane voltage determination impossible. The technique assumes that pipette voltage equals membrane voltage, as voltage commands are imposed on the pipette, not on the cell. However, recording pipette and access resistance (due to potential clogging at the electrode tip) are in series with the current recording and the voltage command. This series resistor in conjunction with the membrane resistance acts as voltage divider to all imposed voltages. Consequently, only in cases where the membrane resistance greatly exceeds the series resistance is an adequate voltage clamp (point clamp) assured. Under experimental circumstances, series resistance accounts for at least $5\,M\Omega$, and more typically 10 to $15\,M\Omega$. To keep the voltage error below 1%, membrane resistance has to be two orders larger than the series resistance, in our example $\sim1\,G\Omega$. This is hardly the case, and certainly does not hold true during activation of ionic currents! To ensure the best possible recording conditions, the following steps are absolutely necessary:

1. Electrode resistance has to be minimized as much as possible, depending on the size of the cells to be studied. In our experience, and dependent on the solutions used, cells from 8 to $40\,\mu m$ in size can be successfully patched with electrodes in the 1.5- to 3-$M\Omega$ range. However, once the whole-cell recording configuration has been achieved, it is important to frequently check for adequate compensation. If all compensation mechanisms are turned off, one may see a dramatic decrease in the magnitude of initial capacitance transients observed. The most likely explanation for such a change is the clogging of the electrode tip with cell membrane, which directly interferes with clear access to the cell's interior. This phenomenon of membrane healing around the electrode tip can be prevented by buffering $[Ca^{2+}]_i$ using high concentrations of ethylene glycol tetracidic acid (EGTA) or 1,2-bis(o-aminophenoxy) ethane-N,N,N′,N′-tetra acidic acid (BAPTA). In an acute situation, slight positive or negative pressure can reverse electrode clogging. In our experience, it is possible to achieve access resistances of $5\,M\Omega$ prior to series resistance compensation.
2. Series resistance ($R_s$) needs to be compensated for. Most patch-clamp amplifiers provide a positive feedback series resistance compensation circuit, in which a signal proportional to the measured current is added to the command potential. $R_s$ is determined by adjusting $R_s$ and $C_p$ controls to square out a command voltage. Subsequently, $R_s$ compensation is activated. Although theoretically near 100% compensation is possible, real experiments hardly allow gain setting of more than 50% to 80%. $R_s$ compensation scales the command input to account for the voltage loss across $R_s$. $R_s$ compensation is very sensitive to changes in the fast (pipette) capacitance compensation. It is

essential to adjust this compensation properly to achieve maximum percent compensation settings. Note that continuous bath perfusion may result in some oscillations of the bath fluid level. As a consequence, this would also change the effective capacitance of the pipette and would make the fast capacitance compensation unstable and thereby $R_s$ compensation prone to ringing. Effective compensation under those circumstances thus requires a stable bath perfusion level. The problem can be reduced by the use of heavily Sylgarded electrodes, which have a much reduced capacitance.

In an ideal case, with an $R_s$ of 10 M$\Omega$, a voltage step of 100 mV results in a current flow of 1 nA, and an apparent input resistance $R_{cell} + R_s$ of 100 M$\Omega$. A 1-nA current flow across $R_s$ generates a 10-mV voltage drop across $R_s$ and thus a 10% error. Using $R_s$ compensation, assuming an 80% compensation, the error is reduced to 2 mV or 2% of command voltage, a tolerable error. The errors for uncompensated $R_s$ become larger as current amplitude increases. For example, suppose activation of voltage-dependent channels gives rise to a 10 nA current, with an uncompensated $R_s$ of 10 M$\Omega$. This results in a 100-mV steady-state voltage error. Even with compensation dialed in at 80%, this still results in a 20 mV error!

3. Cells with low input resistances are almost impossible to record from. Should the membrane impedance be <100 M$\Omega$ it is advisable to increase it by inclusion of ion channel blockers to block conductances that are not of immediate interest. Thus, K$^+$ channel blockers could be included and Cl$^-$ replaced by acetate to allow resolution of small Na$^+$ currents.

Uncompensated $R_s$ has two additional detrimental effects on current recordings. It affects the time response to a voltage change, and it results in increased signal noise. The transmembrane voltage resulting from a step change in voltage is described by $V_m = V_c [1 - \exp(-\tau/(R_s * C_m))]$ with an effective time constant of $\tau = RC$ (if $R_s \ll R_m$). Assuming an uncompensated $R_s$ of 10 M$\Omega$ and $C_m$ of 100 pF, charging of the membrane will be slowed with an effective time constant ($\tau$) of 1 ms. Unfortunately, $R_s$ will also filter any current flow recorded with this arrangement. In the absence of compensation, this results in a single pole $RC$ filter with a $-3$ dB frequency described by $F = 1/(2\pi * R_s * C_m)$, resulting in a cutoff frequency ($F$) of 159 Hz for the above example.

## 5.2. Voltage-Clamp Errors

The voltage clamp is prone to error, as it makes numerous assumptions that may not be valid under the given experimental conditions. It assumes that the cell is equipotential and that the voltage measured at any one point across the membrane is the true membrane voltage. Analogously, current injection, which imposes change to the membrane voltage, is thought to be uniformly realized in all parts of the membrane, including distant processes. This, however, is not the case. As a result, two sources of error exist, namely, voltage (point) and space-clamp errors. As illustrated below, whole-cell patch-clamp recordings are even more susceptible to error than classical two-electrode recordings, and as such, the experimenter needs to be constantly aware of possible sources of this error.

### 5.2.1. Space Clamp

Space-clamp limitations are intrinsic to voltage-clamp and do not differ in their principles between different voltage-clamp techniques. Current injected into the cell to maintain or establish a change in membrane voltage spreads radially from the injection site and decays across distance with the space (length) constant ($\lambda$). In small-diameter spherical cells, this is a minimal concern. However, in a process-bearing cell, the current signal may have been distorted by the time it reaches distant processes hundreds of micrometers away from the injection site. In the best of circumstances, the signal will be attenuated and the voltage-changes imposed will be smaller. In the worst case, distant membranes may not experience any voltage change at all.

Double-electrode voltage-clamp methods are somewhat advantageous in that they facilitate detecting space-clamp problems more readily. In this case, the voltage electrode can be inserted at a distance from the current electrode, and closer to the site of interest. Whole-cell recordings clamp the voltage of the electrode tip, and thus provide no means to establish any true recording at a site distant from the electrode. Some investigators have chosen to insert a second patch-clamp or sharp microelectrode into cells to monitor the true voltage changes observed. Space-clamp problems can only be reduced by recording from small cells with simple morphology, ideally spherical cells. Nonspherical cells can sometimes be rounded up by exposure to trypsin or treatment with dibutyrl cyclic adenosine monophosphate (dBcAMP). However, often the most interesting cells bear extensive arborized processes. A second way to ensure that injected current can travel further is to increase the cell's impedance. Thus, it is often possible to block parts of the cell's conductance (e.g., $K^+$ conductance) pharmacologically to effectively increase the length constant ($\lambda$) of the cell.

## 5.2.2. Voltage (Point) Clamp

The whole-cell recordings technique effectively voltage-clamps the electrode tip, and, by assuming that its resistance is small relative to the cell's resistance, the cell's voltage is assumed to be clamped. As described above, this is only the case if series resistance is small and well compensated for. Unlike space-clamp errors, point-clamp errors can be readily detected and often eliminated. After canceling series resistance error (at least partially, see above) the experimenter can calculate the voltage error, which now is a linear function of current flow. Point-clamp errors produce primarily two distortions to the recorded signal: (1) slowing of current kinetics and (2) apparent attenuation of true current amplitudes due to uncompensated voltage error. Errors due to slow activation can be readily identified from the current traces.

## 5.2.3. Determining Quality of Clamp from I-V Plots

The *I-V* plots are a very sensitive way to evaluate point-clamp errors. Figure 11 demonstrates $Na^+$ current recordings from three different cells of a neuronal cell line (BlO4). Current recordings were obtained under appropriate and poor voltage control, and peak current values were plotted as a function of membrane potential and superimposed in Fig. 11D. Under the imposed ionic gradients of the recordings, the theoretical equilibrium potential for $Na^+$ was ~40 mV. The current traces in Fig. 11A yield an *I-V* plot that reverses close to sodium reversal potential (ENa), indicative of proper voltage-control, whereas the traces in Fig. 11B indicate a 20-mV more positive reversal potential, and in Fig. 11C the currents do not reverse at all. Thus the *I-V* plot directly indicates the severity of the voltage errors in Fig. 11B,C. While the recording in Fig. 11B still yielded the same peak in the *I-V* relationship (at −20 mV) as the recording in Fig. 11A, the voltage step to −40 mV (arrows in

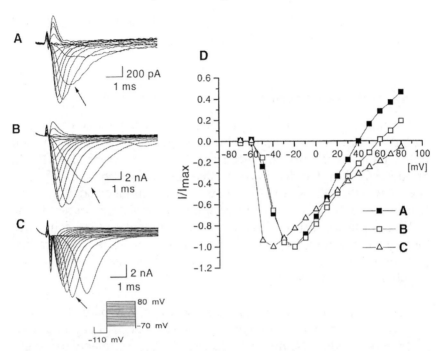

Fig. 11. *I-V* curves used to judge quality of point-clamp. (**A–C**) Families of current recordings from three different B104 cells (neuronal cell line) using the same step protocol indicated in inset. Peak Na⁺ currents at each potential were plotted in **D**. Arrows in **A–C** point to current trace in response to a –40 mV voltage step. Only the recording in **A** is under appropriate point clamp. **B** and **C** are distorted by a delay in current activation at threshold (arrows) and by a shift in the *I-V* relationship (**D**), overestimating the true current reversal potential. ENa was 40 mV.

Fig. 11A–C) was significantly delayed. In Fig. 11C, voltage control is lost at the threshold of current activation (–50 mV), and currents do not activate in a graded fashion but rather "escape" to reach near-peak amplitude with a severe delay. Similar *I-V* plots of transmitter-induced currents for which the major ion carrying the response is known (such as GABA) facilitate utilizing the reversal potential as an indicator for proper voltage control.

We have found that by minimizing $R_s$ and utilizing *I-V* curves, currents of up to 10 nA can be properly voltage-clamped over a wide voltage range. However, under all these circumstances, potential space-clamp errors remain, and most likely currents activated in remote processes are not recorded at all.

## 6. Special Applications

A number of specialized applications utilize the whole-cell patch-clamp recording technique. Three of these, namely perforated patch, patch-slice, and single-cell polymerase chain reaction (PCR), are described in detailed in other chapters of this book. One application worth mentioning is the use of patch-electrodes for dye loading. As whole-cell recordings allow low resistance access to the cell's cytoplasm, it provides a convenient means to load cells with biological markers such as fluorescence indicators, horseradish peroxidase, or biotin. Loading of cells can be used to address a number of questions:

1. As fluorescence markers may diffuse through gap junctions, dye diffusion can be used to search for the existence of gap junctions between cells. Numerous fluorescent low-molecular-weight compounds that are aldehyde-fixable can be utilized for this purpose. Lucifer yellow (LY, excitation 425 nm) has been a long-time favorite. However, Invitrogen (Carlsbad, CA), offers fluorescent hydrazide salts that are available at the most commonly used wavelengths for mercury-arc lamps and lasers. Like Lucifer yellow, these dyes are fixable. The inclusion of these fluorescent fixable dyes in the patch pipette solution allows visualization during recording, assuming a fluorescent light source is available. These dyes may also be used as a cell marker, allowing the localization of a cell from which recordings were obtained. After fixation, cell-specific antibodies can be utilized to antigenically identify the cell. Alternatively, some investigators prefer the inclusion of biocytin (~0.2–0.5%) in their patch pipette solutions.
2. Although membrane-permeable acetoxymethyl (AM) esters exist for most ratiometric fluorescent indicator dyes, their characteristics can differ depending on the method of cell loading (Almers and Neher, 1985). Dyes can be readily loaded through a patch pipette through which electrophysiological recordings can be obtained simultaneously while imaging an ion of interest ratiometrically.

# 7. Conclusion

What started out as a spinoff from single-channel patch recordings has resulted in a technique more useful than its inventors had anticipated (Sigworth, 1986). Its ease of use makes whole-cell recordings the most widely used intracellular recording technique to date; in many instances it has replaced sharp microelectrode recordings. However, whole-cell patch-clamp recordings are notoriously prone to error and may not always generate accurate recordings. It is thus important to understand the limitations of the technique. If proper care is taken, a whole-cell patch clamp allows the study of almost any small cell of interest, and has opened the field of single-cell electrophysiology. Modifications (perforated patch) and applications of the technique to more intact preparations (patch-slice) have provided invaluable insight into nervous system function.

# Acknowledgments

During preparation of this manuscript, the first author was supported by grants RO1-NS31234 and RO1-NS36692 from the National Institutes of Health.

# References

Almers, W. and Neher, E. (1985) The Ca signal from fura-2 loaded mast cells depends strongly on the method of dye-loading. *FEBS Lett.* **192**, 13–18.

Bezanilla, F. and Armstrong, C. M. (1977) Inactivation of the sodium channel: I. Sodium current experiments. *J. Gen. Physiol.* **70**, 549–566.

Connor, J. A. and Stevens, C. F. (1971) Voltage clamp studies of a transient outward membrane current in gastropod neural somata. *J. Physiol.* **213**, 21–30.

Hamill, O. P., Marty, A., Neher, E., Sakmann, B., and Sigworth, F. J. (1981) Improved patch-clamp techniques for high-resolution current recording from cells and cell-free membrane patches. *Pflügers Arch.* **391**, 85–100.

Hodgkin, A. L. and Huxley, A. F. (1952) A quantitative description of membrane current and its application to conduction and excitation in nerve. *J. Physiol. (Lond.)* **117**, 500–544.

Hodgkin, A. L., Huxley, A. F., and Katz, B. (1952) Measurement of current-voltage relations in the membrane of the giant axon of loligo. *J. Physiol. (Lond.)* **116**, 424–448.

Neher, E. (1982) Unit conductance studies in biological membranes. *Tech. Cell Physiol.* **P121**, 1–16.

Neher, E., Marty, A. (1982) Discrete changes of cell membrane capacitance observed under conditions of enhanced secretion in bovine adrenal chromaffin cells. *Proc. Natl. Acad. Sci.* **79**, 6712–6716.

Pusch, M. and Neher, E. (1988) Rates of diffusional exchange between small cells and a measuring patch pipette. *Pflügers Arch.* **411**, 204–211.

Sigworth, F. J. (1983) Electronic design of the patch clamp, in *Single-Channel Recording* (Sakmann, B. and Neher, E., eds.), Plenum Press, New York, London, pp. 3–35.

Sigworth, F. J. (1986) The patch clamp is more useful than anyone had expected. [Review]. *Fed. Proc.* **45**, 2673–2677.

## Recommended Readings

For a more in-depth description of the patch-clamp technique, the following publications are highly recommended:

Ferreira, H. G. and Marshall, M. W. (eds.) (1985) *The Biophysical Basis of Excitability.* Cambridge University Press, London.

Hille, B. (2001) Ionic Channels of Excitable Membranes, 3$^{rd}$ ed. Sinauer, Sunderland, MA.

Jones, S. (1990) Whole-cell and microelectrode voltage clamp, in *Neuromethods*, vol 14 (Boulton, A. A., Barker, G. B., and Vanderwolf, C. H., eds.), Humana Press, Clifton, NJ.

Rudy, B. and Iverson, L. E. (1992) Ion channels, in *Methods in Enzymology* (Rudy, B. and Iverson, L. E., eds.), Academic Press, San Diego (1992).

Sakmann, B. and Neher, E. (1995) *Single-Channel Recording*, 2nd ed., Plenum Press, New York.

Sherman-Gold, R. (ed.) (1993) The Axon Guide for Electrophysiology and Biophysics Laboratory Techniques, Axon Instruments, Inc., Sunnyvale, CA.

# 3

# Single-Channel Recording

## David J.A. Wyllie

## 1. Introduction

The ability to record the real-time movement of a protein is quite enthralling; in effect, this is what we achieve when we record the activity of an individual ion channel. Clearly we do not "see" the amino acids that make up the ion channel undergo the conformational changes that must occur, but we observe the consequences—the current flow through the ion channel pore. The currents that flow are very small, a few picoamperes ($10^{-12}$ A), and, in the case of the nicotinic acetylcholine receptor found at the muscle end plate, for example, they represent the net movement of about 25,000 monovalent ions per millisecond through the channel pore. These currents are recorded using an electrophysiological technique termed *patch-clamp recording*. The word *patch* is self-evident—we record ion channel activity from a small area of membrane. The word *clamp* refers to the fact that we maintain this membrane at a constant potential (for a description of the electronics that achieve the voltage clamp, see Chapter 2).

Many texts have been written on single-channel recording. Given space limitations, this chapter cannot treat the subject in depth; rather, it discusses the fundamentals of single-channel recording, the data that can be obtained from the various recording configurations, and some of the theory and concepts involved in the analysis of Markov processes, and it provides a brief overview of the implementation of analysis of single-channel data records using both empirical fitting methods and reaction scheme-based methods. For a detailed treatment of the subject of practical and theoretical considerations as applied to single-channel recording and data

From: *Neuromethods, Vol. 38: Patch-Clamp Analysis: Advanced Techniques, Second Edition*
Edited by: W. Walz @ Humana Press Inc., Totowa, NJ

analysis, the reader is referred to the six excellent texts listed at the end of the chapter (see Recommended Reading).

## 2. Single-Channel Recording

### 2.1. Resolving Single-Channel Currents Requires Low Noise

   To resolve membrane currents that are only a few picoamperes (or in some cases <1 pA) in amplitude requires that the background noise of the recording is as low as possible. In their seminal paper, Neher and Sakmann (1976) demonstrated that it was possible to observe individual ion channel activity if an electrode was pressed against a muscle fiber and the current flowing in the patch monitored. In 1981, Hamill et al. documented that further improvements in background noise levels were possible if one was able to form a "gigaseal." Such a high-resistance seal was obtained by applying suction (negative pressure) to the glass patch-pipette when it contacted the surface of the muscle. This reduced the background noise levels by a further order of magnitude (compared to that achieved previously) and consequently increased the temporal resolution of the recording. The formation of a gigaohm seal ($>10^9 \Omega$) is therefore the key to obtaining high-quality single-channel current recordings. The noise associated with a current source is determined by the resistance of that source and is described by the following equation:

$$\sigma_n = \sqrt{\frac{4kTf_B}{R}},$$

where $\sigma_n$ is the root mean square (rms) of the current noise, $k$ = Boltzmann's constant, $T$ = absolute temperature, $f_B$ = the bandwidth of the recording, and $R$ = the resistance of the source. Noise of many forms, and from a variety of sources, contributes to the final level of a patch recording; however, the main sources are the feedback resistor in the head stage, the resistance of the patch, the pipette capacitance, and the seal resistance itself. The total rms noise resulting from $n$ uncorrelated sources is given by the following relationship:

$$\sigma_{total} = \sqrt{(\sigma_1^2 + \sigma_2^2 + \ldots + \sigma_n^2)},$$

Thus, because of the nature of this relationship, the source with the highest rms noise level will contribute most significantly to the final noise levels. For example, imagine that three noise sources have rms values of 0.1, 0.2, and 0.4 pA, respectively. The final noise is 0.46 pA, and as such, noise from the third source would dominate the recording.

Clearly even the best gigaohm seal does not permit the resolution of ion channel activity if other precautions to reduce noise levels have not been taken, for example, use of a Faraday cage, good "earthing" and "shielding" of equipment, and use of antivibration air tables. As a guide the resistance between any two (metal) points in the setup should be less than $1\,\Omega$. To achieve this, earth all equipment to a common point on the electrophysiological setup; the use of a metal rod that has holes drilled in to accept 4-mm in-line plugs is particularly convenient. Take a single earthing cable from this point to the "signal ground" point on the back of the patch-clamp amplifier. Ensure that any power cables are routed well away from the cable that connects the head stage to the amplifier. Additionally, it is often necessary to construct a shield that can screen the head stage of the patch-amplifier. This can be made from either aluminium foil or fine copper mesh, its essential feature being that it should be easily removable to permit the experimenter to check that all is well with the recording chamber, perfusion system, etc., during the course of a recording. This shield should also be grounded.

To summarize, a high-resistance seal between the patch-pipette and membrane together with a low-noise current-to-voltage amplifier with the recording made in a well-grounded and shielded environment should permit the high-resolution recording of single-channel currents.

## 2.2. Patch-Clamp Amplifiers

Single-channel currents are recorded using patch-clamp amplifiers that use either a resistive feedback head stage or a capacitive feedback head stage. Amplifiers such as the Axopatch-1D (Molecular Devices, Union City, CA) (together with its predecessors), which unfortunately is no longer manufactured, and HEKA's EPC7 and EPC8 (Lambrecht/Pfalz, Germany), each use resistive feedback. The value of the resistor in the head stage is switched to $50\,G\Omega$ (from $500\,M\Omega$) when the experimenter selects the $\beta = 100$ setting on the Axopatch, or a gain of $50\,mV/pA$, or above, on the HEKA

amplifiers. The theory of the operation of the amplifier is similar
to that described for whole-cell recording in Chapter 2, but now
the maximum current that can be recorded is 240 pA (12 V ÷ 50 GΩ).
The increase in the resistance of the feedback resistor reduces the
noise level of the recording (see above). However, the use of resistors with such large resistances introduces two main problems:
noise levels associated with these resistors are considerably higher
than those expected from theoretical considerations, and the resistors possess not insignificant capacitance. This latter feature limits
the bandwidth of the recording. For example, a 50-GΩ resistor with
a stray capacitance of 0.1 pF results in a bandwidth of only 31.8 Hz
(obtained from $f_c = 1/2\pi RC$). As this would be useless for recording
the fine detail of channel openings and closings, an additional
circuit is built into the amplifier to increase the bandwidth to around
20 kHz (for details see Sherman-Gold, 1993; Sigworth, 1995).

Patch-clamp amplifiers employing a capacitive feedback head
stage (e.g., Axopatch 200B) are popular for the recording of single-
channel currents because the noise levels that can be achieved with
such amplifiers are less that those obtained with traditional resistive feedback technology. In this sort of head stage a capacitor
replaces the resistor in the feedback circuit of the current to voltage
converter. Such a circuit acts as an integrator. By differentiating the
slope of the voltage that charges the capacitor, the amount of current
passed is determined. The bandwidth of a capacitive current to
voltage converter is wide (70 kHz), and although such scope is not
needed, given the fact that most single-channel currents are filtered
with low-pass filters with an $f_c$ of 40 kHz (−3 dB) or less, the lower
noise levels that can be achieved are very valuable. The drawback
of capacitive feedback head stages is that, with time, the voltage
across the capacitors reaches that of the supply voltage. When this
occurs the capacitor needs to be discharged (the discharge itself
only takes about 50 µs). The frequency of such discharges depends
on the amount of current passed. For example, if the discharge is
triggered when the voltage reaches 10 V, this would occur every 2
seconds if the capacitor has a value of 1 pF and the current passed
is 5 pA (from $V = Q/C$). Such reset rates are acceptable. Clearly, if
the current passed is in the order of several hundred picoamperes,
the resets become more frequent (every 20 ms for a current of
500 pA). Therefore in whole-cell recording, the capacitive feedback
system is less valuable; indeed, the lower inherent noise level of
the capacitive feedback system is not much use since noise levels

are considerably higher in whole-cell recording (for further details see Sherman-Gold, 1993; Sigworth, 1995). Finally, more advanced amplifiers such as the HEKA EPC10 are available that minimize recording noise levels by using software control and fiber optic data cables. Nevertheless, even with the inherently noisier resistive-feedback amplifiers, it is possible, with careful shielding, to obtain noise levels that allow for the resolution of single-channel events with amplitudes <1 pA.

### 2.3. Patch-Pipettes

Patch-pipettes (electrodes) used to study single-channel activity are fabricated usually from thick-walled glass. This type of glass yields electrodes with lower noise properties than can be achieved from thin-walled glass. Thin-walled glass has a lower access resistance and hence is preferred for whole-cell recording. Borosilicate glass (e.g., GC150F-7.5, supplied by Harvard Apparatus, Holliston, MA) is well suited to making patch-pipettes. This glass has an outside diameter of 1.50 mm and an inside diameter of 0.86 mm. The "F" refers to the fact that the capillary contains a fine-glass filament that makes for the easier filling of the pipette with recording solution, and the capillary length of 7.5 cm means that two pipettes of usable length can be pulled from each tube. Quartz glass can also be used to make patch-pipettes and has the advantage that it allows for much lower noise recordings but the disadvantage that a laser is required to heat the glass in order that pipettes can be fabricated. Pipettes used for making isolated patch recordings tend to have higher resistances than those used for whole-cell recording, with resistances of between 5 and 20 MΩ being common.

The capacitance of patch-pipettes, and hence the noise from this source (see above), can be reduced by coating the patch-pipette with a coating of Sylgard No. 184 resin (Dow Corning, Midland, MI). This is a hydrophobic compound that prevents the external recording solution from creeping up the outside of the patch-pipette. Care needs to be taken to avoid blocking the end of the patch-pipette with the resin, and therefore it is advisable (although not essential) to apply the resin while viewing under a dissecting microscope. Try to get the coating within 100 μm of the electrode tip and extend it to beyond where the electrode will exit the solution in the recording chamber. The resin is "cured" by passing the coated patch-pipette into a heated wire coil. Rotating the patch-pipette during the curing process avoids the buildup of a large amount of Sylgard on one side of the electrode.

Prior to using the patch-pipettes for making channel recordings, their tips are fire-polished. It is a good idea to polish a few electrodes at a time, use them over the space of a couple of hours, and then polish another batch, since small particles of debris can accumulate on the tips over time. Fire-polishing involves heating a platinum wire in close proximity to the tip of the patch-pipette. When the wire is heated, it will expand and move upward by several tens of micrometers; therefore, it is important not to have the patch-pipette too close to the wire before heating, as one can easily break the pipette with the deflection of the wire. Additionally, it is important to melt a small amount of glass in the form of a bead onto the wire to avoid any platinum vapor finding its way onto the pipette tip. Commercial fire polishing units are manufactured by several companies and generally have one low-powered objective that is useful for locating the patch-pipette and placing it near to the wire. A second, higher-powered objective is then used to visualize the actual fire-polishing. During the polishing it should be possible to observe a small amount of shrinking (melting) of the tip of the electrode. When using thick-walled glass, this is not as dramatic as can be seen with thin-walled glass, and indeed the purpose of the polishing is simply to clean the surface of the pipette tip, rather than change its overall shape. All this being said, it is entirely possible to obtain high-quality single-channel recording with pipettes that have not been fire-polished, although the success rate of obtaining gigaohm seals may be increased with additional polishing. More details regarding the fabrication of and noise properties of patch-pipettes are given in Chapter 1.

Most electrophysiologists have their preferences for internal and external recording solutions, and therefore there are no universal standards. To begin with, stick with solutions that have been "tried and tested," and modify these basic recipes, as necessary, once you are confident that you are able to record routinely the channel activity of interest to you.

## 2.4. Current Conventions in Electrophysiology

Electrophysiologists have agreed on a convention to describe currents recorded in patch-clamp experiments. Positive ions flowing out of the patch-pipette are measured as positive currents and are illustrated as upward deflections on the current trace. Positive ions flowing into the patch-pipette are measured as negative currents and are illustrated as downward deflections on the current trace. The words *inward* and *outward* refer to the direction of movement of ions flowing across the membrane. Thus the net movement of positive ions from the outside of the cell to the inside would be referred to as an inward current, while the net movement of positive ions from the inside of the cell to the outside would be referred to as an outward current. However in the case of negative ions [e.g., Cl⁻ ions in the case of the γ-aminobutyric acid (GABA$_A$) receptor], movement from the inside of the membrane to the outside also generates negative currents (and would, by convention, be displayed as a downward deflection). Similarly, the net movement of anions from the outside of the cell to the inside of the cell would generate a positive current (upward deflection). These conventions can at times become a little confusing. For example, in both cell-attached and inside-out membrane patches, the net movement of positive ions in the direction of the outer to inner membrane surface is, by definition, an inward current, but as positive ions are leaving the patch-pipette, this would be recorded as a positive (or upward) current. Thus it is common for single-channel currents that have been recorded in these configurations to be inverted so that the inward movement of ions is represented as a downward deflection. Things are much simpler with outside-out membrane patches as the net movement of positive ions from the outer to the inner membrane surface generates a negative patch current (as, here, positive ions flow *into* the electrode).

## 2.5. Configurations Used to Record Single-Channel Currents

Single-channel recordings are most commonly made from visualized cells, for example, in cell culture, acutely dissociated cells, or cells in thin tissue slices. The critical point is that the surface of the cell to be patched is devoid of any material that could prevent the patch-pipette from making a high-resistance seal with the cell membrane. When recording from isolated cells, this is not generally a problem, but if one records from tissue slices, a certain amount of cleaning may be required (Edwards et al., 1989). A good-quality microscope is a real advantage; the choice of an inverted versus an upright microscope if one is recording from acutely isolated or cultured cells is a matter of personal preference to a large extent. However, if one plans to record from cells in thin tissue preparations, then an upright microscope is essential. The working distance of water-immersion lenses has improved dramatically over the last few years (with little compromise of their numerical apertures), and it is common to have more than 3 mm of working distance with a 40× water-immersion lens and approximately 2 mm with a 63× objective. When recording from isolated cells, phase-contrast optics are generally all that is required; however, for recording from cells in tissue, the use of Nomarski (differential interference contrast) optics is preferable. The use of infrared optics adds considerably more expense, but allows for visualization of cells located deeper in the slice. When considering micromanipulators, there are no additional requirements for single-channel recording over and above those needed for whole-cell recording. Indeed, it may well be possible to use an old micromanipulator that drifts too much to be of use for whole-cell recording if one is mainly interested in recording from either inside-out or outside-out membrane patches, so check those cupboards before buying an expensive new one.

There are three basic patch-clamp configurations that can be used to record the activity of single-channel current, and each has advantages and disadvantages. One needs to decide the nature of the experiment and the type of information required from it before selecting the appropriate configuration. The following subsections describe the configurations and cite selected examples of data that illustrate the sorts of experiments that can be carried out with each configuration.

### 2.5.1. The Cell-Attached Patch

This configuration is the easiest to achieve and causes no disruption to the intracellular composition of the cell under investigation. The cell-attached patch is the starting point for all patch configurations (including whole-cell recordings). The protocol for obtaining the cell-attached configuration is as follows:

1. Apply a small amount of positive pressure to the patch-pipette and then place the pipette tip in the recording bath. Take care not to apply too much pressure, as this will possibly prevent the formation of a gigaohm seal.
2. Offset the pipette current to zero.
3. Monitor the electrode resistance while approaching the cell. Voltage steps of about 5 to 10 mV can be applied via the patch-clamp amplifier and the current deflection monitored on an oscilloscope. Patch amplifiers such as the Axopatch-1D have an

internal oscillator that can be set to 100 Hz to produce to voltage steps. The polarity and magnitude of the voltage step can be set by the Step command. The Axopatch 200B amplifier supplies a +5 mV seal test at line frequency; however, the polarity and magnitude of this pulse cannot be altered.

4. Immediately before approaching the cell, check that there has been no drift in the pipette current (correct if necessary).
5. Observe a small indentation to the cell as the pipette touches the membrane. At this point the size of the current deflections should decrease, indicating an increase in the resistance of the pipette. Remove the positive pressure and observe a further increase in the seal resistance.
6. Monitor the seal resistance. Sometimes a gigaohm seal forms spontaneously without any further intervention on the part of the experimenter. If the gigaseal does not form, apply some gentle negative pressure. This is most easily done by mouth, which gives greater control; however, a syringe may be used.
7. Seal formation is sometimes aided by applying a negative voltage to the pipette. In my experience there is no prescribed voltage that routinely works; somewhere in the range of −20 to −40 mV is suggested.
8. When the gigaseal forms, the test pulses should cause no detectable deflection on the current trace; all that remains are the capacitive transients indicating the onset and offset of the voltage pulse. Cancel these transients using the controls on the patch-clamp amplifier. You now should have a cell-attached patch.

An immediate question one asks when recording in the cell-attached configuration is, What is the transmembrane potential of the patch of membrane under the patch-pipette? This is not trivial, since in order to answer the question we need to know the resting potential of the cell. If the resting membrane potential ($V_{mem}$) is known, then the patch potential ($V_{patch}$) can be obtained from

$$V_{patch} = V_{mem} - V_{cmd},$$

where $V_{cmd}$ is the command potential set on the patch-clamp amplifier. Estimates of the resting potential can be obtained from the Nernst equation; however, this is unlikely to be very accurate unless a high $K^+$ concentration is present in the external recording solution. Although the Goldman-Hodgkin-Katz (GHK) equation may give a better approximation of the resting potential, it is unlikely that one will have accurate values of the necessary permeability ratios. The Nernst equation is

$$V_{mem} = \frac{R \cdot T}{z \cdot F} \cdot \ln \frac{[ion]_{out}}{[ion]_{in}},$$

and the GHK equation is

$$V_{mem} = \frac{R \cdot T}{z \cdot F} \cdot \ln \frac{P_K \cdot [K^+]_{out} + P_{Na} \cdot [Na^+]_{out} + P_{Cl} \cdot [Cl^-]_{in}}{P_K \cdot [K^+]_{in} + P_{Na} \cdot [Na^+]_{in} + P_{Cl} \cdot [Cl^-]_{out}},$$

where $R$ = the gas constant, $T$ = absolute temperature, $F$ = Faraday's constant, $z$ = ion valency, $[ion]_{out}$ = extracellular concentration of ion, $[ion]_{in}$ = intracellular concentration of ion, and $P_K$, $P_{Na}$, and $P_{Cl}$ are the permeabilities of $K^+$, $Na^+$, and $Cl^-$ ions, respectively.

One method that is sometimes employed with cell-attached recording is to raise the extracellular $K^+$ concentration to depolarize the membrane potential of a neuron to approximately $0\,mV$. The membrane potential of the patch is then equal to the negative of the command potential set by the patch amplifier. One problem with this approach is that the depolarized state of the neuron may alter intracellular processes that regulate the activity of the channels under investigation. In addition, altering the external solution will affect all the other cells in the recording chamber. An alternative method is to record the unitary amplitudes of the channel currents of interest at a variety of applied command potentials. If the unitary single-channel conductance of the channels is known, then it is simple to estimate the patch potential that gives the observed single-channel current amplitude; if one knows the command potential set by the patch-clamp amplifier, one can estimate the membrane potential of the cell. It is important to realize that in the cell-attached patch configuration, the patch-clamp amplifier controls only the potential of the membrane patch under the patch-pipette; the remainder of the cell membrane is at the resting potential for the conditions used in the experiment. Apart from the difficulty in estimating the membrane potential of the cell, another disadvantage of this recording method is that if one is studying the activity of ligand-gated ion channels, the agonist needs to be included in the patch pipette. Clearly it is important that the agonist concentration selected gives the appropriate amount of activity; it is difficult (though not impossible) to alter the agonist concentration in the pipette solution once the recording has been established.

## APPLICATION EXAMPLE

As mentioned above the great advantage of the cell-attached patch configuration is that we do not disturb the intracellular environment of the cell. Thus, this configuration allows us to study the actions of drugs that can modulate the activity of ion channels via intracellular signalling pathways. An example of such a use is shown in Fig. 1 (adapted from Selyanko and Sim, 1998). The upper current trace (Fig. 1A) shows the activity of sustained $K^+$ channels under a cell-attached patch-pipette. Openings of the channel are shown as upward deflections to indicate that these are outward currents (the traces have been inverted from what would actually have been observed). Addition of oxotremorine-M (a muscarinic acetylcholine receptor agonist) to the extracellular bathing solution results in a dramatic reduction of $K^+$ channel activity (Fig. 1B). Since the agonist was not applied, directly, to the $K^+$ channels under the patch-pipette (and it could not have diffused into this isolated membrane area because of the high-resistance seal between the patch-pipette and the cell membrane), we must assume that the activation of muscarinic receptors results in the production of an intracellular messenger(s) that causes the inactivation of the $K^+$ channels (middle trace). This effect is reversible since when oxotremorine-M is removed, the channel activity returns (Fig. 1C). The plot in Fig. 1D documents the probability of the channel being open ($P_{open}$) throughout the course of the entire experiment.

## 2.5.2. The Inside-Out Patch

As the term suggests, in this configuration the intracellular face of the lipid bilayer faces the external bathing solution in the recording chamber. To obtain the inside-out patch configuration, it is first necessary to obtain a cell-attached patch. Following the formation of the cell-attached patch, pull the patch-pipette away from the cell membrane rapidly; an inside-out patch may form directly. However, it is not uncommon for a membrane vesicle to form. If this is the case, an inside-out patch can be made by lifting the patch-pipette from the recording solution for a short period of time; this should burst the vesicle, leaving an inside-out membrane patch. The

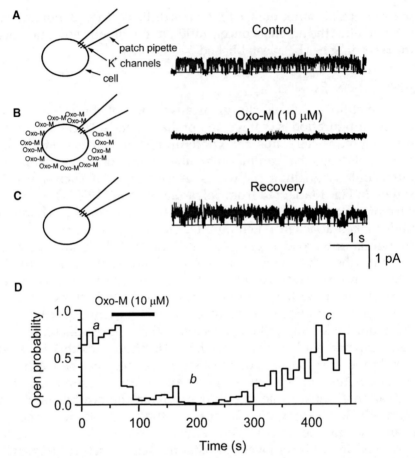

Fig. 1. Oxotremorine-M inhibits the activity of single sustained $K^+$ channels in cell-attached patches. (A) Control level of $K^+$-channel activity recorded in a cell-attached patch. The solid line indicates the closed level. (B) Addition of oxotremorine-M (10 μM) to the external bathing solution (and therefore not directly to the area of membrane under the patch-pipette) results in the inhibition of the $K^+$-channel activity. (C) Removal of oxotremorine-M allows the $K^+$-channel activity to recover. (D) Quantification of the Popen of $K^+$-channel activity recorded throughout the experiment. (Adapted from Selyanko and Sim (1998), with permission. © Blackwell Publishing.)

membrane potential of the inside-out patch is simply the negative of the command potential set by the patch-clamp amplifier. As mentioned above, it is common to invert the current traces recorded in this configuration so that, for example, inward currents are displayed as downward deflections.

APPLICATION EXAMPLE

The inside-out mode of patch-clamp recording is ideally suited to the study of agents that modulate the activity of ion channels by interacting with their cytoplasmic domains. Figure 2 shows an experiment where the inside-out configuration has been utilized to study the modulation of large conductance $Ca^{2+}$-activated $K^+$ channels (BK channels). $K^+$ channels of this type are activated by increases in the intracellular $Ca^{2+}$ concentration and therefore are well suited to being studied using the inside-out patch configuration where one can alter the concentration of $Ca^{2+}$ in the external bathing solution. BK channels normally run down in the absence of adenosine triphosphate (ATP) in the solution bathing the intracellular side of the membrane; however, Fig. 2 illustrates that channels carrying the point mutation S689A do not show this phenomenon. In either the absence or presence of ATP (applied to the cytoplasmic face of the channel), there is a "run-up" of channel activity (unpublished data of Martin Hammond and Mike Shipston, Centre for Integrative Physiology, University of Edinburgh).

## 2.5.3. The Outside-Out Patch

This is the preferred configuration for the study of ligand-gated ion channels since the external face of the lipid bilayer (and hence the ligand binding site of the channel) faces the external recording solution, permitting the exchange of control and agonist solutions with the membrane patch. The membrane potential of an outside-out patch is simply the value of the command potential set on the patch-clamp amplifier. The procedure for making an outside-out patch is as follows:

1. Obtain a cell-attached patch and keep applying the small voltage steps that were used to monitor the increase in seal resistance during the formation of the cell-attached patch.
2. Set the command potential somewhere between –20 and –40 mV; the precise value will be determined, empirically, by the success of obtaining the whole-cell configuration (see below).

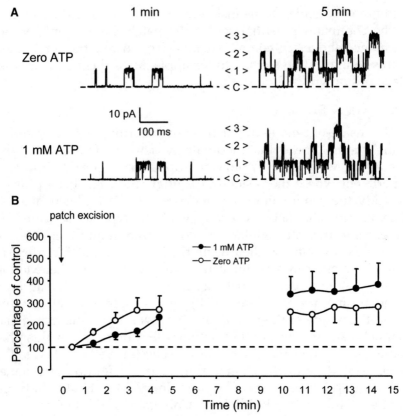

Fig. 2. Inside-out patch recording of BK channels. (**A**) Representative traces of recordings of S869A BK channels in the absence ($P_{open}$ increases from 0.067 at 1 min to 0.467 at 5 min) and presence of adenosine triphosphate (ATP) ($P_{open}$ increases from 0.045 at 1 min to 0.315 at 5 min). $\langle C \rangle$ indicates the closed state, and $\langle n \rangle$ the number of open channels. (**B**) Plot of the change in the activity of S869A BK channels in the presence and absence of ATP. In 1 mM ATP the activity of the channel increased and after 15 minutes has increased to 380% ± 98% ($n = 25$, $p < .05$). In zero ATP the activity of the channel also increased and after 15 minutes has increased to 279% ± 80% ($n = 10$, $p > .05$). There is no a significant difference between the activity of S869A BK channels in 1 mM ATP and zero ATP after 15 minutes ($p > .05$). All data are mean ± standard error of the mean (SEM). Dotted line represents the level of control activity observed during the first minute of recording. (Courtesy of M. Hammond and M. Shipston, Centre for Integrative Physiology, University of Edinburgh.)

3. Apply gentle suction to the patch (suction via mouth gives far better control than is achieved via a syringe). The object here is to rupture the membrane under the patch-pipette without compromising the integrity of the gigaohm seal formed between the patch-pipette and the cell membrane. Successful "breakthrough" is most often achieved by applying the suction in brief gentle "kisses."
4. Breakthrough is immediately apparent as the current response to the voltage steps reappears on the oscilloscope but now has large capacitive transients associated with them. The magnitude of the steady-state current trace associated with the voltage-steps gives an indication of the input resistance of the cell. This is now the whole-cell patch-clamp recording configuration.
5. Slowly pull the patch electrode away from the cell. One can generally observe a neck of membrane being pulled away from the cell. Depending on the type of cell, it is sometimes necessary to back the patch-pipette many tens of micrometers away from the cell. As the outside-out patch forms, there is a reduction in the DC holding current, the noise of the recording, and the magnitude of the capacitive currents associated with the voltage-pulses being applied to the patch-pipette.
6. A successful outside-out patch forms if the membrane pinches off the cell, and there is no loss in the seal resistance. This takes a little practice, and some cells give rise to good quality outside-out patches more readily than others.
7. To minimize the noise levels further, the volume of the bathing solution can be reduced so as to leave only the tip of the patch-pipette in the recording solution (this reduces the pipette capacitance). However, care must be taken to avoid the tip leaving the bathing solution. If one is recording the steady-state activity of a ligand-gated ion channel, it is possible to switch off the perfusion after exchanging completely the control bath solution for one that contains agonist. This eliminates noise associated with the perfusion system, and there is no danger of the solution level in the bath changing.

APPLICATION EXAMPLE

As mentioned above, the outside-out patch configuration is widely used in the study of ligand-gated ion channels. In addition to being used for studies of the steady-state functioning of ion

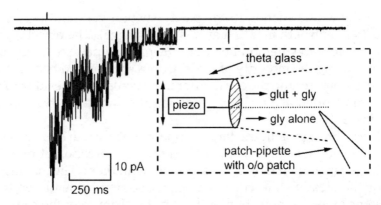

Fig. 3. Use of a concentration jump to activate *N*-methyl-D-aspartate (NMDA) receptor channels in an outside-out (o/o) patch. The current trace shows an individual response obtained from an outside-out membrane patch excised from a *Xenopus laevis* oocyte expressing recombinant NR1/NR2A NMDA receptor channel. Channel activity was evoked by exposing the patch to a solution containing 1 mM glutamate for 1 ms. Unitary events are clearly discernible in the sweep. The inset shows a cartoon illustration of recording method used in such concentration jump experiments. [Adapted from Wyllie et al. (1998), with permission. © Blackwell Publishing.]

channels, solutions can be applied rapidly to the outside-out patch in an attempt to mimic more closely the agonist concentration profile that an ion channel might experience following the synaptic release neurotransmitter. Thus outside-out patches are used in concentration-jump experiments. Figure 3 shows an example of channel activity recorded from an outside-out patch in response to a 1-ms application of 1 mM glutamate. The patch was isolated from cell expressing recombinant NR1/NR2A *N*-methyl-D-aspartate (NMDA) receptor channels and was held at −100 mV. The inset shows a sketch of the concentration-jump technique. Two solutions flow from either side of a piece of theta glass; one contains the control solution (glycine alone), while the other contains the test solution (in this case glycine +1 mM glutamate). The movement of the theta glass is controlled by a piezo crystal. Applying a high voltage to the crystal stack causes its expansion. The patch-pipette is placed a few micrometers away from the sharp interface that forms between the control and test solutions. Applying a high voltage to the crystal for 1 ms causes the theta glass to move, and the patch-pipette is exposed to the glutamate-containing solution. In the trace shown, 11 individual channels are open simultaneously at the peak of the response. With time, channels close at random, leaving the receptor in a lower affinity state from which agonist dissociation may occur.

Figure 4 illustrates another outside-out patch recording. In this case, single-channel events have been recorded at different holding potentials to determine the conductance of the channel under investigation. This property is discussed below.

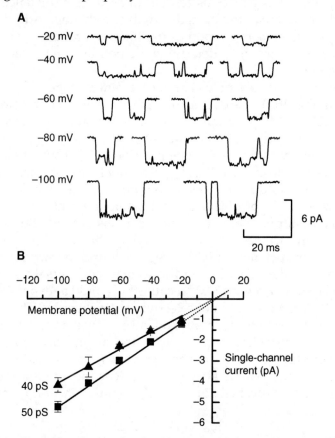

Fig. 4. Current-voltage plot for NMDA receptor-channels. (**A**) Examples of NMDA NR1/NR2A(T671A) single-channel currents recorded in an outside-out membrane patch at the holding potentials indicated. As the holding potential approaches the reversal potential of the current, the driving force decreases and hence the amplitudes of the unitary events become smaller. (**B**) Current-voltage plot indicating the existence of two (50 pS and 40 pS) prominent conductance levels. Extrapolating the least-squares fitted lines to the data gives an estimated reversal potential of +2 mV. [Adapted from Anson et al. (2000), with permission. © Blackwell Publishing.]

## 2.6. Single-Channel Conductance — One of the Signatures of an Ion Channel

Let us assume that we have selected the patch-clamp recording configuration best suited to our channel of interest. Perhaps the most straightforward parameter to measure is the amplitude of the single-channel current. Measuring amplitudes at a variety of holding potentials allows us to determine the conductance of the channel, one of the parameters that defines a channel. Additionally, altering the ionic composition of the internal and external recording solution allows us to determine the permeation characteristics of a channel.

An example of a single-channel current-voltage (I/V) plot is shown in Fig. 4B. These data are taken from a study of recombinant NR1/NR2A NMDA receptors carrying a point mutation in the region of the glutamate-binding site, where an alanine residue replaced the threonine residue at position 671 (Anson et al., 2000). As can be seen from both the single-channel current traces and the I/V plot, two conductance levels are present. The main level has a conductance of 50 picoSiemens (pS), while the sublevel has a conductance of 40 pS. These conductance levels are indistinguishable from the conductances of wild-type NR1/NR2A NMDA receptors and high-conductance NMDA receptors found in central neurons. Extrapolating the fitted lines indicates that the reversal potential of these currents is around +2 mV, a value that would be expected given that NMDA receptors are nonselective cation channels. The fact that the data sets for both the main and subconductance level are well fitted with straight lines indicates that the channel is ohmic, that is, the amplitude of the single-channel current, $I_{sc}$ at any holding potential, $V_m$, can be determined if we know the channel's unitary conductance, $g_{sc}$, and its reversal potential, $V_{rev}$. Thus from Ohm's law:

$$I_{sc} = g_{sc}(V_m - V_{rev})$$

## 2.7. Single-Channel Amplitudes Are Described by Gaussian Distributions

Figure 5A,B shows examples of single-channel currents recorded in outside-out membrane patches excised from a *Xenopus laevis* oocyte expressing either recombinant NR1/NR2A (Fig. 5A) or NR1/NR2D NMDA (Fig. 5B) receptors. In both cases the patch

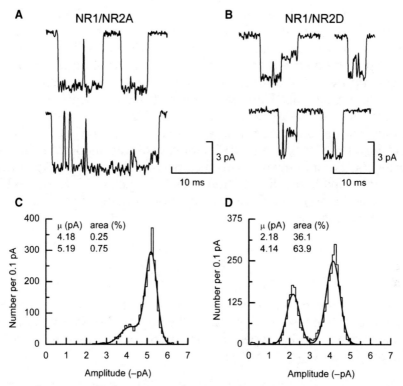

Fig. 5. Single-channel currents and amplitude histograms. (**A,B**) Examples of recombinant NR1/NR2A (**A**) and NR1/NR2D (**B**) NMDA receptor single-channel currents recorded from outside-out patches held at –100 mV. (**C,D**) Amplitude histogram for NR1/NR2A and NR1/NR2D NMDA receptor single-channel events with durations greater than 2.5 $T_r$ (415 μs). In each case the histograms are fitted with a mixture of two Gaussian components with means and standard deviations as indicated. **A, C**: unpublished data. [**B,D** adapted from Chen et al. (2004), with permission. © Blackwell Publishing.]

potential was –100 mV, and it can be assumed that both channel types have reversal potentials that are around 0 mV. Even without detailed analysis, visual inspection of the two types of channel events indicates that (1) NR2A-containing NMDA receptors have larger single-channel amplitudes than NR2D-containing NMDA receptors, and (2) these latter events possess a very prominent subconductance level that is around half the amplitude of the main conductance level; NR2A-containing NMDA receptors also possess a subconductance level, but this is not as common. Thus, for both channel types, fits of amplitude distributions are best described by the sum of two Gaussian (or normal) distributions (Fig. 5C,D). In principle, the distribution of amplitudes that have been measured by taking the mean amplitude of the current when the channel is open is not actually Gaussian. This is because events with long durations give a more accurate measure of the true amplitude than do shorter duration events (more details can be found in Colquhoun and Sigworth, 1995, Appendix 2). Indeed, for the histograms shown in Fig. 5C,D, only events with durations greater than 415 µs have been included in the distribution, which is equal to 2.5 $T_r$ of a 2-kHz filter (see below). The Gaussian function equation for a Gaussian distribution is

$$f(x) = \frac{1}{\sigma\sqrt{2\pi}} \exp\left(\frac{-(x-\mu)^2}{2\sigma^2}\right)$$

where µ is the mean of the distribution and $\sigma^2$ is the associated variance. In the next section we shall compare the Gaussian distribution with two other types of distribution that are used to describe the properties of ion channels.

## 3. Some Background Theory and Practical Issues of Single-Channel Data Analysis

This section of the chapter discusses the elementary basics and concepts involved in the analysis of single-channel data. For a more rigorous discussion of the topics mentioned here, see the Recommended Readings at the end of the chapter, for example, Conn (1998), Neher and Sakmann (1995), and Ogden (1994). These texts, in addition to providing excellent insight into the interpretation of single-channel data, give considerably more detail about the mathematical background and theory of single-channel analysis than is given here.

### 3.1. Translating the Single-Channel Data Record in Dwell Time Intervals

Channel data are best stored on videotape or on digital audiotape for subsequent digitization. It is best to record single channels with the minimum of filtering, for example at around 10 to 20 kHz

(–3 dB or half-power), as it is a simple matter to filter the data more heavily when it comes to the analysis. Most patch-clamp amplifiers have a built-in filter unit, which can be set from around 500 Hz to 50 kHz. It is common to filter data with a low-pass filter of between 2 and 5 kHz (–3 dB) when it comes to the final analysis; however, be sure to use a Bessel-type filter to prevent the "ringing" that occurs if one filters channels with Tchebychev or Butterworth filters. As a guide it is best to digitize data at 10 times the filter rate; for example, digitize at 20 kHz if filtering at 2 kHz. If for any reason this is not possible or practical, then one can digitize as low as five times the final filter level. It is important to be aware of the cascade effect of filters. For Bessel and digital Gaussian filters, the final cut-off frequency ($f_c$) is given by the following relationship:

$$\frac{1}{f_c} = \sqrt{\frac{1}{f_1^2} + \frac{1}{f_2^2} + \frac{1}{f_3^2} + \dots \frac{1}{f_n^2}},$$

where $f_1$, $f_2$, etc. are the –3 dB values of the various filters used in the cascade series. For example, if one has filtered single-channel data at 10 kHz using the Bessel filter on the patch-clamp amplifier and then filters the data using another filter set at 5 kHz, then the final cut-of frequency is not 5 kHz but 4.47 kHz. To achieve a final cut-off frequency of 5 kHz, one would need to filter the data at 5.77 kHz (–3 dB). However, if one has filtered the data using a filter setting of 20 kHz on the patch-clamp amplifier and then filtered at 5 kHz, the final cut-off frequency would be 4.85 kHz, which would probably be acceptable for most applications. Thus the take-home message is to use a minimum amount of filtering when recording the data onto tape.

From the channel data record we need to obtain (1) the amplitude of the event and (2) the time that the channel spends at that amplitude level. Thus for a channel that opens to only one conductance level our ideal data record would consist of a series of dwell time intervals with an amplitude of 0 pA (corresponding to closed channel) and a series of dwell times with an amplitude corresponding to the open state of the channel. Optimizing the data can be achieved in several ways:

*The threshold-crossing method.* In this type of analysis an event is detected when the rising phase (opening) of the channel crosses some preset threshold; its duration is measured by the falling phase (closing) recrossing this level. This sounds rather straightforward and in a sense it is; however, many pitfalls await the unwary:

1. The presence of multiple conductance levels in a data record means that determining the 50% threshold values is ambiguous and there is no satisfactory way to measure records with multiple conductance levels by threshold crossing methods.
2. The best resolution that can be achieved is less than that which can be achieved with time-course fitting (see below).
3. A drifting baseline or instability of channel amplitudes (if not corrected or compensated for) can cause havoc with the analysis.

This method of optimizing the data record can be quick, especially if you do not check the validity of the fit to your data! It is essential to *look* at your data beforehand to determine its suitability to this method of analysis. The Strathclyde Electrophysiology Software and the QuB suite of programs (available from http://spider. science.strath.ac.uk/physpharm/ and http://www.qub.buffalo. edu, respectively), contain software for 50% threshold crossing analysis as does the pClamp software supplied by Molecular Devices and Bruxton's TAC single-channel software.

*Hidden Markov models.* The segmental k-means (SKM) algorithm in the QuB suite of analysis programs and a facility in the TAC single-channel software use hidden Markov models (HMM) to detect transitions in the data record. Depending on the number of classes of events contained within the record, simple kinetic schemes are constructed to detect events. For example, if the data record was composed of openings to two conductance levels, it might be possible to detect events with a simple C-O-O (closed-open-open) scheme. Visual inspection of the optimal data record superimposed on the channel recording allows users to satisfy themselves that events have been detected and correctly assigned to the appropriate class. If necessary, a more refined kinetic scheme can be employed to improve the data optimization. Using this sort of detection method (as opposed to simple 50% threshold crossing methods) will allow for the detection of subconductance levels.

*The time-course fitting method.* Only one analysis program allows time-course fitting of single-channel data: SCAN (Single-Channel ANalysis, written by David Colquhoun and available from http:// www.ucl.ac.uk/Pharmacology/dc.html). Be aware that time-course fitting is considerably slower than the 50% threshold-crossing method and HMM methods (if you don't check the fits obtained from these other methods), but in the end not only will you have examined every single-channel transition in your data record, you will be able to fit data records that contain events with multiple conductance levels and impose a significantly better resolution on the ideal record. Each amplitude is fitted as it is seen on the screen with no need for predetermined windows. In time-course fitting, one fits each transition with a square pulse with an amplitude equal to that of the channel transition and that has been filtered at exactly the same $f_c$ as the data record (or, better in principle, with the experimentally determined step response of the entire recording system). More details about the 50% threshold crossing and time-course fitting methods can be found in Colquhoun and Sigworth (1995).

It is important to note that some optimization routines express dwell times as integer multiples of the digitization rate used to sample the data. Thus, if a sample rate of 40 kHz is used, dwell times will be measured in multiples of 25 µs. Therefore, if data records contain a large number of brief events (either openings or closings), then valuable information may be lost (for further discussion and illustration of the consequences of this see Schorge et al., 2005).

## 3.2. Setting the Resolution

Before any analysis of the channel record can be conducted, one needs to impose an acceptable resolution on the data set so that incompletely resolved transitions are clearly distinguished from the baseline noise, but, at the same time, making sure that as few as possible real transitions are missed. It should be said that the presence of multiple conductance levels complicates this further, since incompletely resolved current amplitudes are not always transitions to the same conductance level.

As mentioned above, one needs to determine the final filter cutoff frequency of the data record. Bessel-type filters with four or eight poles are sufficiently close in their response characteristics that they approximate Gaussian filters. In this case such filters have rise times $(T_r)$ that are equal to $0.3321/f_c$, where $f_c$ is the $-3$ dB frequency of the filter. For example, a data record filtered at 10 kHz during the recording and then filtered subsequently at 3 kHz prior to digitization gives an effective overall filtering of 2.87 kHz (−3 dB) and therefore the corresponding $T_r$ is 116 μs. Furthermore, the maximum amplitude, $A_{max}$, of a low-pass filtered square pulse of duration $d$ is given by the equation

$$A_{max} = A_o \text{erf}\left(\frac{0.886d}{T_r}\right),$$

where erf is the error function, which is given by

$$\text{erf}(x) = \int \frac{2\exp(-x^2)}{\sqrt{\pi}}\, dx,$$

and $A_o$ is the amplitude of the original unfiltered pulse.

With ratios of $d/T_r$ of 1.0, 1.5, 2.0, and 2.5, this gives values of $A_{max}/A_o$ of 0.790, 0.940, 0.988, and 0.998, respectively. Thus a low-pass filtered pulse of duration 2.5 $T_r$ reaches 99.8% of its original amplitude. This is why it is common, when fitting amplitude histograms, to include only events that have durations > 2.0 or 2.5 $T_r$.

The probability that an event with amplitude $A_{max}$ arises from random baseline noise is given by the false event rate (Colquhoun and Sigworth, 1995):

$$\text{false event rate} \approx f_c \exp\left[-0.5\left(\frac{A_{max}}{\sigma_n}\right)^2\right],$$

where $\sigma_n$ is the rms of the baseline noise in the channel recording.

Typically, minimum resolvable open and shut times are chosen to give false event rates of less than $10^{-6}$ per second. As a further check of the suitability of a chosen resolution, open and shut histograms can be constructed with a very much unrealistic resolution (e.g., 10 μs), which will give a very high false event rate. When these histograms are displayed on an arithmetic scale (see below), the value at which events start to be missed gives an indication of the upper limit of the resolution that can be obtained in a particular experiment. As a guide, if one analyzes channels with a full amplitude of around 5 pA in a recording with a baseline noise level of around 100 to 130 fA (2 kHz, −3 dB), then it should be possible to set the resolution between 30 and 50 μs.

## 3.3. Channel Life-Times Are Distributed Exponentially

In the analysis of single-channel events, we attempt to understand the average functioning of a channel by studying many hundreds or thousands of *individual* events; there is no point in trying to estimate the mean open time of a channel by simply measuring half a dozen event durations. The random nature of the dwell times (both shut times and open times) means that we need to use probability distributions to describe their functioning. In addition to the Gaussian distribution that we encountered when discussing the distribution of single-channel amplitudes, the two most common distributions encountered in single-channel analysis are the exponential distribution and the geometric distribution.

Figure 6A,B compares the features of a Gaussian (normal) distribution with that of an exponential distribution. The Gaussian distribution shown in this figure has a mean equal to 1 and a standard deviation (SD) of 0.2. The mean, median, and mode each has the same value in a normal distribution (in this example these are all 1). The curve is therefore symmetrical around the mean. In addition 95.4% of all values occur ±2 SD of the mean. In contrast, the exponential distribution is very different from the normal distribution. The median occurs at a value equal to 0.693 of the mean value. This is sometimes referred to as the half-life (cf. radioactive decay). Thus 50% of events have a value less than or equal to the median value. The mean value of an exponential distribution occurs when the curve has decayed to 1/e of the initial starting value; this is referred to as an e-fold decay (where e is the universal constant = 2.7182818 . . .). The mean value is also referred to as the time-constant (symbol τ) of the exponential distribution. For a process that is described by a single exponential component, 36.79% of events have a value greater than the mean (63.21% have values smaller than the mean). The mean and median for the curve shown in Fig. 6B are indicated by dashed lines.

Channel dwell times are distributed exponentially (see below). This is a consequence of the fact that single-channel events display a Markov (or memoryless) type of functioning. Two examples of Markov processes are shown in Fig. 6. Figure 6C plots the number of trials (tosses of coin) needed to get a head against the probability. Thus $p = .5$ for a single trial, $p = .25$ for it taking two attempts,

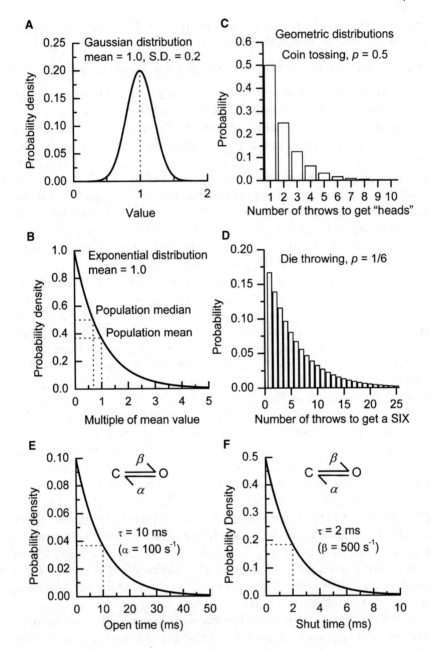

$p = .125$ for it taking three attempts. Figure 6D shows the number of die casts required to obtain a six. In each plot the probability of the $n$th +1 trial being the successful trial is a constant factor (5/6) less than the $n$th trial. In each of these examples, the cumulative probability can be obtained by adding all the individual probabilities that precede a particular trial. For example, when tossing a coin, one will obtain a head three times out of four in a trial of two tosses. What is the analogy with an ion channel? In this case we are interested in the amount of time a channel spends in a particular state (either open or closed). For the sake of argument let us assume that the channel is open. All the bonds in the protein will be bending and stretching until at some point they find themselves in a conformation so close to the transition state that the protein crosses the energy barrier and flips into the shut conformation. There are so many of these bending and stretching events that the protein does not "know" how close any particular set came to causing the channel to close. It is memoryless. The fact that a success occurs (the channel closes) is therefore independent of the number of trials that have taken place. A channel does not "remember" how long it has been open (or closed); therefore, the probability that the channel will close (or open) in the next small time interval is constant. What we find in this situation is that the probability of a successful trial is very small, but the number of trials is very large. When this happens, the geometric distribution approaches the exponential distribution. For a mathematical answer to this question, the interested reader is referred to Colquhoun and Hawkes (1995). The equation describing the geometric distribution is

$$P(r) = p(1 - p)^{r-1} \ (r = 1, 2, 3 \ldots),$$

◄───────────────────────────────────────

Fig. 6. Examples of Gaussian, exponential, and geometric distributions. (**A**) A Gaussian (normal) probability density function (pdf) with a mean equal to 1 and a standard deviation of 0.2. (**B**) An example of an exponential pdf with a mean equal to 1. Dashed lines show where the median and mean occur in such distributions. (**C,D**) Geometric distributions for coin tossing and die throwing. In each histogram the probability given refers to the probability that the $n$th trial is the *successful* trial (i.e., there have been $n - 1$ failures up to this point). (**E**) The pdf of channel open times based on a simple C ↔ O model with the closing rate set to $100 \, s^{-1}$. (**F**) The pdf of channel shut times based on a simple C ↔ O model with the opening rate set to $500 \, s^{-1}$.

where $P(r)$ gives the probability of a success requiring exactly $r$ number of trials given that there has been $r - 1$ failures prior to this, and the success of any given trial is $p$.

As stated above, channel lifetimes are distributed exponentially and these are described by exponential probability density functions (pdf):

$$f(t) = \lambda \exp(-\lambda t) \quad \text{or} \quad f(t) = \tau^{-1} \exp(-t/\tau),$$

where $\lambda$ is the *rate constant* of the distribution ($\tau$ is the *time constant* of the distribution). The term *probability density function* can be interpreted as meaning the curve such that the *area* under the curve between any two values of the variable is the probability of the variable having a value somewhere between these two points. Put another way, the area under the curve up to a point $t$ gives the probability that the lifetime of the channel is less than or equal to $t$. The total area under such curves is unity; this is simply a consequence of the fact that the open time (or closed time) must have a value of between 0 and ∞.

It should be noted that when the number of trials ($r$) is large and the probability ($p$) of a success on any given trial is low, the geometric distribution, which is a discontinuous analogue of an exponential function, approaches an exponential distribution. Thus,

$$(1 - p)^r \rightarrow \exp(-pr),$$

as $p \rightarrow 0$ and $r \rightarrow \infty$ (with the product, $pr$, being finite).

### 3.4. A Simple Two-State Model

Before we consider the analysis of some simulated channel data, let us consider a very simple scheme to allow the introduction of a few terms used in the study of single-channel kinetics. A channel can exist in two states—either open (O) or closed (C). The opening rate constant is $\beta$ and the closing rate constant is $\alpha$, and both have dimensions of reciprocal time (e.g., s$^{-1}$).

$$C \underset{\alpha}{\overset{\beta}{\rightleftharpoons}} O$$

In our two-state model the mean durations of the closed and open states are given by

$$\text{Mean open time} = \tau_0 = \frac{1}{\alpha}.$$

$$\text{Mean closed time} = \tau_c = \frac{1}{\beta}.$$

This result is obtained from a very simple rule that applies to any reaction scheme (no matter how complicated) that can be stated as follows: "The mean duration of an individual state is equal to the reciprocal of the sum of all the rate constants that describe routes for *leaving* that state." These mean values are known as *time constants* and have units of time. The proportion of time that the channel spends in the open state then is given by

$$\frac{\tau_o}{\tau_o + \tau_c} \quad \text{or} \quad \frac{\beta}{\beta + \alpha},$$

and similarly the proportion of time that the channel spends in the closed state is

$$\frac{\tau_c}{\tau_o + \tau_c} \quad \text{or} \quad \frac{\alpha}{\beta + \alpha}.$$

The distribution of open times is described by the following pdf:

$$f(t) = \alpha \exp(-\alpha t) \quad \text{or} \quad f(t) = \tau_o^{-1} \exp(-t/\tau_o),$$

while the distribution of closed times is given by

$$f(t) = \beta \exp(-\beta t) \quad \text{or} \quad f(t) = \tau_c^{-1} \exp(-t/\tau_c).$$

So, in the simple two-state model shown above, if the closing rate $\beta$ is equal to $100\,s^{-1}$ and the opening rate $\alpha$ is equal to $500\,s^{-1}$, then the mean lifetime of the open state would be $10\,ms$, that of the closed state would be $2\,ms$, and the pdfs describing their distributions would be

$$f(t) = 100 \exp(-100t)s^{-1}$$

for openings, and

$$f(t) = 500 \exp(-500t)s^{-1}$$

for closings. These two probability density functions are illustrated in Fig. 6E,F.

## 4. Analysis of Single-Channel Data: Simulated Data Set

The following discussion serves as a brief guide to the analysis of single-channel data. Figure 7 shows 500 ms of a continuous stretch of single-channel activity simulated with the SCSIM program

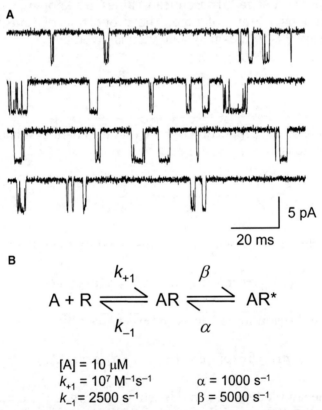

**A**

5 pA

20 ms

**B**

$$A + R \underset{k_{-1}}{\overset{k_{+1}}{\rightleftharpoons}} AR \underset{\alpha}{\overset{\beta}{\rightleftharpoons}} AR^*$$

$[A] = 10\ \mu M$
$k_{+1} = 10^7\ M^{-1}s^{-1}$          $\alpha = 1000\ s^{-1}$
$k_{-1} = 2500\ s^{-1}$          $\beta = 5000\ s^{-1}$

Fig. 7. Simulation of single-channel data using the Castillo-Katz reaction scheme. (**A**) 500 ms of continuous channel activity simulated using the Castillo-Katz reaction scheme and with the rate constants shown in **B**. Note that channel openings tend to occur in trains or bursts of openings that contain relatively short closed periods. Bursts, in turn, are separated from each other by longer closed periods.

(http://www.ucl.ac.uk/Pharmacology/dc.html) and using the del Castillo and Katz (1957) reaction scheme that is shown below the channel traces. The Castillo-Katz scheme is the simplest model that can account for the action of a ligand binding to a receptor resulting in the opening of an ion channel. In this scheme the resting state of the channel is denoted by R, the ligand-bound but shut state by AR, and the open state of the channel by AR*. The association and dissociation rate constants are given by $k_{+1}$ and $k_{-1}$, respectively, while the opening rate constant is denoted by $\beta$ and the closing rate constant by $\alpha$. For the simulation, the values for the rate constants chosen were as follows: $k_{+1} = 1 \times 10^7 \, M^{-1} s^{-1}$, $k_{-1} = 2500 \, s^{-1}$, $\alpha = 1000 \, s^{-1}$, $\beta = 5000 \, s^{-1}$, and the agonist concentration was set at 10 μM. Inspection of the channel activity shows the following:

1. The amplitudes of individual channel openings are very similar. (Note: although these data have been simulated, this can be seen to be true of real channels; e.g., examine the single channel traces shown in Fig. 4A and Fig. 5A,B.
2. The durations of individual openings are not equal.
3. Openings tend to occur in trains or bursts of openings that contain brief closed periods (gaps).
4. Bursts are separated by longer closed periods that again are not of equal duration.

## 4.1. Channel Open Times of the Simulated Data Record

Figure 8A shows the distribution of open times; a rather conservative resolution of 100 μs has been imposed on the data. Since there is only one open state in the Castillo-Katz reaction scheme, open times should be described by a single exponential, and this is indeed the case. Now given the rule regarding how one calculates the mean lifetime of any particular state, we would predict that the time constant of the open time distribution should be 1 ms (since $\tau = 1/\alpha = 1/1000 \, s^{-1}$). The fit of the simulated data gives a time constant of 0.98 ms, which is a good approximation given the fact that we only simulated 2,000 intervals (1,000 openings and 1,000 closings). The continuous line superimposed on the histogram is the exponential pdf. Since the fit of the open time distribution gave an estimate of the time constant of 0.978 ms, this equates to a rate constant, $\alpha$, equal to 1022.5 $s^{-1}$. Thus the open time pdf is

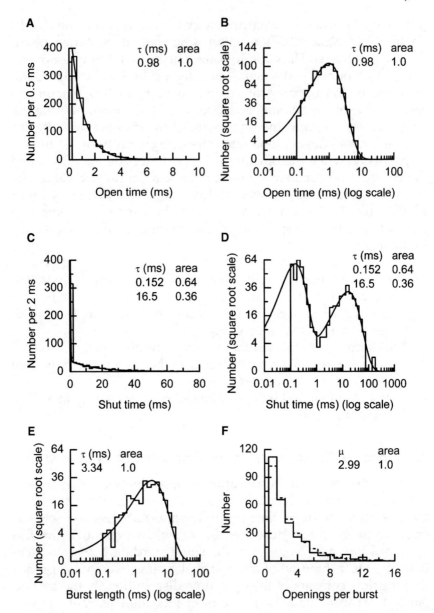

given by $f(t) = 1022.5e^{-1022.5t} s^{-1}$. For illustration, the pdf has been multiplied by the number of events that are predicted, by the fit, to be in the distribution (in this case 1,008) and then expressed in terms of per 0.5 ms (rather than per second). Therefore, the pdf has been multiplied by the factor 1008/2000 (the factor 2000 being obtained from 1 s/0.5 ms). Therefore, the first point on the pdf would be $1022.5 s^{-1} \times 1008/2000 = 515$. The distribution to the right (Fig. 8B) again shows the open time pdf, but in this case the time scale has been plotted on a logarithmic scale and a square root scale has been used for the ordinate (Blatz and Magleby, 1986; Sigworth and Sine, 1987). The use of the square root scale helps to keep the errors associated with the variable bin width constant. This method of displaying data results in the exponential function becoming a skewed bell-shaped curve, where the peak of the curve occurs at the time constant of the distribution. Plotting dwell-time histograms this way has the great advantage that intervals spanning very large time ranges can be plotted on the same graph with multiple exponential components being easily identifiable. It is important to realize that the logarithmic transformation of the abscissa does not simply arise from plotting the exponential distribution semilogarithmically. The pdf that is actually plotted in these sorts of displays is given by

$$\sum a_i \tau_i^{-1} \exp(x - \tau_i^{-1} e^x),$$

Fig. 8. Kinetic properties of simulated data. (**A,B**) Distribution of open times fitted with a single exponential component. The distribution on the left is plotted on a conventional arithmetic scale and distribution has a typical exponential-looking appearance. (**B**) The same data are plotted on the logarithmic-square root scale that is now the preferred method for displaying distributions of single-channel dwell times that span a large time range. (**C,D**) The shut time distribution for the simulated data, where the use of the logarithmic time base makes it easier to see the two components in this distribution. (**E**) Distribution of burst length fitted with a single exponential component. (**F**) Distribution of the number of individual openings contained within each burst. This data is distributed geometrically. In all cases a resolution of 100 µs was imposed on both openings and closings.

where $a_i$ and $\tau_i$ are the relative area and time constant of the *i*th component of the distribution, respectively, and $x = \log_e t$, where $t$ is equal to the length of the interval (more detailed accounts can be found in Blatz and Magleby, 1986; Colquhoun and Sigworth, 1995; and Sigworth and Sine, 1987).

## 4.2. Channel Shut Times of the Simulated Data Record

The shut time distribution for the simulated data is shown in Fig. 8C (arithmetic scale) and Fig. 8D (logarithmic scale). There are two exponential components in the shut time distribution as a result of there being two shut states in the Castillo-Katz scheme. In this example the advantage of the logarithmic display is clear; we can see the two components of the distribution, whereas on the arithmetic display the first component cannot be seen clearly. In fact, in this display it has decayed to practically zero within a single bin width. A question that is frequently asked is, Can we attribute the time constants obtained from the fit of the pdf to particular states in the model? In our model the mean lifetime of AR (liganded but shut state) is equal to 133.3 µs (the reciprocal of the sum of 2500 s$^{-1}$ plus 5000 s$^{-1}$). For the mean lifetime of the R state (unliganded) we need to multiply the association rate constant ($10^7$ M$^{-1}$s$^{-1}$) by the ligand concentration (10 µM) and take the reciprocal of this, thus we obtain a mean lifetime of the R state of 10 ms. From the shut time distribution, we see that the time constants of the pdf are 152 µs (relative area = 0.64) and 16.5 ms (relative area = 0.36). The predicted time constants of the shut time pdf are 132.7 µs (relative area = 0.66) and 15.07 ms (relative area = 0.34). These values were determined using the SCBURST program (http://www.ucl.ac.uk/Pharmacology/dc.html). The 100-µs resolution imposed on the data set means that we will have omitted many events (brief closings) that should have been included in the distribution and thus compromised our ability to determine accurately the time constant of this fast component. In all likelihood we would have obtained an estimate from the fit closer to the predicted time constant if we had set a better resolution. Clearly the predicted value of the fast component of the pdf of *all* shut times (132.7 µs) is very close to the value of the mean lifetime of a single sojourn in the AR state (133.3 µs). Given the simple kinetic scheme used for this simulation, this is not surprising.

However, the same argument cannot be made for the discrepancy between the mean lifetime of the R state (10 ms) and the fit of the data that gives us a value of 16.5 ms. The reason that these two values do not (and would not be expected to) agree is simple. What we observe in the channel record is that the channel is either open or closed; we do not (and cannot) monitor the states that the channel occupies—this can only be inferred (e.g., the brief lifetime of the AR state). In the Castillo-Katz scheme, if the receptor binds to the ligand, it moves to the AR state; in the AR state, it has two options—it can either open or the ligand can dissociate. If the channel opens, the shut time is terminated; however, if the ligand dissociates, the channel enters the unliganded shut state, and the sojourns in three states will be concatenated. As the channel is memoryless, it does not remember that it has visited the AR state, and so the process with the same probabilities of transitions happening starts all over again. With the rate constants chosen in this example, two out of three times that the channel enters the AR state it will open, and on one third of the occasions the ligand will dissociate. Because of this the time constant of the second component in the shut time distribution does not give the mean lifetime of the R state. The predicted mean value of the second component of the shut time distribution using the set of rate constants chosen for the simulation is in fact 15.07 ms (obtained from SCBURST).

Therefore, in general it is not possible to equate the time constants obtained from pdfs with the mean lifetimes of channel states. The possible exception to this statement is when the channel enters a short-lived shut state that occurs within a burst of channel openings. However, even in this case the time constant obtained from the fit of the pdf of all shut times will not give this exactly (see above), and neither is it easy to determine the relationship of such short-lived states to other closed states.

## 4.3. Properties of Bursts of Openings in the Simulated Data Record

It was pointed out above that one of the features of the simulated data was the fact that channel openings tended to occur in quick succession, with short gaps separating the openings within a burst and longer gaps separating bursts from one another. Bursts can be thought of as representing the individual activation of a channel. In the case of a ligand-gated ion channel, bursts represent the series of openings initiated by the binding of ligand to its receptor and are terminated by the dissociation of ligand that causes the channel to enter a state from which it cannot open again without rebinding of ligand (for a mathematical treatment of this problem, see Colquhoun and Hawkes, 1982). Colquhoun and Sakmann (1985) documented the burst structure of nicotinic acetylcholine receptors found at muscle end plates of *Rana temporaria*. The distribution of gaps *within* bursts was described by two exponential components with means of 20 µs and (rarely) 0.51 ms when the ligand used was acetylcholine. However, if we study the individual activations (bursts) of NMDA receptors we see that within a single activation multiple shut time components exist and the values of these range for a few tens of microseconds to several hundred milliseconds depending on the subtype of NMDA receptor, and examples of such activity are shown in the next section. Therefore, it is important to realize that gaps within bursts need not be on the submillisecond time scale.

Returning to our simulated data, to identify the bursts we need to have a method by which we separate shut times that occur *within* a burst from those that occur *between* bursts. Bursts can then be defined as groups of openings that are separated from another group of openings by a shut time interval greater than a defined critical gap length ($t_{crit}$). There is a certain amount of ambiguity as to the optimal value of $t_{crit}$ to select. Three common methods are widely used to estimate the optimal value of $t_{crit}$. All result in a degree of misclassification of events; that is, events are deemed to be *within* bursts when in fact they are *between* bursts and vice versa. These methods are as follows: one can minimize the total number of misclassified events (Jackson et al., 1983), misclassify equal numbers from each component of the shut time distribution (Clapham and Neher, 1984) or misclassify equal percentages from each component of the shut-time distribution (Colquhoun and Sakmann, 1985). Taking the last method, the value of $t_{crit}$ can be obtained by solving, numerically, the following equation:

$$1 - \exp(-t_c/\tau_2) = \exp(-t_c/\tau_1),$$

where $\tau_1$ is the time constant of the longest shut-time component containing intervals thought to occur within a burst, and $\tau_2$ is the time constant of the component that immediately follows this one in the shut-time distribution.

From the shut-time distribution of our simulated data (and also knowing the nature of the kinetic scheme that produced the data), it is safe to assume that gaps contained within the first shut-time component occur mainly within bursts, while gaps in the second component occur mainly between bursts. The critical gap lengths calculated using each of the methods are as follows: 0.81 ms (minimum total number misclassified), 0.60 ms (equal numbers misclassified), and 0.52 ms (equal percentages misclassified). For the analysis of bursts that follows, a $t_{crit}$ of 0.52 ms has been chosen. In our example the estimate of the burst length would be similar if either of the other two values for $t_{crit}$ had been selected. It is important to realize, though, that when the number of events contained within the two components is very different, the first two methods described for obtaining $t_{crit}$ can result in a high proportion of the rarer sort of event being misclassified. Additionally, the method of equal proportions misclassified always results in a greater number of total misclassified events than would be obtained if one used either of the other two methods. Therefore, it is always a good idea to try different methods to see which one gives the most consistent result between different experiments.

With the $t_{crit}$ selected we can now construct the burst length distribution, which is shown in Fig. 8E. Again we have imposed a 100-µs resolution on the data, and the distribution is well described by a single exponential function, with a time constant of 3.34 ms. How does this value compare with what we might expect from our model? First we need to estimate the mean number of brief gaps that would be expected to occur in a burst. This is given by the following equation:

$$\text{Mean number of brief gaps} = \frac{\beta}{k_{-1}}.$$

Thus the ratio of the opening rate constant to the dissociation rate constant gives the mean number of sojourns in the AR state within a burst. Therefore, it follows that the mean number of openings that occur in burst is

$$\text{Mean number of openings per burst} = 1 + \frac{\beta}{k_{-1}}.$$

We can now calculate what the predicted mean burst length should be by

$$\text{Mean burst length} = \frac{1}{\alpha}\left(1 + \frac{\beta}{k_{-1}}\right) + \frac{1}{\beta + k_{-1}}\left(\frac{\beta}{k_{-1}}\right).$$

Therefore, we would expect there to be on average two brief gaps per burst, three openings per burst, and the mean burst length to be 3.27 ms. Thus our estimate of 3.34 ms is a good approximation to the true value. The distribution of the number of openings per burst is not an exponential distribution, but rather it is a geometric distribution. This is plotted in Fig. 8F, where the distribution is best fit with a single geometric component with a mean equal to 2.99 (as predicted).

In general, measurements of burst parameters are less affected by missed events (i.e., events not detected in the analysis because they cannot be differentiated from background noise) than parameters relating to open times and shut times. Furthermore, in the case of ligand-gated ion channels, bursts of channel openings represent single activations of the receptor-channel complex (for example see Béhé et al., 1999; Wyllie et al., 1998). Therefore it is the properties of such events that need to be studied if we want to relate microscopic single-channel activity to the macroscopic activity of ion channels that mediate a synaptic current.

## 5. Analysis of Single-Channel Data: Real Data Set

When it comes to the study of real channels, one might aim to infer what sort of kinetic scheme can account for the data we observe or what effect mutations might have on the type of channel activity observed (see below). In the simulation above, we started with a kinetic scheme and demonstrated the sort of channel activity that we might expect, a luxury not afforded to us normally. Nonetheless, simulations of this nature are useful to demonstrate some basic concepts involved in channel kinetics.

### 5.1. NMDA Single-Channel Currents: Comparison of Two Subtypes

When we carry out an experiment, we find that open and shut time distributions are hardly ever as straightforward as those gen-

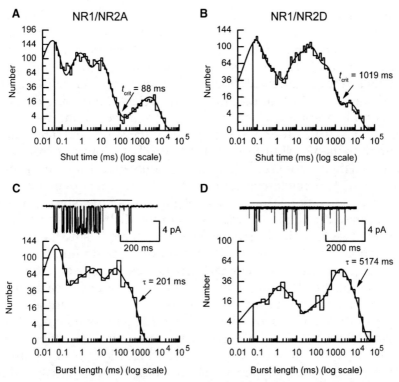

Fig. 9. Example of shut-time and burst-length distribution for two types of NMDA receptor-channels. (**A,B**) Shut-time distributions for recombinant NR1/NR2A and NR1/NR2D NMDA receptor-channels activated by glutamate. Each distribution is fitted with a mixture of multiple exponential components, and the critical shut time that separates gaps occurring within and between separate activations is indicated for each channel type. (**C,D**) Burst length distribution for NR1/NR2A and NR1/NR2D NMDA receptor-channels each fitted with a mixture of exponential components. The most prominent component, in terms of charge transfer, is the last component of each distribution, which has a mean of 201 ms for NR2A-containing receptors and 5174 ms for NR2D-containing receptors. Typical activations from these components are shown above the corresponding burst length distribution. [Adapted from Wyllie et al. (1998), with permission. © Blackwell Publishing.]

erated from the simulation detailed above. The histograms shown in Fig. 9 exemplify this point and serve to indicate that there is much to be done when it comes to understanding the functioning of individual ion channels. Figure 9A,B shows shut-time distributions for recombinant NR1/NR2A and NR1/NR2D NMDA receptors. Each distribution is fitted with a mixture of multiple exponential components and contains dwell times that range from the resolution (40 μs) to several tens of seconds—the advantage of the logarithmic representation of the data is now clearly evident. To determine which of the shut times occur *within* a single activation of the receptor and which occur *between* activation, we need to determine a value for $t_{crit}$ as was done for the simulated data above. Components of a shut-time distribution that show a clear dependence on ligand concentration will generally give a good indication as to which dwell times represent intervals that occur between separate activations. In the case of the histograms shown in Fig. 9A,B, these are the last components in the distribution and therefore the $t_{crit}$ values indicated have been calculated using the method of Clapham and Neher (1984) to produce a minimum, but equal number, of misclassified events from the penultimate and last components of the shut-time distributions. Note that for each of the NMDA receptor subtypes the estimates of $t_{crit}$ are different and suggest that single activations of these receptors contain closed periods that in the case of NR2D-containing NMDA receptors last for up to approximately 1 second. The burst length distributions for the two channel types are shown below their corresponding shut time distributions (Fig. 9C,D). Once again each of these distributions contains several exponential components. Notice that the structure of the bursts for each receptor subtype are quite different—NR2D-containing receptors last for several seconds and have a much lower open probability compared to NR2A-containing receptors (and of course contain some very long *within*-burst shut times). Examples of typical bursts found in the last component of each of the burst length distributions are illustrated. The line above the traces indicates the duration between the first opening and last closing of each event.

One might ask which component in such distributions is the most "physiologically relevant"? The last component (whose time constant is indicated and for which representative examples are shown above each of the histograms) of the shut time distribution carries most of the charge transfer. The burst length distributions illustrated in Fig. 9 were obtained from steady-state recording of NMDA receptor-channel activity using low agonist concentrations to ensure that individual bursts of openings were, temporally, well isolated. Clearly, during synaptic activation of NMDA receptors, the agonist concentration in the synaptic cleft is high (a few millimolar) but decays rapidly (in a few milliseconds). Indeed, using brief concentration jumps to mimic the synaptic activation of NMDA receptors, it has been demonstrated that the deactivation of such currents is dominated by a component with a time constant similar to that observed for the slowest component of the burst length distribution (Wyllie et al., 1998).

## 5.2. NMDA Single-Channel Currents: Example of a Mutation that Affects Kinetic Behavior

As mentioned above, the bursts of openings, if correctly identified, can give an indication of the sorts of single activations that ligand-gated ion channels might experience when activated by synaptically released neurotransmitters. Thus their durations, in part, are determined by the length of time that the ligand occupies its binding site. It is often difficult to determine whether a particular mutation in a ligand-gated ion channel mainly affects binding or gating parameters (Colquhoun, 1998, 2006), although with more advanced reaction mechanism fitting procedures (see below), identification of individual rate constants that are altered can sometimes be achieved. Nevertheless, some predictions about the effects of mutations can be made, and by assessing their effects on single-channel functioning we can test certain hypotheses about the nature of ligand–receptor interactions. An example of such an analysis is shown in Fig. 10.

For the purposes of illustration we will continue using, as an example, single-channel currents mediated by two recombinant NMDA receptor subtypes. Our understanding of structure–function studies of ionotropic glutamate receptors has advanced in recent years through the elegant crystallographic work of Gouaux and colleagues combined with electrophysiological studies (for recent reviews, see Chen and Wyllie, 2006; Mayer, 2006). Amino acids that H-bond with glutamate when it occupies its binding site have been identified. In particular, a conserved threonine residue in the S2 ligand binding domain is found in all NMDA NR2 (glutamate-binding) receptor subunits. Mutation of this residue in both NR2A and NR2D NMDA receptor subunits causes a large reduction in glutamate potency (Anson et al., 1998, 2000; Chen et al., 2004), but from the study of macroscopic studies alone (e.g., concentration response curves), in general it is not safe to conclude that the main effect of such a mutation is to affect ligand binding as opposed to receptor gating reactions that occur following binding

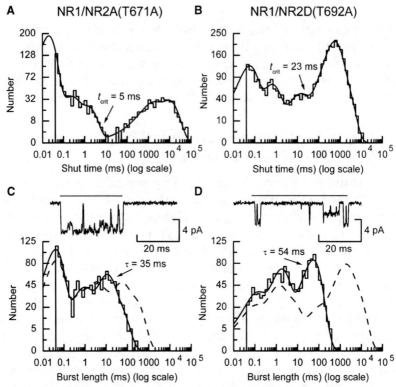

Fig. 10. Example of shut-time and burst-length distribution for two types of NMDA receptor-channels carrying mutations in the ligand-binding site. (**A,B**) Shut-time distributions for recombinant NR1/NR2A and NR1/NR2D NMDA receptor channels carrying a threonine to alanine point mutation in the S2 ligand-binding domain. Each distribution is fitted with a mixture of multiple exponential components and the critical shut time that separates gaps occurring within and between separate activations is indicated for each channel type. (**C,D**) Burst-length distribution for NR1/NR2A(T671A) and NR1/NR2D(T692A) NMDA receptor channels each fitted with a mixture of exponential components. The time constant of the slowest component of each distribution is indicated; the dashed line superimposed on each of the burst length distributions shows the corresponding pdf for wild-type channels. Typical activations from these components are shown above the corresponding burst length distribution. [**A,C** adapted from Anson et al. (2000), with permission. © Blackwell Publishing; **B,D** adapted from Chen et al. (2004), with permission. © Blackwell Publishing.]

events. Studies of the properties of single-channel events can help to address these issues.

Figure 10 shows shut-time and burst-length distributions for NR2A- and NR2D-containing NMDA receptors where this threonine residue has been replaced by an alanine residue. Using the same approach to define which shut times occur *within* and *between* separate activations of receptor-channels, $t_{crit}$ values are calculated to minimize the number of misclassified events. Note that one of the effects of this mutation is to reduce the value of $t_{crit}$ for each receptor subtype (compare Fig. 9A,B with Fig. 10A,B). The resulting burst length distributions for each mutant are shown below the corresponding shut-time distribution (Fig. 10C,D). For both distributions the longest component has a time constant of only a few tens of milliseconds. The values are much shorter [in the case of NR2D(T692A), almost 100-times shorter] than the values seen for wild-type receptors. The dashed line superimposed on each burst length distribution is the corresponding pdf for wild-type receptors (as shown in Fig. 9C,D). The shortening of the burst length is consistent with a mutation in this region of the protein, if we consider that a perturbation in the binding site might affect the duration that a ligand remains bound to its receptor. Thus if the microscopic dissociation rate constant is increased by the mutation, then a shortening of the burst length is not unexpected. By examining other single-channel parameters we can also assess whether this mutation has affected other properties. The examples of single-channel currents shown above each of the burst length distributions show that although the durations of these events are very much shortened (indicated by the line), their single-channel conductances are not altered by this mutation. Further characterization by measuring deactivation rates following concentration jumps can determine whether the burst length from steady-state recording is a good predictor of the macroscopic current time-course (Anson et al., 1998, 2000; Chen et al., 2004; Wyllie et al., 1998, 2006).

## 6. Direct Fitting of Reaction Schemes to Single-Channel Data Records

The dwell-time histograms shown above describing NMDA receptor-channel activity have been analyzed empirically, that is, with no *a priori* knowledge of the underlying reaction scheme for the receptor that generated the results. While such fitting methods still allow us to obtain information about channel activations, and may allow a gross interpretation of the effects of mutations, they cannot, in general, answer questions about whether a mutation affects binding or gating. The relationship between the time constants and areas from the empirical fit are indirectly related to the underlying mechanism, and it is rarely possible to relate the former to the latter. Indeed, in the binding mutation examples given above, we were only able to infer that the mutation affected single-channel kinetic parameters in a manner consistent with an alteration in the dissociation rate constant, but we could not tell by how much.

Recently it has become possible to analyze single-channel data records taking account of all the information they contain—not only the duration of open and shut times but also the sequence in which they occur. Two programs are capable of such global fitting of data records. These are HJCFIT (available from http://www.ucl.ac.uk/Pharmacology/dc.html; Colquhoun et al., 1996; Hawkes et al., 1992) and MIL (available from http://www.qub.buffalo.edu; Qin et al., 1996). Their relative advantages are discussed at http://www.ucl.ac.uk/Pharmacology/dcpr95.html#hjcfit. These powerful approaches to data analysis allow the experimenter to test whether particular reaction schemes are consistent with observed single-channel data records (and vice versa) and moreover allows for the determination of all the rate constants in a reaction scheme (as demonstrated in Colquhoun et al., 2003). Space does not permit a detailed account of how such analysis is implemented. Briefly, and as partly illustrated in Fig. 11, the process works as follows:

1. As for arbitrary fitting of dwell-time histograms, single-channel data sets need first to be optimized to obtain a sequence of open and shut times and a resolution imposed on the data set.
2. For ligand-gated ion channels, simultaneous fitting of data records recorded at several agonist concentrations is very desirable.
3. Using information obtained from the arbitrary fitting of exponential functions, a minimal number of open and shut states required to describe the data can be obtained. This information

can then be used as a starting point in designing reaction mechanisms (Fig. 11A) that might describe the observed data.

4. Dwell times and the sequence in which they occur are then fitted using the method of maximum likelihood with a correction for missed events (i.e., events that are omitted as they have durations that are less than the imposed temporal resolution). For example, if a short shutting is missed, the apparent open time is lengthened.

5. The distributions of apparent dwell times *predicted* from a particular reaction scheme are then superimposed on experimentally observed distributions. It is important to note that the probability density functions shown on the data sets are not fits of experimental histograms, but are the predicted distributions from a specified scheme. Thus, and as illustrated in Fig. 11, the quality of the maximum-likelihood fit can be judged by the number of events that are either over- or underrepresented in the data set.

6. A variety of reaction schemes (and constraints; for example, compare Fig. 11B and Fig. 11C) can be tested and assessed. The resulting maximum-likelihood fit will then provide the experimenter with all the rate constants describing the activation mechanism.

For further details, the interested reader is referred to the documentation supplied on the Web sites mentioned above and recent papers where such methods have been used to investigate, in particular, the activation of ligand-gated ion channels. For some receptor-channels this sort of approach to channel analysis has met with considerable success for the study of both wild-type and mutant receptors (notably nicotinic acetylcholine and glycine receptors; for example, see Beato et al., 2004; Burzomato et al., 2004; Chakrapani et al., 2004; Elenes and Auerbach, 2002; Grosman and Auerbach, 2000; Hatton et al., 2003; Prince et al., 2002; Qin et al., 1996; Shelley and Colquhoun, 2005; Wang et al., 2000; Zhou et al., 2005), whereas for others, such as NMDA receptors where the structure of activations is much more complex, schemes that can account for all of the features observed in single-channel recordings remain more elusive (however, see Auerbach and Zhou, 2005; Banke and Traynelis, 2003; Erreger et al., 2005a,b; Popescu and Auerbach, 2003, Popescu et al., 2004; Schorge et al., 2005). Thus, while considerable progress has been made in recent years, much remains to be done, and it is to be hoped that such direct fitting methods will allow for fuller understanding and interpretations of the effects of different agonists acting at, mutating to, and modulation of ion channels.

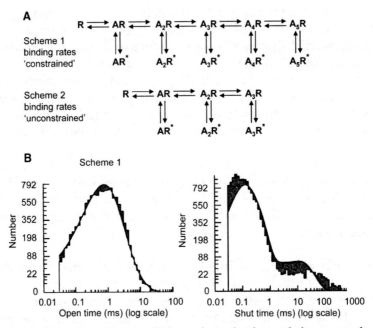

Fig. 11. Direct mechanism fitting of single-channel data records. (**A**) Two reaction schemes used to investigate the action of glycine at $\alpha1$ homomeric glycine receptors expressed in HEK293 cells (Beato et al., 2004). In the upper scheme the rate constants describing binding and unbinding events (i.e., the microscopic association and dissociation rate constants) are constrained to be equal, whereas in the lower scheme these rate constants are unconstrained. Note: the precise values for these rate constants are not illustrated in this example. (**B**) Open- and shut-time distributions obtained from single-channel recording of glycine-evoked ($100\,\mu M$) activity in a cell-attached patch. The solid line superimposed on each of the histograms is the maximum-likelihood distribution *predicted* from scheme 1. Note that this predicted distribution does not describe well the shut-time histogram in particular. This is shown by the shaded gray areas, which highlight areas of the histogram that contain either too many or too few events predicted by the reaction scheme.

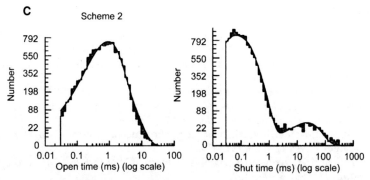

Fig. 11 (*Continued*) (**C**) The same open- and shut-time histograms as in **B**, but this time the maximum likelihood distribution *predicted* by scheme 2 (where binding and unbinding rates are unconstrained) is superimposed on the data sets. Note that there is considerable improvement using this scheme and there are fewer regions in the shut-time histogram where there is either an over- or underrepresentation of events. [Courtesy of M. Beato, D. Colquhoun, and L. Sivilotti, Department of Pharmacology, University College London (UCL).]

## 7. Summary

This chapter provided an overview of techniques and concepts used in the recording and analysis of single-channel data. The reader should not be put off by some of the complexities of the more detailed sorts of analysis described here. Rather, try getting the data first and then seek advice if you find you are struggling. Watching the movement of single proteins is captivating.

The simulation routine and the data analysis programs that have been used, and several others that have been described here, are available, free of charge for academic use, from the following Web sites: http://www.ucl.ac.uk/Pharmacology/dc.html (DCprogs), http://spider.science.strath.ac.uk/physpharm/ (Strathclyde Electrophysiology Software), and http://ww.qub.buffalo.edu (QuB).

## Acknowledgments

Work in my lab is currently supported by grants from the Biotechnology and Biological Sciences Research Council. I would like to thank those colleagues who supplied me with illustrations and unpublished data that have been incorporated into this review. I am grateful to David Colquhoun and Lucia Sivilotti for commenting on, and suggesting improvements to, aspects of this review.

# References

Anson, L. C., Chen, P. E., Wyllie, D. J. A., Colquhoun, D., and Schoepfer, R. (1998) Identification of amino acid residues of the NR2A subunit that control glutamate potency in recombinant NR1/NR2A NMDA receptors. *J. Neurosci.* **18**, 581–589.

Anson, L. C., Schoepfer, R., Colquhoun, D., and Wyllie, D .J. A. (2000) Single-channel analysis of a NMDA receptor possessing a mutation in the region of the glutamate binding site. *J. Physiol.* **527**, 225–237.

Auerbach, A. and Zhou, Y. (2005) Gating reaction mechanisms for NMDA receptor channels. *J. Neurosci.* **25**, 7914–7923.

Banke, T. G. and Traynelis, S. F. (2003) Activation of NR1/NR2B NMDA receptors. *Nature Neurosci.* **6**, 144–152.

Beato, M., Groot-Kormelink, P. J., Colquhoun, D. and Sivilotti L. G. (2004) The activation of α1 homomeric glycine receptors. *J. Neurosci.* **24**, 895–906.

Béhé, P., Colquhoun, D., and Wyllie, D. J. A. (1999) Activation of single AMPA- and NMDA- type glutamate-receptors channels, in *Handbook of Experimental Pharmacology* (Jonas, P. and Monyer, H., eds.), vol 141, Springer-Verlag, Berlin, pp. 175–218.

Blatz, A.L. and Magelby, K.L. (1986) Correcting single channel data for missed events. *Biophys. J.* **49**, 967–980.

Burzomato, V., Beato, M., Groot-Kormelink, P. J., Colquhoun, D., and Sivilotti, L. G. (2004) Single-channel behavior of heteromeric α1β glycine receptors: an attempt to detect a conformational change before the channel opens. *J. Neurosci.* **24**, 10924–10940.

Castillo, J. del and Katz, B. (1957) Interaction at end-plate receptors between different choline derivatives. *Proc. R. Soc. B* **146**, 369–381.

Chakrapani, S., Bailey, T. D., and Auerbach, A. (2004) Gating dynamics of the acetylcholine receptor extracellular domain. *J. Gen. Physiol.* **123**, 341–356.

Chen, P. E., Johnston, A. R., Mok, M. H. S., Schoepfer, R., and Wyllie, D. J. A. (2004) Influence of a threonine residue in the S2 ligand binding domain in determining agonist potency and deactivation rate of recombinant NR1a/NR2D NMDA receptors. *J. Physiol.* **558**, 45–58.

Chen, P. E. and Wyllie, D. J. A. (2006) Pharmacological insights obtained from structure-function studies of ionotropic glutamate receptors. *Br. J. Pharmacol.* **147**, 839–853.

Clapham, D. E. and Neher, E. (1984) Substance P reduces acetylcholine-induced currents in isolated bovine chromaffin cells. *J. Physiol.* **347**, 255–277.

Colquhoun, D. (1998) Binding, gating, affinity and efficacy. The interpretation of structure-activity relationships for agonists and of the effects of mutating receptors. *Br. J. Pharmacol.* **125**, 923–948.

Colquhoun, D. (2006) The quantitative analysis of drug-receptor interactions: a short history. *Trends Pharmacol. Sci.* **27**, 149–157.

Colquhoun, D., Hatton, C. J., and Hawkes, A. G. (2003) The quality of maximum likelihood estimates of ion channel rate constants. *J. Physiol.* **547**, 699–728.

Colquhoun, D. and Hawkes, A. G. (1982) On the stochastic properties of bursts of single ion channel openings and of clusters of bursts. *Philos. Trans. R. Soc. Lond. B* **300**, 1–59.

Colquhoun, D. and Hawkes, A. G. (1995) The principles of the stochastic interpretation of ion-channel mechanisms, in *Single Channel Recording* (Sakmann, B. and Neher, E., eds.), 2nd ed., Plenum Press, New York, pp. 397–482.

Colquhoun, D., Hawkes, A. G., and Srodzinski, K. (1996) Joint distributions of apparent open and shut times of single-ion channels and maximum likelihood fitting of mechanisms. *Philos. Trans. R. Soc. Lond. A* **354**, 2555–2590.

Colquhoun, D. and Sakmann, B. (1985) Fast events in single-channel currents activated by acetylcholine and its analogues at the frog muscle end-plate. *J. Physiol.* **369**, 501–557.

Colquhoun, D. and Sigworth, F. J. (1995) Fitting and statistical analysis of single-channel records, in *Single Channel Recording* (Sakmann, B. and Neher, E., eds.), 2nd ed., Plenum Press, New York, pp. 483–587.

del Castillo, J. and Katz, B. (1957) Interaction at end-plate receptors between different choline derivatives. *Proc. R. Soc. B* **146**, 369–381.

Edwards, F. A., Konnerth, A., Sakmann, B. and Takahashi, T. (1989) A thin-slice preparation for patch-clamp recording from synaptically connected neurones of the mammalian central nervous system. *Plfügers Arch.* **414**, 600–612.

Elenes, S. and Auerbach, A. (2002) Desensitization of diliganded mouse muscle nicotinic acetylcholine receptor channels. *J. Physiol.* **541**, 367–383.

Erreger, K., Dravid, S. M., Banke, T. G., Wyllie, D. J. A., and Traynelis, S. F. (2005a) Subunit specific gating controls rat recombinant NR1/NR2A and NR1/NR2B channel kinetics and synaptic signalling profiles. *J. Physiol.* **563**, 345–358.

Erreger, K., Geballe, M. T., Dravid, S. M., Snyder, J. P., Wyllie, D. J. A., and Traynelis, S. F. (2005b) Mechanism of partial agonism at NMDA receptors for a conformationally restricted glutamate analog. *J. Neurosci.* **25**, 7858–7866.

Grosman, C., and Auerbach A. (2000) Kinetic, mechanistic, and structural aspects of unliganded gating of acetylcholine receptor channels: a single-channel study of second transmembrane segment 12' mutants. *J. Gen. Physiol.* **115**, 621–635.

Hamill, O. P., Marty, A., Neher, E., Sakmann, B., and Sigworth, F. J. (1981) Improved patch-clamp techniques for high-resolution current recordings from cells and cell-free membrane patches. *Plfügers Arch.* **391**, 85–100.

Hatton, C., Shelley, C., Brydson, M., Beeson, D., and Colquhoun, D. (2003) Properties of the human muscle nicotinic receptor, and of the slow–channel syndrome mutant eL221F, inferred from maximum likelihood fits. *J. Physiol.* **547**, 729–760.

Hawkes, A. G., Jalali, A., and Colquhoun, D. (1992) Asymptotic distributions of apparent open times and shut times in a single channel record allowing for the omission of brief events. *Philos. Trans. R. Soc. Lond. B* **337**, 383–404.

Jackson, M. B., Wong, B. S., Morris, C. E., Lecar, H., and Christian, C. N. (1983) Successive openings of the same acetylcholine receptor channel are correlated in open time. *Biophys. J.* **42**, 109–114.

Mayer, M. L. (2006) Glutamate receptors at atomic resolution. *Nature* **440**, 456–462.

Neher, E. and Sakmann, B. (1976) Single channel currents recorded from membrane of denervated frog muscle fibres. *Nature* **260**, 799–802.

Popescu, G. and Auerbach, A. (2003) Modal gating of NMDA receptors and the shape of their synaptic current. *Nature Neurosci.* **6**, 476–483.

Popescu, G., Robert, A., Howe, J. R., and Auerbach, A. (2004) Reaction mechanism determines the NMDA receptor response to repetitive stimulation. *Nature* **430**, 790–793.

Prince, R. J., Pennington, R. A., and Sine, S. M. (2002) Mechanism of tacrine block at adult human muscle nicotinic acetylcholine receptors. *J. Gen. Physiol.* **120**, 369–393.

Qin, F., Auerbach, A., and Sachs, F. (1996) Estimating single-channel kinetic parameters from idealized patch-clamp data containing missed events. *Biophys. J.* **70**, 264–280.

Schorge, S., Elenes, S., and Colquhoun, D. (2005) Maximum likelihood fitting of single channel NMDA activity with a mechanism composed of independent dimers of subunits. *J. Physiol.* **569**, 395–418.

Selyanko, A. A. and Sim, J. A. (1998) $Ca^{2+}$-inhibited non-inactivating $K^+$ channels in cultured rat hippocampal pyramidal neurones. *J. Physiol.* **510**, 71–91.

Shelley, C. and Colquhoun, D. (2005) A human congenital myasthenia–causing mutation (epsilon-L78P) of the muscle nicotinic acetylcholine receptor with unusual single channel properties. *J. Physiol.* **564**, 377–396.

Sherman-Gold, R. (ed.) (1993) *The Axon Guide for Electrophysiology and Biophysics Laboratory Techniques.* Axon Instruments, Union City, California.

Sigworth, F. J. (1995) Electronic design of the patch clamp, in *Single Channel Recording* (Sakmann, B. and Neher, E., eds.), 2nd ed., Plenum Press, New York, pp. 95–127.

Sigworth, F. J. and Sine, S. M. (1987) Data transformations for improved display and fitting of single-channel dwell time histograms. *Biophys. J.* **52**, 1047–1054.

Wang, H. L., Ohno, K., Milone, M., et al. (2000) Fundamental gating mechanism of nicotinic receptor channel revealed by mutation causing a congenital myasthenic syndrome. *J. Gen. Physiol.* **116**, 449–462.

Wyllie, D. J. A., Béhé, P., and Colquhoun, D. (1998) Single-channel activations and concentration jumps: comparison of recombinant NR1A/NR2A and NR1A/NR2D NMDA receptors. *J. Physiol.* **510**, 1–18.

Wyllie, D. J. A., Johnston, A. R, Lipscombe, D., and Chen, P. E. (2006) Single-channel analysis of a point mutation of a conserved serine residue in the S2 ligand binding domain of the NR2A NMDA receptor. *J. Physiol.* **574**, 477–489.

Zhou, Y., Pearson J. E., Auerbach, A. (2005) Phi-value analysis of a linear, sequential reaction mechanism: theory and application to ion channel gating. *Biophys. J.* **89**, 3680–3685.

## Recommended Reading

Aidley, D. J. and Stanfield, P. R. (1996) *Ion channels—Molecules in Action*, Cambridge University Press, Cambridge.

Ashley, R. H. (ed.) (1995) *Ion Channels—A Practical Approach*, IRL Press, Oxford.

Conn, P. M. (ed.) (1998) *Methods in Enzymology—Ion Channels*, Part B, Academic Press, New York.

Hille, B. (1992). *Ionic Channels of Excitable Membranes*, 2nd ed., Sinauer Associates, MA.

Hille, B. (2001). *Ion Channels of Excitable Membranes*, 3rd ed., Sinauer Associates, Sunderland, MA.

Neher, E. and Sakmann, B. (eds) (1995) *Single Channel Recording*, 2nd ed., Plenum Press, New York.

Ogden, D. C. (ed.) (1994) *Microelectrode Techniques: The Plymouth Workshop Handbook*, 2nd ed., Company of Biologists Limited, Cambridge.

# 4

# Combined Fluorometric and Electrophysiological Recordings

## Hartmut Schmidt and Jens Eilers

## 1. Introduction

Combined electrophysiological and fluorometric recordings have proven to be a powerful tool, especially in the field of neurobiology. With this combination it became possible to overcome three major limitations of pure electrophysiological measurements. First, high-resolution recordings from cellular compartments distant to the recording electrode became feasible, allowing, for example, the quantification of the occupancy of postsynaptic receptors (Mainen et al., 1999) and the density of voltage-gated $Ca^{2+}$ channels (Sabatini and Svoboda, 2000) at the level of single dendritic spines. Second, the analysis of cellular responses not directly associated with electrical signals became possible. Examples include synaptically evoked $Ca^{2+}$ release from intracellular stores (Takechi et al., 1998; Finch and Augustine, 1998) and $Ca^{2+}$ buffering by endogenous $Ca^{2+}$-binding proteins (Zhou and Neher, 1993). Third, the spatiotemporal extent of second messenger signals, for example, during subthreshold synaptic activity (Eilers et al., 1995a) and during the induction of synaptic plasticity (Eilers et al., 1997b), can be monitored by means of fluorescence imaging.

In recent years, several technical advances have extended the range of possible applications of fluorescence microscopy. Thus, two-photon microscopy (Denk et al., 1990), green-fluorescent proteins (Chalfie et al., 1994), and indicators that use the Förster resonance energy transfer (FRET) effect (Adams et al., 1991) opened exciting new avenues for studying cell physiology. In view of these ongoing developments, combined fluorometric and electrophysio-

From: *Neuromethods, Vol. 38: Patch-Clamp Analysis: Advanced Techniques, Second Edition*
Edited by: W. Walz © Humana Press Inc., Totowa, NJ

logical recordings will continue to be the method of choice for high-resolution in vitro and in vivo studies.

This chapter provides a detailed overview of the special requirements for performing fluorometric measurements in combination with standard patch-clamp recordings. The emphasis is on measurements of dendritic $Ca^{2+}$ signals and the determination of protein mobility in neurons in brain slice preparations since such recordings can encompass significant difficulties arising from the fast time course or compartmentalization of the fluorescence signals. However, the methodology is also applicable for other preparations and applications, such as dendritic sodium signals (Rose et al., 1999).

## 2. Methods

### 2.1. Recording Setup

#### 2.1.1. Vibration Control

As for standard electrophysiological recordings, good isolation from vibration is also critical for fluorometric experiments. While vibrations affect the stability of the patch-clamp recording, they also lead to fluctuations in fluorescence signals, especially when recording from small cellular compartments, such as spines, that have to be kept in focus with high precision. Thus, any vibrations of the setup need to be prevented by mounting it on a high-quality air table. In addition, any movements of the fluorescence detector (i.e., camera or confocal system) relative to the microscope body need to be restricted by a rigid connection between the microscope and the detector system.

If the fluorescence system incorporates a laser, severe vibrations may be introduced by its fan (for air-cooled lasers) or the water flow through its cooling pipes (for water-cooled lasers). Such vibrations are difficult to dampen. Fortunately, most modern laser-based imaging systems use fiber optics to connect the laser to the scanning unit. Here, the laser does not have to be located directly on the air table and, therefore, vibrations cause no problem. However, in some older systems the laser is rigidly connected to the imaging system. If critical vibrations occur in these systems, an experienced service engineer could take out the laser and reconnect it via an optical fiber (e.g., from Melles Griot, Irvine, CA).

## 2.1.2. Microscope

Several important considerations concern the selection of the microscope and its components. Whether the setup is based on an inverted or an upright microscope is determined by the preparation under study. For work on cultured cells, inverted microscopes are preferable. Here, the objective can be positioned quite close to the specimen since the patch-clamp electrode is above the focal plane. Thus, objectives with a high numerical aperture (NA) can be selected that have a good optical resolution and a high collection efficacy. For work on thick samples (e.g., brain slices), upright microscopes (Fig. 1) are preferable because they allow the establishment of the electrophysiological recording under visual control. Suitable objectives, however, have a slightly reduced NA due to the long working distance required to be able to position the electrode under the objective.

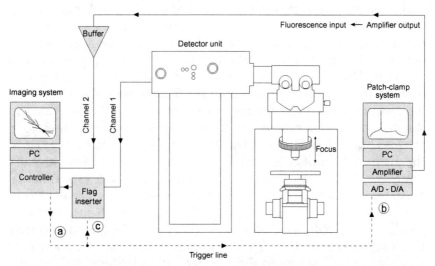

Fig. 1. Schematic drawing of the experimental setup. The system consists of an upright microscope (center) equipped with a fluorescence detection unit (left) and an electrophysiology unit (right). Focusing moves the nosepiece and the attached objective only. The imaging system can trigger the patch-clamp system (a to b) or the flag inserter (c, see Fig. 4), which electronically places markers into the fluorescence signal (channel 1) before it is recorded by the imaging system. The electrophysiological signal can be amplified (buffer) and fed into a free channel of the imaging system (channel 2) to allow accurate correlation of electrical and fluorometric responses (see Fig. 5).

The z-stage should be "fixed," and focusing should only move the nosepiece with the attached objective (Fig. 1). With a fixed stage, the specimen is not moved during focusing, and the manipulators can be mounted on rigid columns surrounding the microscope (see, for example, the mechanical setup from Luigs and Neumann, Ratingen, Germany). For two reasons, the focusing should also not move the trinocular, to which, in most cases, the imaging system is coupled (Fig. 1). First, some systems (especially confocal laser-scanning systems) have a considerable weight, which may lead to slow defocusing or may not be supported by the focusing unit at all. Second, as stated above, the imaging system should be mounted rigidly on the air table to prevent vibrations. If the system is only attached to the trinocular, the fluorometric recordings are prone to movement artifacts.

Some patch-clamp experiments may require the use of infrared differential interference contrast (IR-DIC) optics to visualize somata and dendrites in thick samples (Dodt and Zieglgänsberger, 1990). While this technique generates high-contrast images, even deep in brain slices, it necessitates the addition of two optical elements between the specimen and the trinocular, namely a prism and a polarizer. During fluorometric recordings, these two elements block a substantial amount of the emitted fluorescence. However, the light emitted from the fluorescent probes is of ultimate value in fluorescence measurements of living preparations and, therefore, the loss of photons induced by DIC optics is not acceptable. Thus, the distinct DIC optics (especially the polarizer) should be carefully removed during the fluorescence recording. If the experimental arrangement does not allow this maneuver, Dodt's gradient contrast illumination (Schiefer et al. 1999; Luigs and Neumann) or a condenser for oblique illumination (Olympus, Tokyo, Japan) may be an attractive solution. These components are based on a simple modification of standard bright field illumination and do not require any special optical elements in the emission light path. In many preparations the resolution is as good as that obtained with IR-DIC.

The most critical optical element is the objective. Its selection is primarily based on the requirements of the electrophysiology. Thus, its immersion medium, working distance, and geometry should easily permit patch-clamp recordings. Beside this, the two most important parameters for fluorometric recordings are the wavelength-dependent transmittance and the numerical aperture.

### 2.1.2.1. TRANSMITTANCE

Standard objectives transmit light efficiently in the visible spectral range but may have serious limitations in the ultraviolet (UV) and infrared (IR) range. Since most available dyes emit light in the blue to yellow range, transmittance is no problem for *emission*. However, *excitation* of the dyes may require UV (<400 nm) or IR (>800 nm) light. Typical examples include quantitative $Ca^{2+}$ measurements using the fluorescent indicator *Fura-2* (excitation 340–380 nm) and in vivo two-photon microscopy (excitation 760–900 nm). Thus, depending on the experimental conditions (available imaging system, dye to be used, etc.), a UV- or IR-optimized objective may be required. While all major microscopy companies provide such lenses, a comparison between objectives from different companies is difficult. This is due to the fact that most companies do not inform their customers about the spectral transmission curves of their objectives. In that respect, a demonstration and direct comparison of suitable objectives may provide useful information on which objective to buy.

### 2.1.2.2. NUMERICAL APERTURE (NA)

For fluorometric recordings, the NA is of the utmost importance. A thorough explanation of the optical and numerical basis of the NA is beyond the scope of this chapter; see Lanni and Keller (2005) for details. Generally speaking, the NA describes how much of the emitted fluorescence is collected by the objective. A higher NA corresponds to a higher collecting efficacy. Note, however, that the NA is not linearly related to the efficacy. For water immersion objectives, a NA of 0.5 corresponds to a collecting efficacy of 5%, a NA of 0.7 to 7.5%, and a NA of 0.9 to 13.5%. Objectives with as high an NA as possible should be chosen to collect a maximum of the emitted fluorescence.

### 2.1.3. Imaging System

Today, there is a vast and somewhat confusing diversity in fluorescence detection systems available from commercial suppliers. Choices range from standard camera-based fluorescence systems to confocal laser-scanning microscopes and, more recently, to two-photon microscopes. All these systems have their specific pros and cons, so no single favorite can be recommended. The following subsections cite specific requirements that should be considered when selecting a system for combined electrophysiological and fluorescence recordings.

### 2.1.3.1. TEMPORAL RESOLUTION

Depending on the application, one may have to cope with extremely rapid fluorescence signals. Dendritic $Ca^{2+}$ transients, for example, can have decay time constants well below 100 ms (see Fig. 5). To reliably quantify such signals, the fluorescence system has to be capable of recording at a comparable speed. Several approaches can be used to fulfill this demand. Thus, some cameras as well as laser-scanning confocal systems and two-photon microscopes (Fan et al., 1999) can acquire full images at video rate (see Fig. 4). Other systems allow the use of binning, or the selection of regions of interest to sample at rates of tens of Hertz. An even higher time resolution is obtained by restricting the acquisition to a single line (the so-called line-scan mode; Fig. 2) or a single point (see Fig. 5).

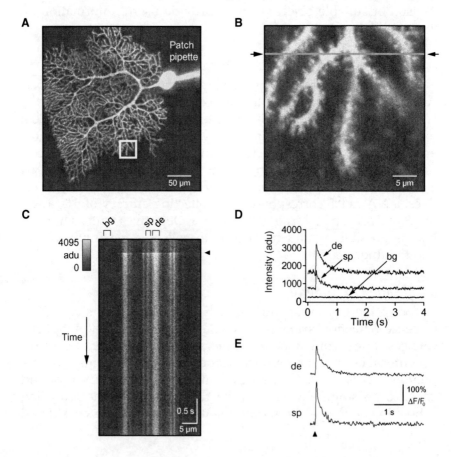

Sample rates of around 500 Hz (line scan) and 0.5 MHz (point mode) can be achieved. The ultimate temporal resolution is achieved by random-access scanners (Bullen et al., 1997; Salome et al., 2006) that enable recording fluorescence signals in the kilohertz range from several selected points of interest. In any case, when deciding on a system, special care should be taken to test the functionality of the hardware and software. Systems may perform brilliantly as long as standard functions are carried out, but may crash, more or less systematically, as soon as one tries to push the temporal resolution to its limits.

### 2.1.3.2. SPATIAL RESOLUTION

Fluorescence signals may be restricted to small cellular compartments such as submembrane shells (Eilers et al., 1995b; Hernández-

---

Fig. 2. Line-scan recordings of $Ca^{2+}$ transients in spines and dendrites. (A) Confocal fluorescence image of a Purkinje neuron in a cerebellar slice preparation. The cell was loaded with the $Ca^{2+}$ indicator Oregon Green BAPTA-1 via the patch pipette. The white rectangle outlines the region that is shown at higher magnification in (B). The image represents a maximum projection of a stack of 16 $x$-$y$ images taken at 1-μm $z$ intervals. (B) High-resolution image of spiny dendrites taken with a 60× objective (0.9 NA) and an additional 10× magnification provided by the confocal imaging system. The arrows and the gray line indicate the area in which activity-induced $Ca^{2+}$ transients were monitored in the line-scan mode (see panel C). Note that dendrites that were out of focus appear dim and blurred. (C) Line-scan image ($x$ versus time) of the region outlined in B during activation of the afferent climbing fiber. In this and the following figures the time point of stimulation is denoted by arrowheads. The stimulation resulted in a suprathreshold excitatory postsynaptic potential (EPSP) (not shown) and a transient increase in fluorescence in spines and dendrites, which reflects a transient increase in the intracellular $Ca^{2+}$ concentration. Three regions [background (bg), spine (sp), and dendrite (de)] are demarked, in which the time course of the fluorescence changes were analyzed in detail (see D). The temporal resolution of the acquisition was 2 ms per line. The pixel intensities of the 12-bit images range from 0 (black) to 4095 (white) in arbitrary digital units (adu; left inset). (D) Averaged pixel intensity from the three regions outlined in (C) plotted against time. Note that all values are well above 0 and below 4095 adu, indicating a correct setting of gain and contrast of the imaging system (see text for details). (E) Fluorescence transients in the spine and the dendrite are converted to background-corrected $\Delta F/F_0$ values (see text).

Cruz et al., 1990) or dendritic spines (Müller and Connor, 1991; Takechi et al., 1998). For studying such signals, standard fluorescence techniques are of limited use due to image blurring by out-of-focus light. Although such stray light can be mathematically removed from the images off-line (Carrington et al., 1995), this procedure is rather complicated and time-consuming. A simpler, though more expensive alternative is given by confocal laser-scanning microscopy (CLSM; Minsky, 1961) and two-photon microscopy (TPM; Denk et al., 1990). Here, out-of-focus light is either prevented from reaching the detector (CLSM) or is not generated at all (TPM), yielding high-resolution images even in opaque tissue. The examples shown in Figs. 2, 4, and 5 were generated with commercial CLSM systems, and the example in Fig. 6 was generated with a custom-built TPM.

### 2.1.3.3. EXCITATION LIGHT

Different fluorescence systems are equipped with different light sources. While camera-based systems usually employ arc lamps equipped with filter wheels or monochromators, confocal or two-photon systems use lasers for excitation. In both cases severe vibration may occur either due to the cooling device of the laser or when the filter wheel or monochromator is activated during selection of different wavelengths. Thus, whenever possible, the light source should be connected to the fluorescence system via an optical fiber to prevent these vibrations from affecting the electrophysiological recordings or the optical resolution. Furthermore, the fluorescence system should permit simple and vibration-free exchange of its optical elements, such as barrier filters and dichroic mirrors. Thus, switches and turrets should operate smoothly and no strong solenoids should be used for shuttering the excitation light.

### 2.1.3.4. SOFTWARE

Fluorescence responses may have small amplitudes and fast kinetics and, thus, may be hard to detect by eye. For the detection of such signals the acquisition software should provide simple measurement tools, such as a quick brightness versus time analysis from selected regions of interest. Also, a basic macro language that allows the programming of simple acquisition protocols (Fig. 3) is extremely helpful in detecting minute fluorescence signals. The

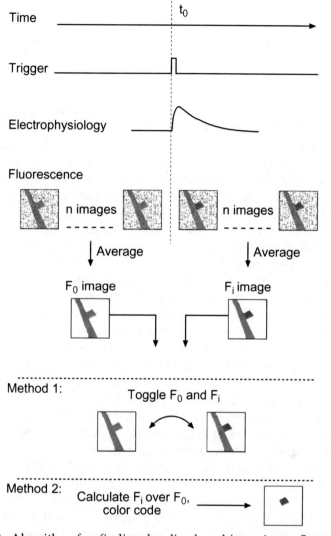

Fig. 3. Algorithm for finding localized and/or minute fluorescence transients. $t_0$ denotes the time point at which the fluorescence system sent a trigger signal that started the stimulation of afferent fibers. The electrophysiology trace shows an example of EPSP. The rectangles represent simulated images of a dendrite and an adjacent spine with a fluorescence signal restricted to the spine (right images). In addition, the top row of images contains synthetic noise to illustrate the low signal-to-noise ratio of raw recordings. Before and after the stimulation several images are averaged to obtain the $F_0$ and $F_i$ image, respectively. Differences in the fluorescence intensity can easily be identified either by alternating the display of the $F_0$ and $F_i$ image at about 1 to 4 Hz (Method 1) or by calculating the ratio of the images and displaying the result using a color code that emphasizes values different from 1 (Method 2).

idea is to be able to judge during or immediately after a recording whether it was successful or not.

## 2.1.3.5. HARDWARE

The fluorometric system should be able to start the acquisition of the electrophysiology system. This will facilitate recordings in which either small and fast responses need to be identified or the electrophysiological response needs to be correlated with the fluorescence signal with a high temporal precision. For this purpose a trigger signal (as in Fig. 1) from the fluorescence unit should be available that either signals the acquisition status (active/inactive) or can be set by macro commands in the fluorescence software. This output will in most cases be a TTL (transistor-transistor logic, 0 V and 5 V representing "off" and "on," respectively) line that can directly be connected to a corresponding input of the electrophysiological unit (b in Fig. 1). Most patch-clamp acquisition software packages allow acquisition to be triggered by an external stimulus; thus, the described trigger line can be used to have the fluorescence unit control the exact time point of, for example, a somatic depolarization or synaptic stimulation. Figures 2 to 5 illustrate recordings in which this capability was of critical importance.

The trigger output can also be used to automatically place a marker into the stream of the fluorescence data (c in Fig. 1). For this function, an appropriate electronic device (flag inserter in Fig. 1) needs to be added that, when triggered, electronically inserts a marker, the flag, into the fluorescence data (see Fig. 3B). During the recording as well as during playback of the data, the flag will be a clearly recognizable indicator of the time point of stimulation. While electronic circuits for this purpose are not commercially available, they can easily be constructed. For video-rate fluorescence imaging, an appropriate circuit has been described (Eilers et al., 1997a) that could be adapted to different fluorescence systems.

For the highest demands in temporal correlation, a free fluorescence input of the imaging system can be used for recording electrophysiological in parallel with fluorometric data (channel 2 and channel 1, respectively, in Fig. 1). An example of this application is shown in Fig. 5. Note that the output impedance of the electrophysiological signal (usually in the range of $500\,\Omega$) must match the input impedance of the fluorescence input (usually about $50\,\Omega$). Also, the fluorescence input will most likely expect positive

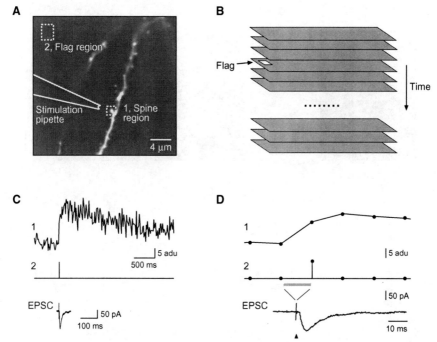

Fig. 4. Synchronizing electrical and fluorometric recordings using the flag inserter. (**A**) Confocal image of apical dendrites of a CA1 pyramidal neuron loaded with Calcium Green-1. The dashed boxes delineate the regions that were used for measurement of the fluorescence in the spine and the flag signal generated by the flag inserter circuit (see Fig. 1 and text). Afferent fibers were activated with an extracellular stimulation pipette. (**B**) Schematic illustration of the acquired image sequence. The gray rectangles indicate individual images. The white mark represents the flag that is inserted electronically by the flag inserter. The flag is drawn at a fixed position into the image generated immediately after the time point of synaptic stimulation. (**C**) Activation of the afferent fibers produced an excitatory postsynaptic current (EPSC) (bottom trace) and an associated fluorescence increase in the spine (trace 1). The flag signal (trace 2) shows the single data point of maximal intensity (reduced to 5% for clarity) that corresponds to the image into which the flag was inserted. Image acquisition rate was 60 Hz. (**D**) The fluorometric data from **C** plotted on the same time scale as the EPSC. Individual data points are indicated by closed circles. The flag signal (third data point in trace 2) allowed the EPSC and the fluorescence signal to be correlated with an accuracy of 16.5 ms, indicated by the gray line. (Modified from Eilers et al., 1997a.)

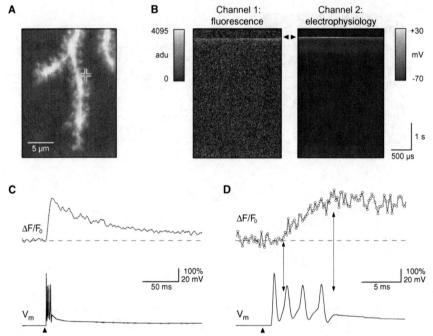

Fig. 5. Simultaneous electrophysiological and fluorometric recordings in the point mode. (**A**) Confocal image of spiny dendrites of a Purkinje neuron loaded with Oregon Green 488 BAPTA-1. The cross indicates a single spine from which the fluorescence was sampled in the point mode (see **B**), channel 1. (**B**) Point recording of the electrical response (channel 2, recorded with a somatic patch pipette) and the fluorometric response (channel 1) in the spine marked in **A** evoked by activation of the afferent climbing fiber. The confocal system allowed the recording of two channels (sampling rate 0.5 MHz). While the fluorescence signal was recorded in the first channel, the cell's membrane potential ($V_m$) was fed into channel 2 (see text for details). The arrowhead denotes the time point of stimulation. Note that both the $x$ and the $y$ axis of the images represent time. (**C**) Signals from **B**, covering the first 200 ms of the climbing fiber response, plotted against time. The fluorescence was binned to 0.5 kHz and converted to $\Delta F/F_0$ values. The electrical signal ($V_m$) includes a so-called complex spike, the typical response to climbing fiber stimulation. (**D**) Horizontal expand of **C**. Note that the start and the end of the rising phase of the $Ca^{2+}$ transient can be correlated to distinct components of the complex spike (arrows).

voltage signals while the electrophysiological signal (membrane voltage or current) will be of negative polarity. Thus, an amplifier (buffer in Fig. 1) needs to be incorporated that inverts and amplifies the electrophysiological signal before it is fed into the fluorescence system.

Grounding of electrophysiological setups is somewhat complicated by incorporating fluorescence equipment. Cameras and lasers, for example, generate high-frequency electrical noise. Furthermore, since any modification of the ground connections of these devices is not recommended for safety reasons, ground loops will also be a problem. In this respect it may be helpful to electrically isolate critical equipment by connecting it to the microscope with insulating adapters only.

Finally, it should be possible for the user to easily align the fluorescence detection unit. Both the overall transmittance and the optical resolution should be carefully checked on a regular basis. For this purpose mixed stained pollen grains (No. 204264 Carolina Biological Supply Co., Burlington, NC) can be used as a simple but reliable test specimen. These pollen grains are stained with different fluorescent dyes and can easily be visualized with most fluorescence optics. They possess small protrusions that allow the optical resolution of fluorescence systems to be checked and adjusted.

## 2.2. Recordings

As stated above, this chapter focuses on the methodology for fluorometric $Ca^{2+}$ measurements and the quantification of the mobility of dye-labeled proteins in neurons. We start by addressing topics that equally apply for both types of recordings, like solutions and seal formation. Thereafter, special requirements for the two types of experiments are discussed. The guideline is also applicable to fluorometric measurements of other signaling molecules, such as $Cl^-$, $Na^+$, and cyclic adenosine monophosphate (cAMP).

### 2.2.1. Solutions

Solutions used in combined fluorometric and patch-clamp recordings are similar to those used in standard patch-clamp recordings. Our standard extracellular saline contains (in mM) 125 NaCl, 2.5 KCl, 1.25 $NaH_2PO_4$, 26 $NaHCO_3$, 1 $MgCl_2$, 2 $CaCl_2$, 20 glucose, and pH 7.3 to 7.4 when gassed with 95% $O_2$ and 5% $CO_2$. Depending on the experimental requirements, various intracellular solutions can be used. As an example, one of our standard pipette solutions contains (in mM): 150 K-gluconate, 10 NaCl, 3 Mg-ATP, 0.3 guanosine triphosphate (GTP), 10 4-(2-hydroxyethyl)-1-pipe-razineethanesulfonic acid (HEPES), and 0.1 Oregon Green 1,2-bis(o-aminophenoxy)ethane-N,N,N',N'-tetraacetic acid (BAPTA-1); adjusted to pH 7.3 with KOH; osmolarity ~310 mOsm. Note that no extra $Ca^{2+}$ buffers like ethyleneglycotetraacetic acid (EGTA) or BAPTA are included since these would substantially buffer $Ca^{2+}$ transients and reduce the fluorometric signal. However, when low-affinity $Ca^{2+}$ indicator dyes like magnesium green are used in prolonged recordings, EGTA or BAPTA may be included to compensate for possible washout of endogenous $Ca^{2+}$ buffers (Zhou and Neher, 1993). For measuring the mobility of proteins, the $Ca^{2+}$ indicator dye is replaced by the dye-labeled protein of interest. Depending on whether the protein is intrinsically $Ca^{2+}$ binding or not, EGTA or BAPTA may be included in the intracellular solution to maintain a physiological $Ca^{2+}$ buffering capacity during the whole-cell recording (Helmchen and Tank, 2005).

For efficient use of the expensive indicator dyes, we prepare a concentrated (125%) solution of the intracellular solution from which the dye/protein has been omitted as well as a concentrated (about 2 mM) stock solution of the indicator dye dissolved in double-processed water (Sigma, No. W-3500, ST. Louis, MO). Both solutions should be kept at −20°C for no longer than 2 weeks. Suitable quantities of these two solutions are mixed (and sonicated) on a daily basis. Before use, the solution should be filtered with syringe filters (Nalgene, No. 180-1320, Carolina Biological Supply Co., Burlington, NC). Metal needles must not come into contact with the indicator dye or the pipette solution. They release trace amounts of metals that interfere with various cellular functions. For the same reason, the intracellular solution should be prepared using double-processed water only.

## 2.2.2. Patch-Clamp Recordings

Seal formation and whole-cell recordings are similar to standard patch-clamp recordings. Low-resistance pipettes should be used to allow fast dye/protein loading. We routinely use 4- to 5-MΩ pipettes (measured with the intracellular solution mentioned above) for somatic recordings from cerebellar Purkinje neurons. When approaching the cell, the application of strong positive pressure should be avoided. Otherwise, the surrounding tissue will be stained by dye ejected from the pipette, which may reduce the optical resolution during the fluorometric recordings. We use the following protocol for patching cells located close to the surface of the slice preparation: when entering the bath and when approaching the cell, weak positive pressure is applied to the pipette to prevent extracellular solution from entering the pipette; immediately before the cell is contacted, a strong but brief (<1 second) pressure pulse is given to clean the surface of the cell; then the pressure is relieved and weak suction is applied to facilitate seal formation (Eilers and Konnerth, 2005).

After the whole-cell configuration is established, equilibration of the cytosol with the pipette solution is determined by three factors: the series resistance ($R_s$; Pusch and Neher, 1988), the mobility of the dye/protein (Pusch and Neher, 1988), and the cell geometry (Rexhausen, 1992). In cells that possess an elaborate dendritic tree, such as cerebellar Purkinje neurons (Figs. 2 and 6), the time constant for loading remote dendrites may be as slow as 30 to 45 minutes (Rexhausen, 1992). Under such conditions, $R_s$ should be optimized to accelerate dye loading. Thus, the initial pipette resistance should be as low as tolerated by the cells under investigation. Furthermore, after breaking into the cell, $R_s$ should be monitored and kept as low as possible. We rather aggressively try to keep $R_s$ low during the first 15 minutes of the whole-cell recordings. While we lose some cells by this approach, we mostly obtain recordings that stay stable for at least 60 to 120 minutes. Indicator dyes as well as proteins of up to 30 kDa are easily loaded with standard pipettes. Larger proteins may require low-resistance pipettes (1–2 MΩ).

## 2.2.3. Fluorometric Calcium Measurements

### 2.2.3.1. CALCIUM INDICATOR DYES

The selection of an appropriate indicator dye is governed by several parameters, the most obvious one being its excitation spectrum. Depending on the fluorescence system in use, only certain wavelength may be available. Confocal laser-scanning microscopes, for example, are usually equipped with ion lasers that emit only at a few distinct wavelengths and, in most cases, do not allow excitation in the UV range. In contrast, a broad spectrum of different dyes can be excited with monochromators (Uhl, 2005) or two-photon microscopes (Xu et al., 1996). Note that excitation as well as emission spectra of most commonly used dyes are available online (for example, from Molecular Probes/Invitrogen, Eugene, OR). The

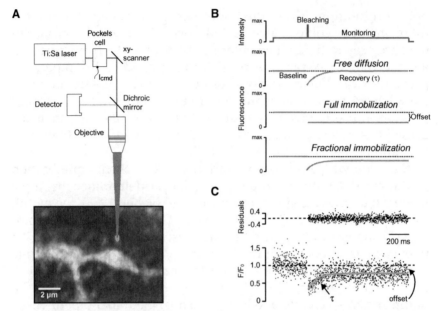

Fig. 6. Quantification of molecular diffusion in dendritic spines by fluorescence recovery after photobleaching (FRAP). (**A**) Equipment diagram (top) and spiny dendrites of a Purkinje neuron loaded with dye-labeled calbindin D28k (CB, 200-μM pipette concentration) for ~1 hour via a somatic whole-cell patch pipette. The pulsed light generated by a mode locked Ti:sapphire laser was intensity-modulated ($I_{cmd}$) by a Pockels cell. For FRAP recordings, the laser light was restricted to a single spine (point mode). The fluorescence light generated in the focal volume was detected by a photomultiplier tube. (**B**) FRAP principle: following a baseline period at low laser intensity, a brief high-intensity laser pulse is applied to irreversibly bleach the fluorophores within the focal volume (top panel). Thereafter, the fluorescence recovery (τ) is monitored. Depending on the mobility of the protein under study, the fluorescence either shows a rapid and complete recovery to baseline (2nd panel), no recovery at all (offset, 3rd panel), or a partial recovery (4th panel). (**C**) Normalized fluorescence signal ($F/F_0$, bottom) from the spine shown in **A**. After bleaching, the fluorescence did not fully recovered to baseline, indicating a fractional immobilization of CB. The upper panel shows the residuals (i.e., data fit). (Modified from Schmidt et al., 2005.)

second critical parameter concerns the expected magnitude of the $Ca^{2+}$ signal to be studied and the affinity of the $Ca^{2+}$ indicator. As a rule of thumb, $Ca^{2+}$ indicator dyes allow measurements in the range of 0.1 to 10 times their dissociation constants ($K_d$). Thus, for example, a high-affinity $Ca^{2+}$ indicator with a $K_d$ of 300 nM allows measurements in the range from 30 to 3000 nM. Note that the exact value of the $K_d$ depends on several parameters, including the pH, osmolarity, temperature, and the concentration of other divalent ions such as $Mg^{2+}$. Thus, for quantitative measurements, the $K_d$ should be determined specifically for the intracellular solution, in which it is dissolved. For this purpose, ready-made calibration kits (e.g., C-3273, Molecular Probes) and the public domain software Max-Chelator (available at www.stanford.edu/~cpatton/maxc.html), which calculates the equilibrium concentration of $Ca^{2+}$, $Mg^{2+}$, and EGTA, are available.

Additional criteria for selecting an indicator include its resting fluorescence, that is, the brightness of the dye at the intracellular calcium concentration ($[Ca^{2+}]_i$) at rest. For recordings from fine cellular structures, such as dendritic spines, dyes that exhibit a substantial resting fluorescence (such as Oregon Green BAPTA-1) are preferable. Otherwise, fine structures may be too dim to be seen under resting conditions, or no reasonable baseline recording may be obtained (see below). Finally, if possible, a dye that allows the quantification of $Ca^{2+}$ signals by means of ratiometric recordings should be chosen. Unfortunately, most such dyes either require UV excitation (Fura-2 and Indo-1) or are inherently dim (Fura Red). In conclusion, no single dye can be recommended as the gold standard. A good starting point, however, is the high-affinity indicator Oregon Green BAPTA-1 (Molecular Probes), which can be excited by most light sources and has a substantial fluorescence at resting $[Ca^{2+}]_i$.

Once an indicator dye has been selected, the next question is at what concentration it should be used. On the one hand, the dye concentration needs to be high enough to allow fluorescence measurements with a reasonable signal-to-noise ratio. On the other hand, for most applications, the concentration should be as low as possible for the following reasons: (1) any $Ca^{2+}$ indicator will act as a buffer that reduces the amplitude and prolongs the duration of $Ca^{2+}$ transients (Neher, 2005); (2) $Ca^{2+}$ indicators may have unexpected pharmacological side effects. Several commonly used $Ca^{2+}$ indicators, for example, block inositol 1,4,5-triphosphate ($IP_3$) receptors at submillimolar concentrations (Morris et al., 1999). We routinely use a dye concentration of 20 to 50 µM for somatic and 100 to 200 µM for dendritic recordings. However, these values should be optimized for the required spatiotemporal resolution, the sensitivity of the fluorescence detector in use, and the endogenous buffering capacity of the cell under study (Helmchen and Tank, 2005).

## 2.2.3.2. ACQUISITION OF CALCIUM IMAGING DATA

High-resolution fluorescence recordings from living preparations require delicate tuning of different parameters, most notably of the excitation intensity. This is due to the fact that the generation of fluorescence is inevitably associated with phototoxic damage and dye bleaching. Photodamage may be evident from local swelling (blebbing) of the cell or from a decrease in the membrane resistance, with a resulting increase in holding current and reduced excitability (altered waveform or reduced number of spikes). More subtle signs of photodamage include $Ca^{2+}$ transients that, in contrast to the normal case (Fig. 2D,E), do not fully return to baseline. Thus, the excitation intensity should be reduced as much as possible, and an appreciable level of noise in the fluorescence needs to be accepted. This is especially important for line-scan and point recordings (Fig. 3 and 5, respectively), in which the excitation is restricted to small cellular compartments. For these recordings, the excitation intensity should be reduced by a factor of at least 10 (line scan) and 100 (point mode) compared to the intensity used to scan the whole field of view.

The signal-to-noise ratio should not be improved by simply increasing the dye concentration because of the side effects discussed in the previous section. Rather, the detection efficacy should be improved by using objectives with a high NA and by optimizing gain and offset of the fluorescence detector. Figure 2D shows an example where the dynamic range of the detector was used properly. The fluorescence signal was sampled by a 12-bit analog/digital (A/D) converter. Thus, values between 0 and 4095 arbitrary digital units (adu) could be obtained. Note that the peak of the fluorescence transient did not saturate the A/D converter (<4095 adu) and that the background signal was well above 0 adu. The latter point is critical for a correct quantification of the fluorescence data. Unfortunately, some fluorescence systems perform an automatic offset calibration that may set the offset too high. This may lead to an overestimate of the background intensity and to errors in the quantification of the fluorescence signals. To avoid saturation of the detector, so-called look-up tables (LUTs), available in most fluorescence systems, should be used to inspect the data. The LUTs use distinct colors to highlight fluorescence intensities that are either close to 0 adu or to saturation (4095 adu in our system) and, thereby, provide an easy-to-use tool for adjusting gain and the offset correctly.

Especially if high-magnification is employed to focus the excitation light on small cellular regions (see Fig. 2B), the periods of excitation should be kept as short as possible in order to effectively reduce photodamage. The trigger lines described in Section 2.1.3.5 and in Fig. 1 are designed to fulfill this purpose. The line-scan recording illustrated in Fig. 2 represents an example where the fluorescence detection unit triggered the electrophysiology system, which, in turn, stimulated an afferent excitatory fiber after a fixed interval of 300 ms (Fig. 2C). Thus, the length of the baseline recording was short but well defined. Manual triggering of the stimulation may have resulted in a baseline recording that is too long or too short.

When fluorescence responses are spatially restricted, for example to single dendritic spines, they might be hard to detect by eye. Photodamage occurs if the cell is screened too long during the search for the active site. Figure 3 displays a simple search algorithm that aids the identification of localized or fast responses. The implementation of the algorithm critically depends on the capabilities of the fluorescence system. While some systems allow the use of macro commands (e.g., MetaFluor, Molecular Devices, Sunnyvale, CA), others may be limited to predefined commands that cannot be modified. However, even in such systems it may be possible to implement the algorithm by acquiring two images (each representing an average of $n$ images) with the stimulation occurring after the first averaged image.

The timed stimulation illustrated in Fig. 2 already provides a rough estimate of the temporal relationship of the electrical and fluorometric signals. In the case of video-rate fluorescence microscopy (Inoué and Spring, 1997), a more accurate correlation can be obtained by a special electronic circuit (Eilers et al., 1997a; see also Fig. 1) that electronically marks the time point of stimulation. This method achieves a resolution of 16.5 ms (Fig. 4) or 60 µs (not illustrated; see Eilers et al., 1997a), which is sufficient for most applications. This approach has the additional advantage (e.g., for failure analysis) that the occurrence of stimulation is indicated even if no cellular response was evoked. If a perfect correlation between the electrical and fluorometric signals is required, the regimen illustrated in Fig. 5 could be used. Here, the electrical signal was fed into a free input channel of the fluorescence system, which was sampled in parallel to the standard fluorescence signal (see also Fig. 1). Thus, both signals were acquired simultaneously by the same instrument. This approach allows a correlation, for example, between the cell's spike pattern and the onset and duration of dendritic $Ca^{2+}$ transients (Schmidt et al., 2003b).

### 2.2.3.3. ANALYSIS OF CALCIUM DATA

The first step in the analysis of fluorescence data is the definition of so-called regions of interest (ROIs), in which the amplitude and time course of responses is to be quantified (see Figs. 2C and 4A). The software of the fluorescence detection unit allows the definition of ROIs of arbitrary shape and size. (Note that in line-scan recordings, a ROI is a line segment.) It is important that the ROIs are set correctly. If a ROI is drawn too small, that is, if it does not cover the entire cellular region in which the signal is to be analyzed, part of the fluorescence signal is lost and the signal-to-noise ratio is decreased. If the ROI is set too large, that is, if it also covers inactive parts of the cell or unstained areas, the amplitude of the signal will be reduced due to the averaging of active and inactive areas. Furthermore, the data should be inspected using LUTs (see above), and parts of the cell in which the signal was saturated should not be included in the analysis. Note that for a quantitative analysis of the fluorescence data, the background signal needs to be measured in an unstained region of the preparation. If the software of the fluorescence system does not allow the analysis of ROIs, the public domain software NIH Image (available from rsb.info.nih.gov) may be used for this purpose.

Once the ROIs are defined, the software enables the averaged intensity from individual ROIs to be plotted versus time and the data to be exported to external programs. Since the following analysis steps are well defined and stereotyped, they may be automated. In our hands, the analysis program Igor Pro (Wavemetrics, Lake Oswego, OR) provides a simple but powerful programming environment. With this software, all of the analysis steps outlined below can be programmed, allowing the analysis of even hundreds of data sets with a single keystroke.

We now describe the quantification of fluorescence data using the simple $\Delta F/F_0$ method, which can be applied to any fluorescence recording. The more complex quantification of ratiometric recordings (e.g., Fura 2, Indo-1, or a mixture of Fluo 3/Fura Red) or fluorescence lifetime recordings (Wilms et al., 2006) is beyond the scope of this chapter; see Grynkiewicz et al. (1985) and Helmchen (2005) for details. The $\Delta F/F_0$ method corrects for variations in excitation intensity, dye concentration, and thickness of the cellular structures. Let's consider the example shown in Fig. 2D. Fluorescence signals were recorded from a dendrite, a spine, and the background. The signal recorded from the dendrite is larger (in absolute numbers) than that recorded from the spine. This is probably due to the fact that the dendrite is thicker than the spine and, thus, more dye molecules were excited when the dendritic area was scanned. Alternative explanations could be that the dye concentration in the spine was lower than that in the dendrite or that somehow the illumination of the field of view was inhomogeneous, directing less excitation light onto the spine than onto the dendrite. Whatever the correct explanation, we would like to know whether the response, that is, the $Ca^{2+}$ transient, was larger in the dendrite than in the spine. To answer this question we have to

1. subtract the background from both the dendritic and the spine signal;
2. average the baseline of the two background-corrected traces (this yields two $F_0$ values);
3. subtract each $F_0$ value from the corresponding fluorescence traces (to obtain $\Delta F$ traces); and
4. divide both $\Delta F$ traces by the corresponding $F_0$ value.

In the resultant $\Delta F/F_0$ ratios, any of the uncertain parameters mentioned above (thickness of the structure, dye concentration, and excitation intensity) are canceled out and only the relative change in fluorescence remains (see Fig. 2E). Note that this is only true if the cell did not change its size (important for recordings from muscle cells), the dye concentration did not change significantly, and the excitation intensity was constant during the recording period (4 s in Fig. 2). If we finally assume that both compartments had the same $Ca^{2+}$ concentration during the baseline recording, the amplitude of the $\Delta F/F_0$ signals is solely dependent on the amplitude of the $Ca^{2+}$ transient in the spine and the dendrite. Thus, we can directly compare the two signals and find, in our example, that the spine signal was considerably larger than the dendritic one. The

$\Delta F/F_0$ method also allows the comparison of successive recordings from the same cell and even from different cells. It should be kept in mind, however, that only minor differences in the dye concentration are canceled out. Larger difference and the associated difference in $Ca^{2+}$ buffering may affect the amplitude and time course of the signals (Neher, 2005).

## 2.2.4. Measuring Protein Mobility

In this section we discuss how the diffusional mobility of dye-labeled proteins can be quantified by fluorescence recovery after photobleaching (FRAP) (Axelrod et al., 1976; Edidin et al., 1976). The underlying principle of this approach is that proteins of interest, labeled with a fluorescent dye, are introduced into single cells via a patch pipette. After equilibration, the labeled proteins are bleached in defined cellular compartments and the recovery of fluorescence is recorded. Subsequently, the diffusion coefficient ($D$) and the fractional immobilization of the protein is calculated from the recovery. The FRAP recordings provide an easy-to-use method for studying protein mobility and interaction in situ.

### 2.2.4.1. PROTEIN LABELING

Most proteins can easily be labeled using commercially available kits (e.g., No. A-10235, Molecular Probes, which randomly labels primary amines). We typically achieve a ~1:1 ratio of label to protein (Schmidt et al., 2003a, 2005), which we consider optimal in terms of brightness and increase in molecular weight (MW) of the protein. Since the Stokes-Einstein relation predicts that $D$ is proportional to the inverse cubic root of the MW, diffusion of the labeled protein may be considerably retarded compared to that of the native, unlabeled protein if the labeling ratio is too high. Consequently, the MW of the label should be significantly smaller than the MW of the protein under study (making GFP an inappropriate choice for small proteins). In our experience, Alexa dyes (MW ~0.6 kDa) work well for 10- to 30-kDa proteins (see, for example, Schmidt et al., 2003a, 2005). Compared to fluorescein-based labels, Alexa dyes show less bleaching, which is a critical parameter in diffusion measurements. For in situ quantification of diffusion, the cells should be loaded with protein concentrations similar to the assumed physiological value. Otherwise, potential intracellular targets may be saturated by overly high protein levels, which may lead to a severe overestimation of the protein mobility.

### 2.2.4.2. ACQUISITION OF FRAP DATA

We focus here on FRAP recordings from small cellular compartments such as dendritic spines, in which the derivation of $D$ values is rather straightforward. Since diffusional equilibration in such small compartments can occur within less than 100 ms, FRAP recording are typically performed in the line-scan (Svoboda et al., 1996) or point mode (Schmidt et al., 2003a). The baseline fluorescence of dye-labeled molecules is monitored at a low ("monitoring") laser intensity before a brief, high-intensity laser pulse is applied to irreversibly bleach a substantial fraction (30–60%) of the fluorophores in the focal volume. Subsequently, the recovery of the fluorescence, which reflects molecular diffusion, is recorded again at the monitoring intensity.

A prerequisite of FRAP recordings is the ability to modulate the excitation light in the submillisecond range. This is easily achieved by passing laser light through a Pockels cell. Thus, FRAP recordings can readily be performed on CLSM or TPM systems in which the Pockels cell is placed between the laser output and the scanning system. While Pockels cells are faster than mechanical shutters, they show some nonlinear switching behavior that has to be taken into consideration. It requires recording the laser intensity during the FRAP recording in parallel to the fluorescence signal. This is achieved by a second detector, toward which a small fraction of the excitation light is directed or by a second recording, in which the FRAP experiment is performed on a reflecting surface (e.g., aluminum foil; for details see Schmidt et al., 2005). For both recordings (intensity and fluorescence), the background intensity should also be recorded to allow a quantitative analysis (see next section).

Due to the fact that the laser light is directed onto a small cellular compartment and that a strong bleach pulse is applied, phototoxic damage is a major concern in FRAP recordings. Apart from the obvious signs of injury discussed above, a more subtle indication for phototoxic damage, which requires special attention, is a systematic prolongation of recovery during successive recordings (Koester et al., 1999). To overcome these problems, the monitoring laser intensity needs to be carefully adjusted. This will also minimize undesired bleaching of fluorophores during the baseline, which, if present, may downsize the effect of immobilization of proteins (Schmidt et al., 2005). For the bleach pulse, a delicate balance between bleaching efficiency and phototoxic damage has to be established. However, careful adjustment of the laser intensity allows reliable recordings that can be repeated many times even in fine structures such as dendritic spines (Schmidt et al., 2003a, 2005).

### 2.2.4.3. ANALYSIS OF FRAP DATA

The first step in analyzing FRAP data is to calculate background- and intensity-corrected normalized fluorescence curves. This is achieved by (1) subtracting the background curves from the corresponding intensity and fluorescence curves, and subsequently (2) dividing the fluorescence curve by the intensity curve (for one photon excitation) or by the square of the intensity curve (for two-photon excitation; Brown et al., 1999), and (3) normalizing the fluorescence to its baseline value ($F/F_0$). As illustrated in Fig. 6B, the time course of the decay and the degree of recovery depend on the mobility and fractional immobilization of the protein (Reits and Neefjes, 2001; Sprague et al., 2004). Subjecting the $F/F_0$ curve to an appropriate fitting function reveals the $D$ value as well as the fractional immobilization of the protein.

The exact form of the fitting function depends on the cellular compartment under study. For a dendritic spine, diffusional coupling to the parent dendrite is governed by the geometry of the spine and its neck. Diffusion can be described as diffusion *through* a pipe (i.e., the spine neck) and a median $D$ value can be derived as

$$\tilde{D} = \frac{\bar{l}\,\overline{V}}{\tilde{\tau}\,\pi \overline{r^2}}$$

where $\bar{l}$ is the mean length of the spine neck, $\overline{r^2}$ is the expectation of its radius squared, $\overline{V}$ is the mean volume of the spine head, and $\tau$ is the time constant of fluorescence recovery (Schmidt et al., 2003a). The immobilized fraction can be computed as the steady-state post-bleach offset (Fig. 6B) divided by the maximal decrease in fluorescence after the bleach pulse (Schmidt et al., 2005). Figure 6C shows an example experiment with partial protein immobilization of ~30%.

For FRAP experiments in thin cylindrical structures such as axons, diffusion can be described as diffusion *within* a pipe (Schmidt et al., 2007). The fitting equation can be deduced according to the formalism for three-dimensional (3D) diffusion in the bulk cytoplasm described by Brown et al. (1999). Finally, FRAP experiments can be combined with synaptic or pharmacological stimulation protocols to identify activity-dependent interaction partners in subcellular compartments (Schmidt et al., 2005). The femtoliter resolution of FRAP helps to uncover and study highly localized protein interactions in live cells.

## List of Selected Suppliers

| | |
|---|---|
| www.carolina.com | Carolina Biological Supply Co. (test slides) |
| www.moleculardevices.com | Molecular Devices (MetaFluor, acquisition software) |
| www.luigs-neumann.com | Luigs and Neumann (Dodt gradient-contrast illumination) |
| www.olympus-global.com | Olympus (oblique condenser) |
| www.mellesgriot.com | Melles Griot (fiber optics) |
| probes.invitrogen.com | Molecular Probes (fluorescent probes) |
| rsb.info.nih.gov | NIH Image and ImageJ (image analysis software, public domain) |
| www.stanford.edu/~cpatton | Chris Patton, MaxChelator (program for determining the free metal concentration in the presence of chelators, public domain) |
| www.wavemetrics.com | Wavemetrics, Igor (programmable graphing and analysis software) |
| www.conoptics.com | Conoptics (Pockels cell) |

## Acknowledgments

This work was supported by grants from the Bundesministerium für Bildung und Forschung to Jens Eilers.

# References

Adams, S. R., Harootunian, A. T., Buechler, Y. J., Taylor, S. S., and Tsien, R. Y. (1991) Fluorescence ratio imaging of cyclic AMP in single cells. *Nature* **349**, 694–697.

Axelrod, D., Koppel, D. E., Schlessinger, J., Elson, E., and Webb, W. W. (1976) Mobility measurement by analysis of fluorescence photobleaching recovery kinetics. *Biophys. J.* **16**, 1055–1069.

Brown, E. B., Wu, E. S., Zipfel, W., and Webb, W. W. (1999) Measurement of molecular diffusion in solution by multiphoton fluorescence photobleaching recovery, *Biophys. J.* **77**, 2837–2849.

Bullen, A., Patel, S. S., and Saggau, P. (1997) High-speed, random-access fluorescence microscopy: I. High-resolution optical recording with voltage-sensitive dyes and ion indicators, *Biophys. J.* **73**, 477–491.

Carrington, W. A., Lynch, R. M., Moore, E. D., Isenberg, G., Fogarty, K. E., and Fay, F. S. (1995) Superresolution three-dimensional images of fluorescence in cells with minimal light exposure. *Science* **268**, 1483–1487.

Chalfie, M., Tu Y., Euskirchen, G., Ward, W. W., and Prasher, D. C. (1994) Green fluorescent protein as a marker for gene expression. *Science* **263**, 802–805.

Denk, W., Strickler, J. H., and Webb, W. W. (1990) Two-photon laser scanning fluorescence microscopy. *Science* **248**, 73–76.

Dodt, H. U. and Zieglgänsberger, W. (1990) Visualizing unstained neurons in living brain slices by infrared DIC-videomicroscopy. *Brain Res.* **537**, 333–336.

Edidin, M., Zagyansky, Y., and Lardner, T. J. (1976) Measurement of membrane protein lateral diffusion in single cells. *Science* **191**, 466–468.

Eilers, J., Augustine, G. J., and Konnerth, A. (1995a) Subthreshold synaptic $Ca^{2+}$ signalling in fine dendrites and spines of cerebellar Purkinje neurons. *Nature* **373**, 155–158.

Eilers, J., Callewaert, G., Armstrong, C., and Konnerth, A. (1995b) Calcium signaling in a narrow somatic submembrane shell during synaptic activity in cerebellar Purkinje neurons. *Proc. Natl. Acad. Sci. USA* **92**, 10272–10276.

Eilers, J., Hof, D., and Konnerth, A. (1997a) The Flaginserter: a reliable event marker for video recordings. *J. Neurosci. Methods* **78**, 151–156.

Eilers, J. and Konnerth, A. (2005) A practical guide: dye loading with patch-pipettes, in *Imaging in neuroscience and development: a laboratory manual* (Yuste, R. and Konnerth, A., eds.), Cold Spring Harbor Laboratory Press, New York, pp. 277–281.

Eilers, J., Takechi, H., Finch, E. A., Augustine, G. J., and Konnerth, A. (1997b) Local dendritic $Ca^{2+}$ signaling induces cerebellar LTD. *Learning and Memory* **4**, 159–168.

Fan, G. Y., Fujisaki, H., Miyawaki, A., Tsay, R. K., Tsien, R. Y., and Ellisman, M. H. (1999) Video-rate scanning two-photon excitation fluorescence microscopy and ratio imaging with cameleons. *Biophys. J.* **76**, 2412–2420.

Finch, E. A. and Augustine, G. J. (1998) Local calcium signalling by inositol-1,4,5-trisphosphate in Purkinje cell dendrites. *Nature* **396**, 753–756.

Grynkiewicz, G., Poenie, M., and Tsien, R. Y. (1985) A new generation of $Ca^{2+}$ indicators with greatly improved fluorescence properties. *J. Biol. Chem.* **260**, 3440–3450.

Helmchen, F. (2005) Calibration of fluorescent calcium indicators, in *Imaging in neuroscience and development: a laboratory manual* (Yuste, R. and Konnerth, A., eds.), Cold Spring Harbor Laboratory Press, New York, pp. 253–263.

Helmchen, F. and Tank, D. W. (2005) A single-compartment model of calcium dynamics in nerve terminals and dendrites, in *Imaging in neuroscience and development: a laboratory manual* (Yuste, R. and Konnerth, A., eds.), Cold Spring Harbor Laboratory Press, New York, pp. 265–275.

Hernández-Cruz, A., Sala, F., and Adams, P. (1990) Subcellular calcium transients visualized by confocal microscopy in a voltage-clamped vertebrate neuron. *Science* **247**, 858–862.

Inoué, S. and Spring, K. R. (1997). *Video microscopy—The fundamentals*, 2nd ed., Plenum Press, New York.

Koester, H. J., Baur, D., Uhl, R., and Hell S. W. (1999) $Ca^{2+}$ fluorescence imaging with pico- and femtosecond two-photon excitation: signal and photodamage. *Biophys. J.* **77**, 2226–2236.

Lanni, F. and Keller, H. E. (2005) Microscopy and microscope optical systems, in *Imaging in neuroscience and development: a laboratory manual* (Yuste, R. and Konnerth, A., eds.), Cold Spring Harbor Laboratory Press, New York, pp. 711–765.

Mainen, Z. F., Malinow, R., and Svoboda, K. (1999) Synaptic calcium transients in single spines indicate that NMDA receptors are not saturated. *Nature* **399**, 151–155.

Minsky M. (1961) Microscopy apparatus. U.S. patent No. 3013467.

Morris, S. A., Correa, V., Cardy, T. J., O'Beirne, G., and Taylor, C. W. (1999) Interactions between inositol trisphosphate receptors and fluorescent $Ca^{2+}$ indicators. *Cell Calcium* **25**, 137–142.

Müller, W. and Connor, J. A. (1991) Dendritic spines as individual neuronal compartments for synaptic $Ca^{2+}$ responses. *Nature* **354**, 73–76.

Neher, E. (2005) Some quantitative aspects of calcium fluorimetry, in *Imaging in neuroscience and development: a laboratory manual* (Yuste, R. and Konnerth, A., eds.), Cold Spring Harbor Laboratory Press, New York, pp. 245–252.

Pusch, M. and Neher, E. (1988) Rates of diffusional exchange between small cells and a measuring patch pipette. *Pflügers Arch.* **411**, 204–211.

Reits, E. A. and Neefjes, J. J. (2001) From fixed to FRAP: measuring protein mobility and activity in living cells. *Nat. Cell. Biol.* **3**, E145–147.

Rexhausen, U. (1992) Bestimmung der Diffusionseigenschaften von Fluoreszenzfarbstoffen in verzweigten Nervenzellen unter Verwendung eines rechnergesteuerten Bildverarbeitungssystems. *University of Göttingen*, diploma thesis.

Rose, C. R., Kovalchuk, Y., Eilers, J., and Konnerth, A. (1999) Two-photon $Na^+$ imaging in spines and fine dendrites of central neurons. *Pflügers Arch.* **439**, 201–207.

Sabatini, B. L. and Svoboda, K. (2000) Analysis of calcium channels in single spines using optical fluctuation analysis. *Nature* **408**, 589–593.

Salome, R., Kremer, Y., Dieudonne, S., et al. (2006) Ultrafast random-access scanning in two-photon microscopy using acousto-optic deflectors, *J. Neurosci. Methods* **154**, 161–174.

Schiefer, J., Kampe, K., Dodt, H. U., Zieglgänsberger, W., and Kreutzberg G. W. (1999) Microglial motility in the rat facial nucleus following peripheral axotomy. *J. Neurocytol.* **28**, 439–453.

Schmidt, H., Arendt, O., Brown, E.B., Schwaller, B., and Eilers, J. (2007) Parvalbumin is freely mobile in axons, somata and nuclei of cerebellar Purkinje neurons. *J. Neurochem.* **100**, 735–737.

Schmidt, H., Brown, E. B., Schwaller, B., and Eilers J. (2003a) Diffusional mobility of parvalbumin in spiny dendrites of cerebellar Purkinje neurons quantified by fluorescence recovery after photobleaching. *Biophys. J.* **84**, 2599–2608.

Schmidt, H., Schwaller, B., and Eilers J. (2005) Calbindin D28k targets *myo*-inositol monophosphatase in spines and dendrites of cerebellar Purkinje neurons. *Proc. Natl. Acad. Sci. USA* **102**, 5850–5855.

Schmidt, H., Stiefel, K., Racay, P., Schwaller, B., and Eilers, J. (2003b) Mutational analysis of dendritic $Ca^{2+}$ kinetics in rodent Purkinje cells: role of parvalbumin and calbindin $D_{28k}$, *J. Physiol. (Lond.)* **551**, 13–32.

Sprague, B. L., Pego, R. L., Stavreva, D. A., and McNally, J. G. (2004) Analysis of binding reactions by fluorescence recovery after photobleaching. *Biophys. J.* **86**, 3473–3495.

Svoboda, K., Tank, D. W., and Denk, W. (1996) Direct measurement of coupling between dendritic spines and shafts. *Science* **272**, 716–719.

Takechi, H., Eilers, J., and Konnerth, A. (1998) A new class of synaptic responses involving calcium release in dendritic spines. *Nature* **396**, 757–760.

Uhl, R. (2005) A practical guide: arc lamps and monochromators for fluorescence microscopy, in *Imaging in neuroscience and development: a laboratory manual* (Yuste, R. and Konnerth, A., eds.), Cold Spring Harbor Laboratory Press, New York, pp. 791–803.

Wilms, C, Schmidt, H., and Eilers, J. (2006) Quantitative two-photon $Ca^{2+}$ imaging via fluorescence lifetime analysis. *Cell Calcium* **40**, 73–79.

Xu, C., Zipfel W., Shear, J. B., Williams, R. M., and Webb, W. W. (1996) Multiphoton fluorescence excitation: new spectral windows for biological nonlinear microscopy. *Proc. Natl. Acad. Sci. USA* **93**, 10763–107658.

Zhou, Z. and Neher, E. (1993) Mobile and immobile calcium buffers in bovine chromaffin cells. *J. Physiol.* **469**, 245–273.

# 5

# Combining Uncaging Techniques with Patch-Clamp Recording and Optical Physiology

*Dmitry V. Sarkisov and Samuel S.-H. Wang*

## 1. Introduction

Patch-clamp recording is a powerful approach to monitoring membrane electrical activity with high temporal resolution. However, the spatial resolution of patch-clamp recording in a distributed structure such as a neuron or a brain slice is limited by the fact that each electrode records from just one point, making recording from more than a very small number of points impractical.

An approach that allows biological signals to be monitored and manipulated with high spatial resolution is the use of optical methods. For the monitoring of signals, a powerful approach is the use of activity-dependent fluorescent dyes. In this type of recording, one or more cells are loaded with a fluorescent dye that is sensitive to some change of interest, such as intracellular calcium or membrane voltage. Loading can be done by including the dye in the patch pipette, by bulk loading of many cells at once with acetoxymethyl (AM) ester dyes (Garaschuk et al., 2006; Sullivan et al., 2005; Tsien, 1999) or voltage-sensitive dyes (Djurisic et al., 2003; Grinvald and Hildesheim, 2004), or by expression of an activity-dependent fluorescent protein (Bozza et al., 2004; Wang et al., 2003). Regardless of the means of loading, biological signals lead to variations in fluorescence that can be detected by high-speed fluorescence microscopy. The combination of electrophysiological and fluorimetric recording thus allows monitoring of biochemical and electrical signals simultaneously with high temporal and spatial resolution.

From: *Neuromethods, Vol. 38: Patch-Clamp Analysis: Advanced Techniques, Second Edition*
Edited by: W. Walz @ Humana Press Inc., Totowa, NJ

An additional optical approach, photolysis of caged compounds, can be used to manipulate cellular biochemistry. Uncaging techniques nicely supplement the ability to measure electrical and biochemical signals by providing a means to affect the signals with high spatial and temporal resolution. This chapter discusses the combination of uncaging with patch clamp and fluorescence recording, starting with an overview of caged compounds and applications, a description of the construction of several specific focal uncaging setups, and selected recent technical developments. For additional perspectives we refer readers to other recent reviews (Eder et al., 2004; Kramer et al., 2005; Thompson et al., 2005).

## 2. Principles of Caged Compounds

Before beginning, it is helpful to ask whether uncaging is needed at all. In easily accessible preparations such as isolated cells or ripped-off membrane patches, direct application by pressure ejection through a pipette can achieve rapid application on millisecond time scales (Isaacson and Nicoll, 1991). In semi-intact preparations such as brain slices, another approach is iontophoretic application, which can achieve submicrometer resolution. Uncaging is useful in applications where bringing in a physical electrode is impractical or inadvisable. Examples include the study of intracellular signaling (Adams and Tsien, 1993), dendritic spine physiology (Matsuzaki et al., 2001; Svoboda et al., 1996), and multisite activation of neural circuitry (Gasparini and Magee, 2006; Shoham et al., 2005).

### 2.1. Basics of Uncaging

Caged compounds are biologically active molecules that are made inactive by the addition of a light-sensitive "caging" group (Adams and Tsien, 1993). When illuminated by ultraviolet (UV) light, the cage group absorbs a photon, leading to the breakage of a covalent bond linking the cage group to the rest of the molecule. The molecule is then free to act on its biological target. When the uncaging light is provided at precise times and locations, activation can occur with high temporal and spatial resolution.

Most caged compounds are made by the synthetic addition of cage groups to neurotransmitters, second messengers, and peptides. A few exceptions exist: calcium has been caged by making a chelator whose affinity for calcium is altered by photolysis, and nitric oxide (NO) has been caged by making photolyzable small molecules that release NO upon illumination (Makings and Tsien, 1994; Pavlos et al., 2005). Commercial caged compounds are available for many molecules, including adenosine triphosphate (ATP), glutamate, γ-aminobutyric acid (GABA), N-methyl-D-aspartate (NMDA), and carbachol, and for the second messengers inositol 1,4,5-triphosphate (IP$_3$), calcium, and NO.

In general, photolysis techniques have found great use in biology whenever precise temporal and spatial control is important (Gurney, 1994; Nerbonne, 1996). The first biologically useful caged compound, caged ATP (Engels and Schlaeger, 1977; Kaplan et al., 1978), has been used to study muscle contraction, ATP-dependent channels, and molecular motors. Caged fluorescent markers have been used to track movement of cellular components (Theriot and Mitchison, 1991) and the migration of cells during development (Li et al., 2003), and to measure diffusional coupling between dendrites and spines (Svoboda et al., 1996). The caged second messenger IP$_3$ has been applied to study signal transduction during fertilization (Jones and Nixon, 2000), muscle activation (McCarron et al., 2004), and neuronal signaling (Khodakhah and Armstrong, 1997). Caged compounds are also used in drug discovery (Dorman and Prestwich, 2000).

In addition to the existing approaches to making caged compounds, several recent innovations are worth noting. Molecules have been modified with two groups instead of one to allow chemical two-photon uncaging, a nonlinear two-photon-like effect that especially improves axial resolution (Pettit et al., 1997). Uncaging can be combined with molecular biology-driven approaches, including the expression of light-sensitive ion channels or of nonnative channels that can then be selectively activated using caged ligands (Banghart et al., 2004; Chambers et al., 2006; Tan et al., 2006), opening the possibility of light-based activation of specific cells or cell types. Finally, the recent advent of light-activated inhibitors of protein synthesis or caged proteins opens the possibility of optically probing the role of any protein (Goard et al., 2005; Lawrence, 2005).

## 2.2. Properties of Caged Compounds

Good caged compounds must have several important properties. First, in the inactive state, caged molecules should have minimal interaction with the biological system of interest. Possible interactions with the system are not limited to the receptor of interest, since a caged molecule may be inert at one receptor but still have residual activity at others; for instance, caged molecules with no agonist activity can still act as inhibitors for target receptors (Nerbonne, 1996; Sarkisov et al., 2006). Second, products of the photolysis reaction should not affect the system. Since it seems impossible to test caged compounds for agonist as well as antagonist activity on every putative target, uncaging experiments should be designed with proper controls in mind. A critical third property of caged compounds is that they must release ligands efficiently and quickly in response to illumination, and not at other times. To characterize the ability of a compound to be uncaged, a useful parameter is the *uncaging index*. The uncaging index of a cage group is defined as $\varepsilon^*\varphi$, where $\varphi$ is the quantum yield, or the probability of a group to be photolyzed after it absorbs a photon, and $\varepsilon$ is the extinction coefficient. To estimate the uncaging index, the extinction coefficient $\varepsilon$ needs to be measured at the right wavelength; for a given caged group, $\varepsilon$ is constant. Quantum yield $\varphi$ does not change over a range of wavelengths as long as other absorption bands are avoided, but does vary as a function of the identity of the caged molecule and the caging position. The uncaging index is important because the higher the index, the lower the amount of light needed to achieve uncaging. Light levels used to photolyze caged compounds should not damage or interact in other ways with the biological system. This issue is especially important since most cage groups absorb high-energy UV photons that are more likely to cause damage to the sample than visible or infrared light. Uncaging parameters for some commonly used or otherwise important compounds are shown in Table 1.

## 2.3. Handling of Caged Compounds

In uncaging experiments relatively high concentrations of the caged compound are applied to the system for long periods of time. Under such conditions, even if a small fraction of compound is uncaged, either spontaneously or by stray light, the system will be perturbed. For this reason we recommend storing caged compounds under conditions that minimize their degradation. While exact guidelines vary depending on the stability of the used drug, taking extra precautions is appropriate. Solutions of the caged compound should be kept in the freezer between experiments. A point at which care must be taken is the storage of unused compounds

Table 1
Uncaging Parameters for Selected Compounds

| Compound | $\varepsilon$ in $M^{-1}cm^{-1}$ (wavelength) | $\varphi$ | Uncaging index, $\varepsilon^*\varphi$ |
|---|---|---|---|
| CNB-caged glutamate | ~500 (350 nm) | 0.15 | 75 |
| NPE-caged IP$_3$ | 500 (350 nm) | 0.65 | 325 |
| MNI-caged glutamate | 4,300 (350 nm) | 0.085 | 366 |
| NPE-caged ATP | 660 (347 nm) | 0.63 | 416 |
| CNB-caged carbachol | ~600 (350 nm) | 0.8 | 480 |
| DMNB-caged fluorescein dextran | 4,000 (338 nm) | 0.13* | 520* |
| NDBF-EGTA | 15,300 (350 nm) | 0.7 | 10,710 |

*Estimated.
CNB, α-carboxy-2-nitrobenzyl; NPE, 1-(2-nitrophenyl)ethyl; MNI, 4-methoxy-7-nitro-indolinyl; ATP, adenosine triphosphate; DMNB, 4,5-dimethoxy-2-nitrobenzyl; DMNP, 4,5-dimethoxy-2-nitrophenyl; EDTA, ethylenediaminetetraacetic acid; NDBF-EGTA, nitrodibenzofuran-ethyleneglycoltetraacetic acid (Momotake et al., 2006).

between experiments. Large volumes of solution should be divided into aliquots to minimize freezing/defreezing cycles. If aqueous solutions are unstable (for instance with carboxy-2-nitrobenzyl (CNB)-caged glutamate; see Rossi et al., 1997) or if a compound will be stored for prolonged periods of time, aliquots can be dried in a lyophilizer or SpeedVac™ (Thermo Savant, Holbrook, NY) before freezing.

During experiments, unwanted uncaging may result from exposure to ambient light. Stray uncaging from room light and microscope-transmitted light can be reduced by using UV and yellow filters; the latter can be made from material used for filtering theater lighting. For viewing specimens a good way to minimize uncaging is to visualize the specimen with infrared differential interference contrast (IR-DIC) imaging, which is popular for viewing brain slices. Yet another means of minimizing production of agonist is to use double-caged compounds (Pettit et al., 1997), for which small amounts of light generate products that are predominantly still single-caged.

The expense of caged neurotransmitters dictates conservation of the amount of material used. Conservation is usually achieved by using a recirculating bath with a peristaltic pump. One problem of recirculation is the cumulative buildup of photolysis by-products; unwanted uncaging can be reduced by applying caged neurotransmitters locally through capillary tubing (Furuta et al., 1999).

Commercially available compounds may contain impurities or be partially photolyzed. If doubt exists about the quality of the compound, it may be repurified on a column. Fortunately, commonly used caged compounds are now available from different manufacturers such as EMD Biosciences (San Diego, CA), Invitrogen (Carlsbad, CA), Sigma-Aldrich (St. Louis, MO), and Tocris Bioscience, (Ellisville, MO), so switching to another supplier is sometimes an effective solution. Finally, during experiments, caged compound solutions should be protected from light and kept on ice when not in use.

# 3. Designing an Uncaging Setup

## 3.1. Types of Uncaging Systems

Uncaging systems differ in the way that the uncaging light is delivered to the sample. The particular design affects spatial and temporal resolution of photolysis, cost of construction, and simplicity of construction and maintenance. Ideally, the choice of design depends on the needs posed by studying the biological preparation of interest.

One of the simplest systems delivers a brief pulse of UV light to a whole region of the specimen, such as the full field of view on an epi-illumination microscope (e.g., Brasnjo and Otis, 2004). Among the advantages of such a configuration are low cost and simplicity. A UV flashlamp or even an intense arc lamp can simply be mounted to the optical port of the microscope with appropriate coupling and focusing optics. Although with the use of flash lamps systems configured in full-field mode do not provide high spatial resolution (usually >50 µm), temporal resolution can be submillisecond and spatial resolution can be improved by positioning an aperture in the UV path (Xu et al., 1997).

One means of improving spatial resolution is to deliver uncaging light through a fiber optic light guide introduced into the optical path (Bagal et al., 2005; Diamond, 2005; Dodt et al., 2002; Wang and Augustine, 1995; Yang et al., 2006). Focusing assemblies are available commercially, for instance, from Oz Optics (Carp, Canada), or Rapp OptoElectronic (Hamburg, Germany). Fiber optics can be brought up directly to the sample (Kandler et al., 1998) or even inserted into it (Godwin et al., 1997). With direct introduction of a fiber, resolution of tens of micrometers can be achieved depending on the shape of the fiber ending. Micrometer resolution is possible if tapered quartz fiber is used (Eberius and Schild, 2001).

Resolution of the uncaging system can be improved down to the diffraction limit of light by focusing a laser beam into the specimen through the objective using a system of mirrors (Katz and Dalva, 1994; Sarkisov and Wang, 2006). In both fiber optic and mirror-based systems, power losses are a principal design consideration because of limited acceptance cones and absorption in fibers, and because of losses by mirrors and filters. In both cases, if resolution in all three dimensions is desired, the emerging UV beam should fill as much of the back aperture of the objective as possible. The discussion below focuses on considerations in designing and building a focal uncaging system using a laser and mirrors.

### 3.2. Light Source

Different light sources are used in different types of uncaging systems. The UV light source should be powerful enough to provide sufficient energy for photolysis of caged compounds. The key parameter that determines maximal uncaging efficiency is the amount of light energy delivered per unit area in the specimen. As a rule of thumb, a light density of $\sim 0.5 \, \mu J / \mu m^2$ will be sufficient. In a focused uncaging system, possible widening of the uncaging spot due to scattering by the tissue should be taken into account.

Xenon lamps, flash lamps (Rapp, 1998), and mercury arc lamps (Denk, 1997) can all be used for whole-field uncaging. However, since UV lamps are not intrinsically collimated, significant loss of light energy is unavoidable. For fast uncaging, flash lamps deliver the most light but also generate an electrical discharge that can cause large electrophysiological artifacts.

Lasers are more expensive than lamps but can deliver up to several watts of collimated light. Different types of lasers used for uncaging include nitrogen, frequency-doubled ruby, argon, and neodymium-doped yttrium-aluminium-garnet (Nd:YAG). A convenient and economical solution is the type of laser used in our laboratory, a Q-switched, frequency-tripled neodymium-doped yttrium-vanadate (Nd:YVO$_4$) laser (series 3500, DPSS Lasers Inc, Santa Clara, CA) that provides up to 5 W of 355 nm light at a 20 to 150 kHz repetition rate. The flash duration, 50 to 60 ns, is long enough that multiple excitations of a caged molecule are possible during a single pulse, which is desirable for maximizing the likelihood of uncaging per flash of a given energy. The high power output of Q-switched lasers and the ability to deliver a precise number of flashes (down to a single flash) are important for many applications, and are especially useful for a rapid patterned uncaging system (Shoham et al., 2005) that we will describe later in this chapter.

## 3.3. Ultraviolet Optics

For work using UV lasers, several important issues should be considered. Always use protective eyewear, especially on custom-made systems where the beams are often more exposed. UV and infrared (IR) emission are invisible to the human eye, thus presenting a special hazard. The process of troubleshooting and aligning a system is especially prone to risk.

The concentration of light from pulsed lasers in brief pulses leads to peak power levels that can damage optical components. High-energy pulses can be attenuated using a polarizer, a beamsplitter, and a beam stop (Fig. 1A). An alternate approach is to use reflective neutral density filters that are designed to withstand high-energy UV pulses. A more expensive but popular option is to use a Pockels cell, which allows light to be controlled quickly, within tens of microseconds (Thompson et al., 2005). Finally, to reduce per-area light density, a beam expander can be used to widen the laser beam just after it leaves the laser head.

Measurement of UV power at the sample is very useful and important, but some care must be taken to avoid technical errors. Semiconductor sensors have high sensitivity, but because silicon has a high index of refraction, such sensors require incident light to come in at a near-perpendicular angle, and thus cannot be used to measure a converging beam, as occurs in front of a microscope objective. Thermal sensors do not require collimated light, but may not be sensitive enough to detect light transmitted through the objective. Attempting to increase laser power output to a detectable level may cause permanent damage to the objective. One solution is to measure power at the back of the objective and adjust it by the transmission coefficient of the objective. The transmission coefficient can be estimated from the manufacturer's specification or, better yet, measured directly by passing the converging beam telescopically through a second objective positioned backward.

**A**                                          **Top view**

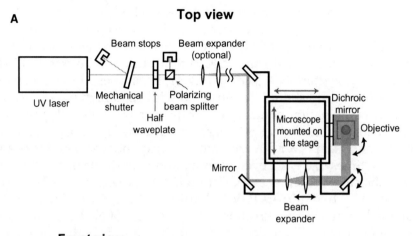

**B**    **Front view**

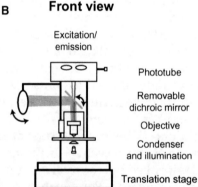

Fig. 1. Optical path of the ultraviolet (UV) uncaging system. After passing through the mechanical shutter and attenuator, the laser beam is directed by two mirrors mounted to the translation stage to the beam expander (**A**). The expanded beam is introduced to the optical path of the system be the dichroic mirror. Black arrows show the optical components used for precise alignment. (Adapted from Sarkisov and Wang, 2006.)

# 4. Construction of a Focal Uncaging Setup

## 4.1. Beam Path

A diagram of a UV laser-based focal uncaging system is shown in Fig. 1. We give specific catalog numbers for the components we use, as well as the general functional criteria for choosing them. Most components have analogous substitutes from other manufacturers.

A beam from a frequency-tripled Q-switched UV laser Nd:YVO$_4$ (model 3501, DPSS Lasers Inc.) is controlled by a mechanical shutter (LS6ZM2 shutter, VMM-D1 driver, Vincent Associates, Rochester, NY) that in the closed state reflects light into a beam trap (BT500, Thorlabs, Newton, NJ). The coating of the shutter's blades as well as the beam trap should be chosen to withstand high power densities. A combination of a zero-order quarter waveplate (WPQ05M-355, Thorlabs) and a calcite polarizer (GL5-A, Thorlabs) is used to attenuate laser power. An optional beam expander widens the beam several times to ensure even illumination over the back-aperture of the objective. After passing through the expander, the beam is directed by two mirrors (UV MAXBRIte, Melles Griot, Irvine, CA) aligned to match the directions of movement of an XY translation stage (XYR-8080, Danaher Corp., Washington, DC) that positions the microscope. Before the beam reaches the objective it is widened by a second, 5× beam expander (BXUV-4.0-5X-355, CVI Laser, Albuquerque, NM) and then introduced to the optical path of the microscope (Fig. 1B) using a dichroic mirror (390DRLP, Omega Optical, Brattleboro, VT). The same beam expander is used to converge the beam to compensate for the focal shift between fluorescence excitation and uncaging light (Sarkisov and Wang, 2006).

For best uncaging resolution, the final UV beam should be flat and approximately fill the back-aperture of the objective. If the back-aperture is overfilled, resolution is improved but the total amount of UV power delivered to the sample is decreased. For instance, when the incoming beam is gaussian with the diameter equal to the size of the back-aperture, 86% of the total incoming light enters the objective, and spatial resolution is made worse by somewhat less than 10%, compared with optimal (Helmchen and Denk, 2005). Diffraction-limited lateral resolution is approximately 0.23 µm full-width half-maximum (Thompson et al., 2005) using a purely mirror-based system. Two-photon uncaging leads to improvements in axial resolution, but because of the longer wavelengths used the lateral resolution is actually worse. A possible solution in the absence of strong scattering is the use of chemical two-photon uncaging.

### 4.2. Aligning the Setup

To achieve the optimal resolution of the uncaging system, the UV beam should follow the optical axis of the system. This is achieved by manual adjustment of the two last mirrors of the UV beam path. Another important requirement is that uncaging light must converge to the same plane of focus as the imaging light. Since objectives bend light differently as a function of wavelength, two collimated beams uncaging and imaging light would usually focus to slightly different planes. To compensate for this focal shift, we make uncaging light slightly diverging (or converging, depending on the objective used) by adjusting the distance between lenses of the beam expander.

For initial alignment a water solution of fluorescein can be used. When illuminated with UV light, the excited fluorescein solution should be visible from the side (Fig. 2A). If the observed cone is not axially symmetric with respect to the optical axis of the system (Fig. 2B) or does not coincide with the point where the excitation light is focused (Fig. 2C), adjustments to the mirrors are necessary. Once the excitation cone appears symmetrical from the side, fluorescein excitation can be observed through the eyepieces and slight adjustments made to the mirror to put the fluorescent spot in the center of the field of view. Then the size of the uncaging spot is minimized by changing the distance between lenses of the beam expander.

Once the size and position of the uncaging spot are optimized under visual control, fine-tuning is then possible using a thin sample of dried caged fluorescein. A dried layer of caged dye greatly simplifies alignment of the uncaging system because it is immobile and can be examined at leisure after an uncaging flash. In our test samples, $10\,\mu L$ water solution of the 1% bovine serum albumin is mixed with $2\,mg/mL$ caged fluorescein dextran and dried on a coverslip. Then the coverslip is positioned under the microscope sample side down as shown in Fig. 2D, and an uncaging flash is delivered. The shape and position of the spot of photolyzed fluorescein are visualized after the flash, and the distance between lenses of the beam expander is adjusted to achieve optimal resolution (Fig. 2E). For a fixed distance between lenses we can measure the focal shift between imaging and uncaging light by finding the axial specimen position at which a light pulse gives a sharply focused uncaging spot. A plot of the focal shift as a function of distance between the lenses of the beam expander is shown in Fig. 2F; zero position is defined as a configuration that does not diverge or converge a collimated UV beam. Detailed instructions on alignment and calibration of the focal photolysis system are given in our recent work (Sarkisov and Wang, 2006).

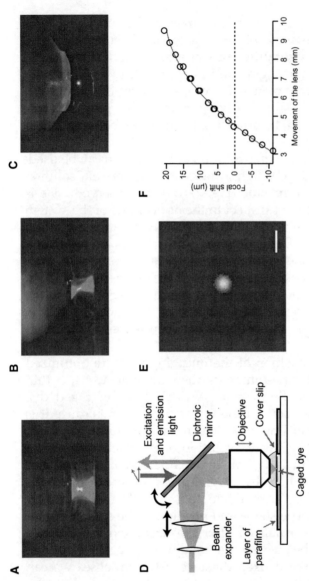

Fig. 2. Optimization of the geometry of the uncaging spot. The shape of the uncaging beam is visualized by the excitation of fluorescein. Optimal resolution is achieved when the cone of excited fluorescein is symmetrical (**A**). If light is asymmetrical (**B**) or does not focus to the location of two-photon excitation (**C**), the shape of the uncaging spot will be suboptimal. (**D**) The focal shift between imaging and the uncaging light is compensated for by changing the distance between the lenses of the beam expander and visualizing the shape of the uncaging spot on the sample of fried caged dye. (**E**) When the focal shift is fully compensated, submicrometer resolution of uncaging is achieved. Image of uncaged fluorescein that was immobilized on the coverslip. Scale bar 1 µm. (**F**) Change of the focal shift between IR and UV light as the distance between lenses increases. (Adapted from Sarkisov and Wang, 2006.)

## 4.3. Control of the System

Uncaging systems can be easily controlled by equipment commonly found in most physiology laboratories. In the simplest case a single transistor-transistor logic (TTL) pulse can be used to trigger a flash lamp or open a shutter. In a system based on a Q-switched laser two signals are necessary: one to open a mechanical shutter, and one to start lasing after a small delay to allow the shutter to open fully. Trigger signals can easily be generated by pulse stimulators (e.g., Master-8, AMPI, Jerusalem, Israel), digitizers (e.g., DigiData, Molecular Devices, Menlo Park, CA), or by a data acquisition board (e.g., National Instruments, Austin, TX) connected to the computer.

Time synchronization among the imaging system, the uncaging system, and the electrophysiological recording system is not hard to accomplish. If the camera or photomultiplier tube (PMT) used for image acquisition is not turned off during the uncaging flash to protect it from high light levels, a flash artifact will be visible on the image. We have not observed deterioration in the performance due to UV flashes in our conventional PMT (R3896, Hamamatsu, Hamamatsu City, Japan) over time. Power supplies of more sensitive gallium-arsenic PMT modules (e.g., H7422P-40, Hamamatsu) are equipped with protective circuits that shut off detectors when they are exposed to high light levels. A time mark that corresponds to the moment of uncaging can be recorded to the unused channel on the physiological system. In many custom microscopy control systems, CfNT (M. Müller, Max Planck Institute for Medical Research, Heidelberg, Germany) and MPScope (Nguyen et al., 2006) marks can also be embedded in the recorded optical signal.

# 5. Recent Developments in Uncaging

## 5.1. Two-Photon Uncaging

Optical two-photon uncaging provides a very effective way to achieve high spatial resolution of uncaging. In this technique a pulsed IR laser is used as the uncaging light source. If a wavelength is chosen such that the cage group can absorb two photons of IR light of similar energy to one UV uncaging photon, the probability of an uncaging event becomes proportional to the second power of light density, thus limiting uncaging to the focal volume where light density is maximal. In recent years cage groups have come into more common use with sufficiently high optical cross-section to allow two-photon uncaging without major specimen damage (Fedoryak et al., 2005).

Aside from confining uncaging to the focal volume, the use of IR photons in true two-photon uncaging allows deeper penetration into the specimen. Since IR light is scattered less than UV light, increases in focal volume are less pronounced as light is focused more deeply into the sample. This advantage of true two-photon uncaging is especially important when working with highly scattering samples as brain slices. Two-photon uncaging is attractive in systems built upon a two-photon microscope since the imaging beam could be used for uncaging and the main necessary additional component is a Pockels cell to gate beam intensity. Use of infrared light for uncaging also simplifies alignment of such a system, since compensation for the focal shift between uncaging and imaging light becomes unnecessary. At present the main disadvantages of two-photon uncaging are the high cost of mode-locked infrared lasers and the limited number of caged compounds with sufficient two-photon uncaging index, a quantity that is closely related to the conventional uncaging index (see Table 1).

### 5.2. Chemical Two-Photon Uncaging

Chemical two-photon uncaging is a simple way to improve characteristics of an existing photolysis system. Improvement is achieved by adding a second inactivating caged group to the molecule of interest. Uncaging a double-caged compound requires absorption of two UV photons, making the probability of photolysis proportional to the second power of light density. Adding a second cage greatly reduces spurious out-of-focus uncaging, since at nonsaturating levels of light most double uncaging events occur in the focal volume where light density is maximal (Pettit et al., 1997). Confinement of photolysis volume is especially important when uncaging of a neurotransmitter is performed in the brain slices, since neurotransmitter action is limited to the cells located above and below the focal volume.

Double caging can also improve the chemical properties of the compound by making it more dissimilar in structure to the native agonist. Dissimilarity reduces the risk of undesired interaction with biological targets (Sarkisov et al., 2007). A third advantage comes from ease of handling; the requirement of two uncaging events makes the production of free agonists due to uncaging by room light or spontaneous degradation less likely.

## 5.3. Patterned Uncaging

Uncaging at multiple locations can be achieved by steering the uncaging beam (Gasparini and Magee, 2006; Matsuzaki et al., 2001; Shepherd et al., 2003; Shoham et al., 2005). Rapid beam steering is usually done using mirrors mounted on scanning galvanometers, as on a confocal or two-photon microscope. A faster means of scanning is the diversion of a UV uncaging beam with acousto-optical deflectors (AODs), which have fast switching times. In an AOD-based system we have achieved stimulation with submicrometer resolution at over 20,000 locations per second (Shoham, O'Connor et al., 2005). Our patterned uncaging system is capable of uncaging in an area of 170 by 170 μm (Fig. 3A), and is integrated with two-photon fluorescence microscopy and patch-clamp recording.

We have used this system to measure scattering in brain tissue of UV light. To perform these measurements fluorescent beads were embedded in different depths in molecular layer of a sagittal cerebellar slice (Wistar rats, P17–21), and the UV beam scanned over the tissue to form an image. Dependence on the size of the reconstructed bead as a function of the depth is shown on Fig. 3B. We found that the focal properties of the UV beam are approximately preserved for the first 25 μm of the slice.

We used this patterned uncaging system to study neuronal and circuit functionality in brain slices. Figure 3C illustrates an experiment in which connections between cerebellar granule cells and Purkinje cells were identified. Caged glutamate was photolyzed at over 20 locations in the granule layer, while electrophysiological responses from the Purkinje cell were recorded by whole-cell patch clamp recording. Out of 20 uncaging locations, seven responding regions were identified (Fig. 3D).

## 5.4. Light-Activatable Proteins and Ectopic Caged Neurotransmitter Receptors

Recent developments in the use of genetically encodable probes provide an approach to controlling cell function that holds great promise (Lawrence, 2005; Miesenböck and Kevrekidis, 2005). Though these methods still require patch-clamp or optical recording, they are useful not only because they are less invasive but also because they have the potential to be cell type–specific. One approach is to express a light-sensitive channel that is normally not

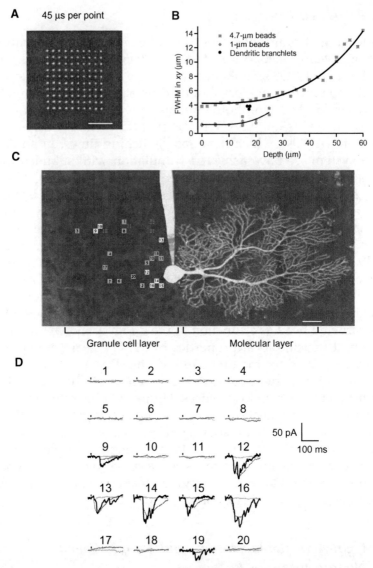

Fig. 3. Patterned uncaging with UV light. In this setup an uncaging beam can be focused laterally to different locations in the focal plane by means of acousto-optical devices. (**A**) Caged fluorescein was photolyzed in 100 locations with 45-μs interpulse time. Scale bar, 50 μm. (**B**) Measurement of the effect of UV scattering on resolution in a cerebellar brain slice. Images of fluorescent beads at different depths were taken by scanning with UV light. (**C**) Cerebellar Purkinje cell filled via patch-clamp recording electrode with fluorescent dye. The squares indicate the positions of the uncaging locations in the granule cell layer. Scale bar, 25 μm. (**D**) Postsynaptic responses to photolysis in the locations shown in **C**. (Adapted from Shoham et al., 2005.)

present in the neurons under a cell type–specific promoter (Boyden et al., 2005; Chambers et al., 2006). When illuminated by the light, the channels open, thus activating a cell of interest. A second approach is to ectopically express a receptor for a transmitter that is not native to the experimental species. Expression in mammals of receptors for the insect transmitter allatostatin has been used to inactivate subpopulations of neurons (Tan et al., 2006). These strategies exemplify the future power of combining protein-based probe design and optical methods.

## Acknowledgments

This work was supported by National Institutes of Health grant NS045193. We thank D. A. DiGregorio and T. S. Otis for helpful comments.

## References

Adams, S. R. and Tsien, R. Y. (1993) Controlling cell chemistry with caged compounds. *Annu. Rev. Physiol.* **55**, 755–784.

Bagal, A. A., Kao, J. P. Y., et al. (2005) Long-term potentiation of exogenous glutamate responses at single dendritic spines. *Proc. Natl. Acad. Sci. USA* **102**(40), 14434–14439.

Banghart, M., Borges, K. et al. (2004) Light-activated ion channels for remote control of neuronal firing. *Nature Neurosci.* **7**(12), 1381–1386.

Boyden, E. S., Zhang, F., et al. (2005) Millisecond-timescale, genetically targeted optical control of neural activity. *Nature Neurosci.* **8**(9), 1263–1268.

Bozza, T., McGann, J. P., et al. (2004) In vivo imaging of neuronal activity by targeted expression of a genetically encoded probe in the mouse. *Neuron* **42**(1), 9–21.

Brasnjo, G. and Otis, T. S. (2004) Isolation of glutamate transport-coupled charge flux and estimation of glutamate uptake at the climbing fiber-Purkinje cell synapse. *Proc. Natl. Acad. Sci. USA* **101**(16), 6273–6278.

Chambers, J. J., Banghart, M. R., et al. (2006) Light-induced depolarization of neurons using a modified Shaker $K^+$ channel and a molecular photoswitch. *J. Neurophys.* **96**(5), 2792–2796.

Denk, W. (1997) Pulsing mercury arc lamps for uncaging and fast imaging. *J. Neurosci. Methods* **72**(1), 39–42.

Diamond, J. S. (2005) Deriving the glutamate clearance time course from transporter currents in CA1 hippocampal astrocytes, transmitter uptake gets faster during development. *J. Neurosci.* **25**(11), 2906–2916.

Djurisic, M., Zochowski, M., et al. (2003) Optical monitoring of neural activity using voltage-sensitive dyes. *Biophotonics* **361**, 423–451.

Dodt, H. U., Eder, M., et al. (2002) Infrared-guided laser stimulation of neurons in brain slices. *Sci STKE* **2002**(120), PL2.

Dorman, G. and Prestwich, G. D. (2000) Using photolabile ligands in drug discovery and development. *Trends Biotech.* **18**(2), 64–77.

Eberius, C. and Schild, D. (2001) Local photolysis using tapered quartz fibres. *Pflügers Arch.* **443**(2), 323–330.

Eder, M., Zieglgansberger, W., et al. (2004) Shining light on neurons—elucidation of neuronal functions by photostimulation. *Rev. Neurosci.* **15**(3), 167–183.

Engels, J. and Schlaeger, E. J. (1977) Synthesis, structure, and reactivity of adenosine cyclic 3',5'-phosphate benzyl triesters. *J. Med. Chem.* **20**(7), 907–911.

Fedoryak, O. D., Sul, J. Y., et al. (2005) Synthesis of a caged glutamate for efficient one- and two-photon photorelease on living cells. *Chem. Commun.* **29**, 3664–3666.

Furuta, T., Wang, S. S.-H., et al. (1999) Brominated 7-hydroxycoumarin-4–ylmethyls: novel photolabile protecting groups with biologically useful cross-sections for two-photon photolysis. *Proc. Natl. Acad. Sci. USA* **96**, 1193–1200.

Garaschuk, O., Milos, R.-I., et al. (2006) Targeted bulk-loading of fluorescent indicators for two-photon brain imaging in vivo. *Nat. Protocols* **1**(1), 380–386.

Gasparini, S. and Magee, J. C. (2006) State-dependent dendritic computation in hippocampal CA1 pyramidal neurons. *J. Neurosci.* **26**(7), 2088–2100.

Goard, M., Aakalu, G., et al. (2005) Light-mediated inhibition of protein synthesis. *Chem. Biol.* **12**(6), 685–693.

Godwin, D. W., Che, D. P., et al. (1997) Photostimulation with caged neurotransmitters using fiber optic light guides. *J. Neurosci. Methods* **73**(1), 91–106.

Grinvald, A. and Hildesheim, R. (2004) VSDI: A new era in functional imaging of cortical dynamics. *Nat. Rev. Neurosci.* **5**(11), 874–885.

Gurney, A. M. (1994) Flash photolysis of caged compounds, in *Microelectrode Techniques* (Ogden, D. ed.), Company of Biologists Limited, Cambridge, pp. 389–406.

Helmchen, F. and Denk, W. (2005) Deep tissue two-photon microscopy. *Nat. Methods* **2**(12), 932–940.

Isaacson, J. S. and Nicoll, R. A. (1991) Aniracetam reduces glutamate receptor desensitization and slows the decay of fast excitatory synaptic currents in the hippocampus. *Proc. Natl. Acad. Sci. USA* **88**(23), 10936–10940.

Jones, K. T. and Nixon, V. L. (2000) Sperm-induced $Ca^{2+}$ oscillations in mouse oocytes and eggs can be mimicked by photolysis of caged inositol 1,4,5-trisphosphate: evidence to support a continuous low level production of inositol 1,4,5-trisphosphate during mammalian fertilization. *Dev. Biol.* **225**(1), 1–12.

Kandler, K., Katz, L. C., et al. (1998) Focal photolysis of caged glutamate produces long-term depression of hippocampal glutamate receptors. *Nature Neurosci.* **1**(2), 119–123.

Kaplan, J. H., Forbush, B. 3rd, et al. (1978) Rapid photolytic release of adenosine 5'-triphosphate from a protected analogue: utilization by the Na-K pump of human red blood cell ghosts. *Biochemistry* **17**(10), 1929–1935.

Katz, L. C. and Dalva, M. B. (1994) Scanning laser photostimulation, a new approach for analyzing brain circuits. *J. Neurosci. Methods* **54**(2), 205–218.

Khodakhah, K. and Armstrong, C. M. (1997) Induction of long-term depression and rebound potentiation by inositol trisphosphate in cerebellar Purkinje neurons. *Proc. Natl. Acad. Sci. USA* **94**(25), 14009–14014.

Kramer, R. H., Chambers, J. J., et al. (2005) Photochemical tools for remote control of ion channels in excitable cells. *Nat. Chem. Biol.* **1**(7), 360–365.

Lawrence, D. S. (2005) The preparation and in vivo applications of caged peptides and proteins. *Curr. Opin. Chem. Biol.* **9**(6), 570–575.

Li, Y. X., Zdanowicz, M., et al. (2003) Cardiac neural crest in zebrafish embryos contributes to myocardial cell lineage and early heart function. *Dev. Dyn.* **226**(3), 540–550.

Makings, L. R. and Tsien, R. Y. (1994) Caged nitric oxide. Stable organic molecules from which nitric oxide can be photoreleased. *J Biol. Chem.* **269**(9), 6282–6285.

Matsuzaki, M., Ellis-Davies, G. C. R., et al. (2001) Dendritic spine geometry is critical for AMPA receptor expression in hippocampal CA1 pyramidal neurons. *Nature Neurosci.* **4**(11), 1086–1092.

McCarron, J. G., MacMillan, D., et al. (2004) Origin and mechanisms of $Ca^{2+}$ waves in smooth muscle as revealed by localized photolysis of caged inositol 1,4,5-trisphosphate. *J. Biol. Chem.* **279**(9), 8417–8427.

Miesenböck, G. and Kevrekidis, I. G. (2005) Optical imaging and control of genetically designated neurons in functioning circuits. *Annu. Rev. Neurosci.* **28**, 533–563.

Momotake, A., Lindegger, N., et al. (2006) The nitrodibenzofuran chromophore: a new caging group for ultra-efficient photolysis in living cells. *Nat. Methods* **3**(1), 35–40.

Nerbonne, J. M. (1996) Caged compounds: tools for illuminating neuronal responses and connections. *Curr. Opin. Neurobiol.* **6**(3), 379–386.

Nguyen, Q. T., Tsai, P. S., et al. (2006) MPScope: a versatile software suite for multi-photon microscopy. *J. Neurosci. Methods* **156**(1-2), 351–359.

Pavlos, C. M., Xu, H., et al. (2005) Photosensitive precursors to nitric oxide. *Curr. Top. Med. Chem.* **5**(7), 637–647.

Pettit, D. L., Wang, S. S.-H., et al. (1997) Chemical two-photon uncaging: a novel approach to mapping glutamate receptors. *Neuron* **19**(3), 465–471.

Rapp, G. (1998) Flash lamp-based irradiation of caged compounds. *Methods Enzymol.* **291**, 202–222.

Rossi, F. M., Margulis, M., et al. (1997) N-Nmoc-L-glutamate, a new caged glutamate with high chemical stability and low pre-photolysis activity. *J. Biol. Chem.* **272**(52), 32933–32939.

Sarkisov, D. V., Gelber, S. E., et al. (2007) Synapse-specificity of calcium release probed by chemical two-photon uncaging of $IP_3$ (in review).

Sarkisov, D. V. and Wang, S. S.-H. (2006) Alignment and calibration of a focal neurotransmitter uncaging system. *Nat. Protocols* **1**(2), 828–832.

Shepherd, G. M. G., Pologruto, T. A., et al. (2003) Circuit analysis of experience-dependent plasticity in the developing rat barrel cortex. *Neuron* **38**(2), 277–289.

Shoham, S., O'Connor, D. H., et al. (2005) Rapid neurotransmitter uncaging in spatially defined patterns. *Nat. Methods* **2**(11), 837–843.

Sullivan, M. R., Nimmerjahn, A., et al. (2005) In vivo calcium imaging of circuit activity in cerebellar cortex. *J. Neurophys.* **94**(2), 1636–1644.

Svoboda, K., Tank, D. W., et al. (1996) Direct measurement of coupling between dendritic spines and shafts. *Science* **272**(5262), 716–719.

Tan, E. M., Yamaguchi, Y., et al. (2006) Selective and quickly reversible inactivation of Mammalian neurons in vivo using the Drosophila allatostatin receptor. *Neuron* **51**(2), 157–170.

Theriot, J. A. and Mitchison, T. J. (1991) Actin microfilament dynamics in locomoting cells. *Nature* **352**(6331), 126–131.

Thompson, S. M., Kao, J. P. Y., et al. (2005) Flashy science: controlling neural function with light. *J. Neurosci.* **25**(45), 10358–10365.

Tsien, R. Y. (1999) Monitoring cell calcium, in *Calcium as a Cellular Regulator* (Carafoli, E. and Klee, C.B., eds.), pp. 28–54, Oxford University Press, New York.

Wang, J. W., Wong, A. M., et al. (2003) Two-photon calcium imaging reveals an odor-evoked map of activity in the fly brain. *Cell* **112**(2), 271–282.

Wang, S. S.-H. and Augustine, G. J. (1995) Confocal imaging and local photolysis of caged compounds, dual probes of synaptic function. *Neuron* **15**(4), 755–760.

Xu, T., Naraghi, M., et al. (1997) Kinetic studies of $Ca^{2+}$ binding and $Ca^{2+}$ clearance in the cytosol of adrenal chromaffin cells. *Biophys. J.* **73**(1), 532–545.

Yang, E. J., Harris, A. Z., et al. (2006) Variable kainate receptor distributions of Oriens interneurons. *J. Neurophys.* **96**(3), 1683–1689.

# 6

# Visually Guided Patch-Clamp Recordings in Brain Slices

## James R. Moyer, Jr. and Thomas H. Brown

## 1. Techniques for Preparing Healthy Brain Slices

Brain slices have become an integral part of synaptic and cellular physiology since the pioneering studies of Henry McIlwain (Li and McIlwain, 1957; Yamamoto and McIlwain, 1966). The hippocampal slice preparation was first brought to the United States from Per Anderson's lab, initially by Tim Teyler and followed shortly after by Phil Schwartzkroin. Over the years the development of the in vitro brain slice preparation has enabled electrophysiologists to study various aspects of the nervous system in an isolated preparation that still retains many of the brain's complement of neuronal connections. Through the use of brain slice preparations, much has been learned about the intrinsic properties and morphology of different populations of neurons, the connectivity between different cell types within or between brain regions, the quantal nature of transmitter release, and the various forms of synaptic plasticity. This section focuses on some of the important aspects of preparing and maintaining healthy brain slices, along with some of the rationales for selecting certain procedures.

### 1.1. Recording and Cutting Solutions

A major value of the in vitro brain slice preparation is the ability to vary in a systematic and controlled manner the extracellular and even the intracellular environment of the neuron(s) under study. To obtain and maintain healthy brain slices, it is necessary to choose the appropriate incubation and recording solutions for a specific set of experiments and to select an appropriate cutting solution in

From: *Neuromethods, Vol. 38: Patch-Clamp Analysis: Advanced Techniques, Second Edition*
Edited by: W. Walz @ Humana Press Inc., Totowa, NJ

which to prepare the brain slices. This section reviews some of the ways to prepare and perhaps modify recording and cutting solutions to obtain healthy brain slices.

### 1.1.1. Recording Solutions for Work with Rat Brain Slices

Selection of an appropriate artificial cerebrospinal fluid (aCSF) for use in brain slice studies depends to some extent on the specific preparation. For example, the composition of aCSF for recording from neurons in rat brain slices should be similar to that found in rat CSF. The following is a listing of compounds found in rat CSF: $Na^+$ 156 mM, $Cl^-$ 126 mM, $K^+$ 2.8 mM, glucose 65 mg/dl (~3.6 mM), lactate 2.8 mM, pH 7.35, 302 mOsm (Sharp and La Regina, 1998). In addition, calcium concentrations in rat cerebrospinal fluid (CSF) have been reported to be about 2.2 mM (see Table 2.6 on page 24 of Davson et al., 1987), which is comparable to other species including cat, dog, goat, and human (see Tables 2.2 through 2.6 in Davson et al., 1987).

For recording, we use the following aCSF (mM): 124 NaCl, 2.8 KCl, 2 $CaCl_2$, 2 $MgSO_4$, 1.25 $NaH_2PO_4$, 26 $NaHCO_3$, 10 D-glucose, 0.4 sodium ascorbate, pH 7.4, 295 mOsm. Historically, glucose concentrations are much higher in brain slice studies than they are in the living brain (typically 10 to 26 mM in slice studies compared with 4 mM in living rat brain). This has its roots in early studies that had difficulty obtaining healthy brain slices using lower concentrations of glucose. Some laboratories use much higher concentrations of glucose (~26 mM) in their aCSF (e.g., Frick et al., 2004; Gulledge and Stuart, 2005). One possible explanation for the benefit of using higher glucose concentrations may be that glucose gets converted to lactate, which has been suggested to be essential during recovery from a hypoxic event (Schurr et al., 1997a,b), such as happens during brain slicing. While we do not use lactate in our own experiments, it may be beneficial in minimizing the anoxic insult inherent in decapitation and preparation of brain slices.

Three different stock solutions are used for making aCSF. They are made approximately every 2 weeks, and we advise that the stocks be replaced if they are more than 1 month old. We use a NaCl stock solution (10×), a sulfate and phosphate stock solution (10× of each of KCl, $MgSO_4$, $NaH_2PO_4$), and a $CaCl_2$ stock solution (100×). The aCSF is made fresh daily by weighing and adding the appropriate amount of glucose, bicarbonate, and ascorbate to the stock solutions. We always add the $CaCl_2$ stock solution last and only after adding about 75% of the total volume of water required. If the $CaCl_2$ stock solution is added earlier or before increasing the volume of water, the solution becomes cloudy due to precipitation of calcium carbonate.

Some researchers supplement their cutting and recording solutions with antioxidants to reduce oxidative cellular damage. Studies have shown that oxidative damage can have detrimental effects on neurons through, for example, mitochondrial damage and membrane lipid peroxidation (Keller et al., 1998; Kovachich and Mishra, 1980; Mattson, 1998). The addition of small quantities of antioxidants such as sodium ascorbate (vitamin C) can ameliorate destructive membrane lipid peroxidation as measured by the accumulation of malonaldehyde as well as edema through water gain (Avshalumov and Rice, 2002; Brahma et al., 2000; Kovachich and Mishra, 1980, 1983; Rice, 1999, 2000). The range of vitamin C concentrations commonly used in aCSF is between 0.4 and 1.0 mM (Borst et al., 1995; Hoffman and Johnston, 1998; Rice et al., 1994). We use 0.4 mM sodium ascorbate in both our cutting and recording solutions. Although glutathione is also an excellent antioxidant, it is not taken up by neurons and does not lead to neuronal preservation (Rice et al., 1994). Some researchers also include an adenosine triphosphate (ATP) regenerative system (e.g., phosphocreatine, creatine phosphokinase, pyruvate, and ATP) in their aCSF solution to offset the rapid depletion of ATP that occurs immediately after decapitation and preparation of brain slices. For example, 2 mM pyruvate has been shown to protect neurons against oxidation from hydrogen peroxide generation (Desagher et al., 1997), so we have used pyruvate when working with aged tissue.

### 1.1.2. Selecting an Appropriate Recording Solution

The specific questions to be addressed dictate the ionic and pharmacological composition of the recording solution. For example, when studying calcium currents it is important to isolate them from other contaminating currents, such as those produced by $Na^+$ and $K^+$ conductances. This is often done pharmacologically by adding to the aCSF antagonists such as tetrodotoxin, which blocks sodium channels (Narahashi, 1974; Narahashi et al., 1964), and tetraethylammonium (TEA) chloride, 4-aminopyridine (4-AP), or internal cesium ions, which block potassium channels (see Chapter 7 of Johnston and Wu, 1995; see also Chapters 2 and 3 of Hille, 2001). It may also be important to conduct ion substitution studies. For example, if a current of interest is suspected to be carried by $Na^+$ ions, it is important to evaluate the current as a function of changes in the concentration of $Na^+$ in the aCSF. To maintain osmolarity and charge balance, this is done by substituting various concentrations of NaCl with an equal concentration of a compound such as choline chloride (where the cation, choline, substitutes for $Na^+$ but is impermeable). Thus, the effects of sodium concentration on the current as well as its reversal potential can be obtained.

In the course of pharmacological studies, certain divalent cations are added to the recording solution to, for example, block certain currents. In these cases, care should be taken to prevent precipitation. This can be accomplished by removing certain phosphates or sulfates and replacing them with appropriate salts to maintain charge and osmolarity balance. Other studies may necessitate changing the ratio of $Ca^{2+}$ and $Mg^{2+}$, which alters not only neurotransmitter release but also the voltage-dependence of $N$-methyl-$D$-aspartate (NMDA) receptor activation.

### 1.1.3. Composition of Different Cutting Solutions

For cutting brain slices, researchers originally relied on using their standard recording solution. Although this has led to some high-quality data in a variety of laboratories, intracellular recordings were difficult to obtain in certain brain regions. For example, obtaining stable, high-quality intracellular recordings from healthy facial nucleus motoneurons from adult rats was not feasible until George Aghajanian's group discovered the benefits of replacing NaCl in their cutting solution with an equimolar concentration of sucrose (Aghajanian and Rasmussen, 1989). This replacement was believed to decrease the neurotoxic effects of passive chloride influx, swelling, and lysis of neurons that occurs during the decapitation and slice preparation processes. We and other laboratories have discovered the value of minimizing excitotoxic damage during slice preparation. In brain slice studies it is now common to use a *neuroprotective* cutting solution.

Sucrose is not the only compound used in place of NaCl during the preparation of brain slices. Some laboratories have also used choline-chloride, with excellent results (e.g., Hoffman and Johnston, 1998). Some laboratories also block glutamate-induced excitotoxic damage by including in the cutting solution one of several glutamate-receptor antagonists, such as kynurenic acid, which is a nonselective blocker of NMDA and alpha-amino-

3-hyroxy-5-methyl-4-isoxazoleproprionic acid (AMPA)/kainate receptors (Christie et al., 1995; Magee et al., 1996).

We have found that a sucrose-based cutting solution works quite well for obtaining stable, long-term, whole-cell recordings (WCRs) as well as synaptically evoked currents. The composition of our sucrose-CSF (in mM) is as follows: 206 sucrose, 2.8 KCl, 1 $CaCl_2$, 1 $MgCl_2$, 2 $MgSO_4$, 1.25 $NaH_2PO_4$, 26 $NaHCO_3$, 10 D-glucose, ascorbic acid 0.4, pH 7.4, 295 mOsm. As with our recording solution (described earlier in Section 1.1.1), our sucrose-CSF is made fresh daily by weighing and adding the appropriate amount of sucrose, bicarbonate, glucose, and ascorbate to the appropriate amount of stock solutions (including a 100× stock solution of $MgCl_2$). Again, if a concentrated stock solution of calcium chloride is used, it should be added last and only after adding a sufficient volume of water (about 50–75% of desired volume). Otherwise the solution will become cloudy, a problem that can sometimes be reversed by oxygenation with carbogen gas (95% $O_2$, 5% $CO_2$).

Our cutting solution is designed to protect neurons in multiple ways. Replacing NaCl with an iso-osmolar concentration of sucrose serves two functions: it reduces sodium influx that occurs when neurons depolarize during the anoxia that accompanies decapitation, dissection, and slicing; and it minimizes excitotoxic cell swelling from passive chloride influx followed by cation and water entry (Rothman, 1985). We also increase the $Mg^{2+}/Ca^{2+}$ ratio from 1:1 to 3:1. This minimizes calcium influx caused by glutamate-induced depolarization (Choi, 1992, 1994, 1995; Feig and Lipton, 1990). Also, our inclusion of 0.4 mM ascorbic acid helps to reduce oxidative damage. We find that this cutting solution yields healthy slices from juvenile, adult, and aged animals (see Sections 1.4 and 1.5 below for additional details).

## 1.2. Making an Incubation Chamber for Maintaining Healthy Slices

One of the most important aspects of in vitro neurophysiology is maintaining a supply of healthy brain slices while one is being used for recording. The older interface-type of recording chamber (Oslo type or Haas type), which was designed for use with a dissecting microscope, had room for several slices. Thus, the recording chamber could also double as a holding chamber. On the other hand, the submerged type of recording chambers that are designed to rest on the stage of a compound microscope typically requires a separate holding chamber. The use of a separate holding chamber also keeps fresh slices free from any drugs that are used during an experiment. In modern holding chambers the slices rest on a nylon mesh, which permits exposure of oxygenated aCSF from both surfaces of the slice (Fig. 1). Several different holding chamber designs can be found in the literature (Edwards and Konnerth, 1992; Moyer and Brown, 1998; Moyer et al., 1992; Sakmann and Stuart, 1995).

### 1.2.1. Making and Assembling an Incubation Chamber

The slice incubation chamber we use results in nearly 100% healthy slices that can be maintained for 6 to 12 hours (Moyer and Brown, 1998). We use a 24-well chamber so that individual slices are kept separate from each other and stored in the order in which they were sectioned. This facilitates identification of the rostrocaudal or dorsoventral level of each section. The design is simple and can be easily made in only a few hours (Fig. 1A). We begin with a standard 24-well culture dish. The bottom is cut off and sanded, and holes are drilled through the walls of each row and column. Thus, each well has four holes for diffusion of aCSF. Next, the dish is then cut in half to yield two 12-well chambers.

Two pieces of standard nylon hosiery are cut from the foot portion, stretched across the bottom of each chamber, and held taught by a rubber band. The nylon is then glued to each chamber using a cyanoacrylate glue (e.g., Krazy Glue pen). The glue is carefully applied onto the nylon hosiery that contacts the bottom of each well and the bottom edges of each chamber. After the chambers dry overnight, excess nylon hosiery is cut away using a sharp razor blade. Eight glass legs, which serve to elevate the chamber and prevent it from floating, are made by cutting pieces from a standard glass stirring rod. Four legs are glued to the four corners of each chamber using a clear silicone-based sealant like that used for aquaria. The chambers are allowed to dry overnight, after which they are submerged in deionized water for several days prior to use.

### 1.2.2. Oxygenation of Brain Slices in the Incubation Chamber

To oxygenate the solutions, we use coarse fritted glass bubblers (Fisher Scientific, Pittsburgh, PA) because they release small bubbles and they are resistant to clogging. The fritted glass bubblers themselves are only ~20 mm long, but they are at joined at the end of a 22-cm Pyrex tube. To remove the Pyrex tube, it is first scored using

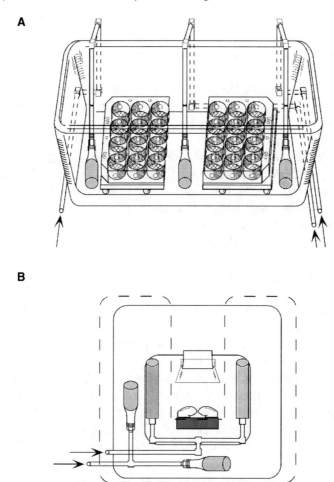

Fig. 1. Design of our brain slice incubation and cutting chambers. (**A**) Schematic diagram of the 24–well slice incubation chamber assembly. Three fritted glass bubblers are used to deliver carbogen gas (95% $O_2$, 5% $CO_2$). Not shown is a stir plate and miniature stir bar used to gently distribute the oxygenated artificial cerebrospinal fluid (aCSF). (**B**) Schematic of the stainless steel Vibratome cutting tray (74 × 74 × 9mm). Small gas dispersion tubes are used to oxygenate the sucrose-CSF during slice preparation. These dispersion tubes are secured by small silver wires that are glued in place by silicone sealant placed on the outer side and bottom surface of the tray. The sucrose-CSF located in the outer chamber of the Vibratome is also oxygenated through two fritted glass bubblers. A 4% (weight/volume) agar block, indicated by the dark gray shaded rectangle, is used to support the brain tissue during cutting. Also shown is a thin Plexiglas lid used to minimize fluid loss by splashing (dashed lines). Arrows in **A** and **B** indicate flow of carbogen gas. (Moyer and Brown, 1998, with permission from Elsevier Science.)

a diamond knife. Then the tube is wrapped in a KimWipe (Fisher Scientific, Pittsburgh, PA), gripped on each side of the score with pliers, and broken along the score line. The bubbler can then be attached to Tygon tubing connected to a multiline manifold that permits regulation of the airflow from the 95% $O_2$/5% $CO_2$ (carbogen) air tank to each of the bubblers. Oxygenation should be just sufficient to maintain a pH of ~7.4 (excess oxygenation can be detrimental to brain slices).

The two 12-well chambers are placed inside a large Pyrex staining dish (the kind used for histological staining). One fritted glass bubbler is placed at either side of the dish and a third is placed between the two incubation chambers (Fig. 1A). A thin Plexiglas lid (with a handle glued to the center) is placed inside the dish to maintain moistened carbogen gas above the slices. The lid is cut to a size appropriate for lowering into the dish, while remaining 2 to 3 cm above the top of the incubation chambers. With the chambers inside the staining dish, the aCSF level is adjusted so that it covers the brain slices and allows fluid exchange between wells via the holes that were drilled through each chamber. The entire apparatus is placed on top of a stir plate (equipped with a heater control), and a small stir bar furnishes gentle circulation of oxygenated aCSF. The slices are also kept at the desired temperature using the heater control. Temperature is monitored using either an external fish tank thermometer (that sticks to the outside of the staining dish) or a small temperature probe (inserted into the aCSF) connected to a portable digital thermometer.

### 1.3. Cutting Brain Slices: Tissue Chopper Versus Vibrating Microtome

Brain slices were first used for electrophysiology by Henry McIlwain (Li and McIlwain, 1957; Yamamoto and McIlwain, 1966). Originally, a tissue chopper was used to cut slices that were typically 300 to 500 μm thick (Alger and Teyler, 1976; Duffy and Teyler, 1975; Schwartzkroin, 1975; Schwartzkroin and Andersen, 1975; Skrede and Westgaard, 1971). The tissue chopper was used for years (and is still used in some electrophysiology labs), but we and others quickly recognized the value of cutting slices using a vibrating microtome (e.g., Keenan et al., 1988; Lipton et al., 1995; Magee and Johnston, 1995a; Moyer and Brown, 1998; Moyer et al., 1992, 2000, 2002).

We specifically compared the visual appearance of slices when they are cut using a chopper versus a vibrating microtome. Comparisons used a 10× air objective or a 40× water objective, infrared-filtered light, differential interference contrast (IR-DIC) optics, and a digital video enhancement device (described in detail in Section 2.1, below). The surface of the slice is much smoother when sectioned with a vibrating microtome (using a slow advance speed and cold cutting saline) and the slices were visibly more healthy (see description of healthy neurons in Section 2.2.2, below). Other scientists have made parallel observations at the metabolic level. For example Charlie Taylor's group stained for mitochondrial activity using 2,3,5-triphenyltetrazolium chloride (Watson et al., 1994) and saw more uniform and greater staining throughout brain slices prepared using a vibrating microtome versus a tissue chopper (see discussion in Aitken et al., 1995). It is not surprising that the vast majority of neurophysiologists now use a vibrating microtome with a cold cutting solution when preparing brain slices for in vitro electrophysiology.

### 1.4. Preparation of Brain Slices from Juvenile Animals

Visually guided, patch-clamp recordings in brain slices are usually done using juvenile rats (2 to 4 weeks of age), for two reasons. First, tissue from young animals is much more resilient to the trauma of the slice preparation than adult tissue. There appears to be a much greater margin of error present at each step in the preparation of slices when using juvenile rats (Moyer and Brown, 1998). Second, it is much easier to see individual cells clearly and at a greater depth within the slice in sections prepared from very young rats (Keenan et al., 1988; Moyer and Brown, 1998), possibly because myelination is not complete in juvenile rats. The use of tissue from juveniles allows cells to be visualized and patched at a greater depth from the surface of the slice. Better imaging facilitates placement of the patch pipette on the soma, dendrite, or axon of the neuron of interest.

Prior to decapitation, the rat pup (or other juvenile animal) is deeply anesthetized, usually with a volatile anesthetic such as halothane or isoflurane. Volatile anesthetics are preferable to injected anesthetics, such as pentobarbital, because the former diffuse out of the brain tissue quite rapidly and are less likely to alter neuronal function. We typically place about 1 to 2 mL of halothane onto a paper towel located within an inhalation narcosis chamber. The rat pup is then placed inside the chamber and removed only after being deeply anesthetized, as evidenced by a lack of movement and slow respiration. The slices are less healthy if decapitation is done after breathing has ceased.

Consistency is one of the most important procedural rules for making healthy brain slices. Consistency is vital not only for finding a set of parameters or procedures that work well, but also for reducing variability among experiments. Day-to-day variability in the slice preparation likely results in variability in the quality of the data recorded. To maintain consistency, we time the dissections with a digital stopwatch, recording the following information onto a data sheet: (1) the time to remove the brain from the animal and place it into a small beaker of ice-cold oxygenated sucrose-CSF; (2) the time when the brain is removed from the beaker and blocked; (3) the time when each hemisphere is glued to the vibrating microtome tray; (4) the time when the first useable slice is cut; (5) the time when the last slice is obtained; (6) the number of slices; and (7) the thickness of the slices. We also record the temperature of the cutting solution at the start and finish of slicing as well as the temperature of the incubation chamber. Figure 2A shows an example of a completed data sheet with entries from the dissection of a juvenile rat. Care and consistency maximize the number of dissections from which usable data are obtained and result in greater consistency among different users within the same lab.

Regarding removal of the brain, faster is better as long as care is taken to prevent damage to the brain. We like to get the brain out in 30 to 45 seconds. With practice, this speed is not difficult to achieve. For rats younger than 6 or 7 weeks of age, we expose the skull by cutting with a scalpel from the nose to the back of the skull. A small pair of scissors (~2.5-cm blade length) is then used to make three cuts. The first is along the occipital suture. The second is a caudal-to-rostral cut along the sagittal suture. The third is across the skull, near the rostral end of the second cut. These cuts leave two plates of skull covering each hemisphere. A small pair of mini–bone rongeurs RS-8310 (Roboz, Gaithersburg, MD), is used to grasp the back of each skull flap and lift firmly upward and outward to remove the skull covering each hemisphere. This must be done carefully, without rotating the wrist. Otherwise the temporal lobe may get damaged by the ventral portion of each plate of skull. Any

| | A DATE: ___ / ___ / ___ | | B DATE: ___ / ___ / ___ | |
|---|---|---|---|---|
| | DISSECTION by: _____ | | DISSECTION by: _____ | |
| Perfusion: | Yes (No) | | (Yes) No | |
| Abdominal Cut: | | | 0:00 | |
| Descending Aortic Clamp: | | | :46 | |
| Cut Right Atrium: | | | :52 | |
| Perfuse Left Ventricle: | | | 1:00 | |
| Fluid from Atrium is Clear: | | | 1:42 | |
| Decapitate: | 0:00 | | 1:51 | |
| Removal of brain: | :33 | | 2:25 | |
| Dissection Start: | 3:25 | | 4:01 | |
| | Left | Right | Left | Right |
| Brain Glued: | 3:56 | 4:05 | 4:32 | 4:42 |
| Removal of 1st Slice: | 6:02 | 6:02 | 6:56 | 6:56 |
| Removal of Last Slice: | 10:16 | 10:16 | 12:28 | 12:28 |
| Number of Slices: | 6 | 6 | 6 | 6 |
| Slice Thickness: | 300 μ / (400 μ) / 500 μ / _____ μ | | 300 μ / (400 μ) / 500 μ / _____ μ | |
| Temperature ( °C) | sucrose aCSF _1_ °C to _1_ °C  normal aCSF _20.5_ °C to _23_ °C | | sucrose aCSF _0_ °C to _-1_ °C  normal aCSF _23_ °C to _24_ °C | |
| Rat Information: | Rec'd: ___ / ___ / ___  weaned: ___ / ___ / ___  Age: _15_ (d) wk / mo / yr  Wt: _42_ (g)/ kg  Sex: (M) / F   Protocol # _____ | | Rec'd: ___ / ___ / ___  weaned: ___ / ___ / ___  Age: _3_ d / wk /(mo)/ yr  Wt: _312_ (g)/ kg  Sex: M /(F)   Protocol # _____ | |
| Additional Comments: | slices done @ ___:___ a.m. / p.m. | | slices done @ ___:___ a.m. / p.m. | |

Fig. 2. Sample dissection time sheet and dissection times. (**A**) Dissection times from a 15-day-old male rat pup. (**B**) Dissection times from a 3-month-old female rat in which an intracardiac perfusion was performed. All times are in a minutes and seconds. Zero time denotes the onset of the first anoxic event. For a standard dissection the first anoxic event occurs at the time of decapitation, whereas when a perfusion is performed the first anoxic event occurs when the diaphragm is cut during the perfusion procedure. The time required to remove the brain is about the same: 33 seconds for the juvenile and 34 seconds for adult. Notice also that the *"removal of first slice"* times are comparable (left and right refer to brain hemisphere).

extra pieces of skull that do not come off cleanly are then quickly removed with the rongeurs to prevent damage during brain extraction.

After the brain is fully exposed, we quickly cut away the cerebellum and insert a periosteal elevator No. 48-1460 (Biomedical Research Instruments, Malden, MA), along the front portion of the brain and slide it forward to cut the olfactory bulbs. We prefer the periosteal elevator because it is slightly curved and very smooth, making it less likely to damage the brain, unlike a narrow spatula. Holding the head nearly upside down, the frontal portion of the brain is gently pushed out of the skull and the optic nerves are severed. The brain is then pushed out of the skull cavity and into a beaker of ice-cold, oxygenated sucrose-CSF.

In older rats, the skull is too thick to cut with scissors, so we use a larger pair of rongeurs (Roboz, RS-803) to remove the skull. We start at the base of the skull and remove the skull behind the occipital suture by gripping it along the suture and lifting upward. Then we gently grip the skull with the tips of the rongeurs and lift the plate of skull covering each hemisphere. If done swiftly and forcefully, the skull separates along the sagittal suture and the entire skull covering one hemisphere comes free. The same approach is used to remove the skull covering the other hemisphere. The brain is removed from the skull as described above for juvenile rats.

All of our dissection tools are chilled in a beaker of ice-cold sucrose-CSF. After the brain is removed it is put into a small beaker (50 mL) containing ice-cold, oxygenated sucrose-CSF from which it is quickly transferred to a larger beaker (250 mL) containing the same solution. The purpose of the intermediate beaker of cold sucrose-CSF is to wash off excess blood that may have dripped into the beaker while removing the brain. We then let the brain chill for around 3 minutes. Of course, if the brain is hemisected prior to removal from the skull, less time is needed to chill each hemisphere thoroughly. The brain is then transferred onto a piece of filter paper, which is placed on top of an ice-packed Petri dish (Pyrex is better than plastic), so that the blocking or dissection occurs *on ice*. The brain is then glued onto a vibrating microtome tray in front of and touching a rectangular agar block (4% agar) that was previously glued to the tray. The agar block supports the tissue during slicing (Fig. 1B). The tray is immediately filled with ice-cold oxygenated sucrose-CSF. The tray and main Vibratome (St. Louis, MO) chamber have bubblers to oxygenate continuously the sucrose-CSF (Fig. 1B).

Using a temperature-controlled Vibratome (No. 3000), 300- to 400-µm thick slices are cut at a temperature of about 1°C (Moyer and Brown, 1998). As each slice is cut, it is removed from the Vibratome tray using the back end of a 3.5-inch glass dropper and placed in a well of the slice incubation chamber (Fig. 1A). Slices are maintained at about 24°C until they are used in an experiment. We find that if slices are removed and placed into the incubation chamber at a temperature below 10°C and allowed to warm slowly to room temperature (or if they are maintained at 10°C or less for 10 to 20 minutes and then transferred to aCSF at room temperature), *nearly every cell in the slice is dead* (see description of dead cells in Section 2.2.1 below; see also Moyer and Brown, 1998).

The slices are therefore placed directly into the incubation chamber in aCSF maintained at room temperature or a few degrees cooler. Slices can be used for recordings after a 45- to 60-minute recovery period. After 45 minutes, the slices look healthy, the cells are easily patchable (Fig. 3), and the recordings typically remain

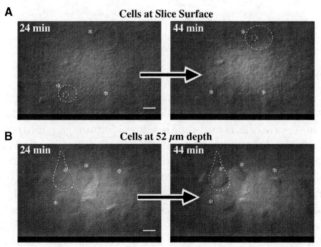

Fig. 3. Recovery of neurons as a function of time after slicing. Infrared differential interference contrast (IR-DIC) video micrographs of perirhinal layer V neurons in brain slices shortly after decapitation (24 minutes, left panels) and about 44 minutes later (right panels). (**A**) Cells located near the cell surface are unhealthy and do not recover. (**B**) Cells located 52 µm beneath the slice surface initially appear unhealthy but recover over time. Asterisks indicate several cells that were monitored over time. (Adapted from Moyer and Brown, 1998, with permission from Elsevier Science.)

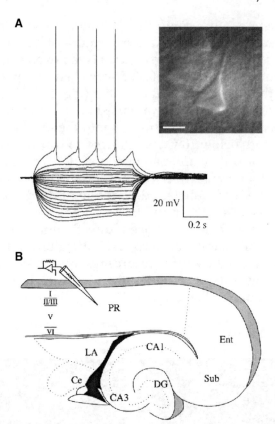

Fig. 4. Visually guided whole-cell recordings from 400–μm-thick brain slices. (**A**) Voltage traces of a regular spiking layer V pyramidal neuron in perirhinal cortex (left) and the IR-DIC video image (right) of the same cell showing the position of the patch pipette (scale bar is 10 μm). This neuron had a resting membrane potential of –78 mV, an input resistance of 405 MΩ, and remained healthy for nearly 2 hours before the experiment was terminated. (**B**) Schematic diagram of a horizontal brain slice containing perirhinal cortex (see Section 3.5). We previously used slices corresponding to plates 98–100 in a rat brain atlas (Paxinos and Watson, 1998) in which recordings are restricted to the rostrocaudal axis from –3.8 to –5.2 mm relative to bregma. PR, perirhinal cortex; Ent, entorhinal cortex; Sub, subiculum; LA, lateral nucleus of the amygdala; Ce, central nucleus of the amygdala; DG, dentate gyrus.

stable for 2 to 3 hours (Fig. 4) and sometimes for as long as 4 to 5 hours. In earlier experiments, recordings were done at room temperature (~22°C), but in more recent experiments, recordings have been done at 31°C or higher (see Section 3.6, below).

## 1.5. Preparation of Brain Slices from Adult and Aged Animals

Preparing brain slices from older animals (adult and aged) requires particular attention to small details regarding both the solutions and the dissection. One obvious difficulty is that the skull of older animals is much thicker and can be more difficult to remove. For example, in young adult rats, the skull can be easily removed by gripping it with rongeurs and lifting upward as described earlier (see Section 1.4). Large sections of skull covering each hemisphere can be removed intact in this manner. However, in aged rats the bone often splinters, so that only small pieces are removed, lengthening the process of brain extraction. Thus it is important to work quickly and efficiently with aged animals. Also, the brain of an older animal is much less resilient. Even a subtle mistake during any step of the dissection process can dramatically affect the ability to obtain high-quality recordings from healthy neurons. The following sections describe our approach to working with tissue from older animals.

### 1.5.1. Osmolarity and Incubation Temperatures

Our procedures have evolved from years of experience in working with adult and aged rabbit hippocampus (Moyer and Disterhoft, 1994; Moyer et al., 1992, 2000) and adult and aged rat perirhinal cortex (McGann et al., 2001; Moyer and Brown, 1998; Moyer et al., 2002). In the earlier studies of hippocampal slices from aged rabbits (36 to 50 months old), sharp electrode recordings (see Section 3.7, below) were obtained from slices prepared using both aCSF and sucrose-CSF as the cutting medium (Moyer and Disterhoft, 1994; Moyer et al., 1992, 1996, 2000; Thompson et al., 1996). In the case of the sucrose-CSF cutting solution, the 124 mM NaCl was substituted with an equimolar amount of sucrose. Since a solution with 124 mM NaCl results in a 124 mM concentration of each $Na^+$ and $Cl^-$ ions, we used 248 mM sucrose (Moyer et al., 1996, 2000). This worked well for cutting isolated hippocampal slices and recording from adult and aged CA1 neurons. The same procedure did *not* work well for WCRs from perirhinal cortex of aged rats (Moyer and Brown, 1998). One problem was that the tissue was difficult to section. We therefore used a much slower cutting speed and began intracardiac perfusions before removing the brain (see Section 1.5.2, below). This improved the quality of the slices. However, after obtaining a WCR, the membrane would often reseal in the electrode tip, thereby reducing the overall success rate.

Noting that the osmolarity of the sucrose-CSF cutting solution was actually much higher than the aCSF, we next reduced the sucrose concentration until the osmolarity matched that of the aCSF (~290 mOsm). We also introduced a preincubation stage in which the slices were first placed in a holding chamber (of the same design) that was maintained at 35° to 37°C. After 30 to 45 minutes at this higher temperature, the slices were then transferred to an incubation chamber maintained at 24°C. Our hunch was that the aged slices may need to return to a warm temperature faster than juvenile tissue. Other laboratories have been using a similar pre-incubation step for juvenile tissue because the cells are easier to patch (Borst et al., 1995; Hestrin and Armstrong, 1996; Jung et al., 2001; Magee and Johnston, 1995a, 1997; Markram et al., 1997a). After adjusting the osmolarity and adding a preincubation stage (typically 30 minutes at ~35°C), the success rate in aged tissue approached that in juvenile tissue (Moyer and Brown, 1998).

### 1.5.2. Benefits of Performing Intracardiac Perfusions

Several researchers have demonstrated the utility of hypothermia for obtaining healthy brain slices (Newman et al., 1992; Okada et al., 1988; Saeed et al., 1993). For example, George Newman's group found that anesthetizing the animal and placing the body on a bed of ice until the rectal temperature was 31°C increased the percentage of healthy CA1 pyramidal neurons in hippocampal slices relative to slices prepared under normal conditions (Newman et al., 1992). When making slices from the isolated hippocampus of adult or aged animals, the brain is hemisected, which facilitates rapid cooling. Also, using a Vibratome to cut from chunks of hippocampus did not present any unusual difficulties. Thus, in past studies of hippocampal slices from adult and aged rabbits, no intracardiac perfusion was performed (Moyer and Disterhoft, 1994; Moyer et al., 1992, 1996, 2000).

When we began preparing perirhinal cortex brain slices from adult and aged rats, it was clear that the cortex with which the blade made initial contact was more likely to compress and tear rather than be cut smoothly through, even at slow cutting speeds (Moyer and Brown, 1998). Thus, the cortex was damaged during the cutting, even though the hippocampus was in relatively good condition. Because of these difficulties, we changed the procedures to include intracardiac perfusions prior to sectioning. This change clearly improved the quality of the brain slices and the success rate in maintaining stable WCRs (McGann et al., 2001; Moyer and Brown, 1998, 2002). Intracardiac perfusion has another obvious benefit; namely, it removes red blood cells, which also stain when tissue is reacted for biocytin (often included in the pipettes for obtaining cellular morphology).

Intracardiac perfusions are performed after deeply anesthetizing the rat using Nembutal (80–90 mg/kg, i.p.). Once the animal is unresponsive to a foot pinch, the chest cavity is opened by cutting across the abdomen, just below the sternum, and then cutting up each side of the ribs. The diaphragm is then carefully cut and the descending aorta is clamped using hemostats to prevent perfusion of the lower body (thereby accelerating perfusion of the brain). The body cavity is exposed by applying hemostats to the cartilage at the bottom of the sternum and lifting the rib cage over the head. If necessary, any connective tissue is cut to free the heart from the chest cavity. A tiny cut is made in the left ventricle using a small pair of scissors, and a blunt-tipped needle (18 to 23 gauge, connected to a perfusion pump) is quickly inserted through the ventricle and into the ascending aorta. The aorta is then clamped with a hemostat to keep the needle in place (note that the needle can be seen through the wall of the aorta).

Next, the right auricle is cut, and ice-cold, oxygenated sucrose-CSF is pumped through the needle using a perfusion pump (Mity-flex, ANKO, Bradenton, FL). After perfusing for about 1 minute (assuming a good clamp on the descending aorta, otherwise the lower body will be perfused), the efflux from the right auricle should be clear. About 20 seconds later the rat is decapitated, the brain is removed, and slices are prepared as described above for juvenile rats. The only difference is that we let the brain chill for only 1 minute or less before starting the dissection. The overall amount of time it takes to get the first brain slice is therefore similar between juvenile and adult or aged animals (compare Figs. 2A and 2B).

## 2. Using Infrared Differential Interference Contrast (IR-DIC) for Visualizing Neurons in Brain Slices

The original extracellular and intracellular studies of hippocampal brain slices used a dissecting microscope to see the cell body layers (such as hippocampal regions CA1 or CA3 or the dentate gyrus) and major fiber pathways (such as the *stratum radiatum*, *stratum oriens*, and *stratum lucidum*). Although individual neurons cannot be seen with a dissecting scope, the optics are suitable for positioning stimulating and recording electrodes into particular cell layers or fiber pathways. For intracellular studies, the high packing density of neuronal somata within the cell body layers of the hippocampus facilitated obtaining intracellular recordings using sharp electrodes. The anatomy of the hippocampus makes this brain region nearly ideal for many types of intracellular and extracellular studies without the need for visualizing individual neurons. In some parts of the hippocampus, the current sinks and sources are spatially organized in a manner that facilitates the interpretation of extracellular recordings from certain populations of synapses.

However, the application of patch-clamp techniques to brain slices (Blanton et al., 1989; Edwards et al., 1989) has multiplied interest in much higher-resolution optics that allow precise positioning of the pipette tip onto specific neuronal targets, typically the cell body or particular parts of the dendritic tree (e.g., Frick et al., 2004; Golding et al., 2005; Magee and Johnston, 2005; Stuart and Sakmann, 1995). Although WCRs can be done "blind" (Blanton et al., 1989; McKernan and Shinnick-Gallagher, 1997; Xiang et al., 1994), visually guided pipette placements are obviously preferable when possible. The blind method is still used for WCRs when IR-DIC optics or other cellular imaging methods are impractical, difficult, or impossible (see Section 4).

## 2.1. Development of IR-DIC Video Microscopy

The first attempts to visualize central nervous system (CNS) neurons were done by Yamamoto in 1975. Using conventional microscopy, he was able to visualize silhouettes of larger cell bodies, but different cells were not easily discerned, and dendrites or small neurons were not observable (Yamamoto, 1975). Subsequently, Yamamoto's group published a paper in which they demonstrated visualization of cerebellar neurons in very thin brain slices (40–120 μm), this time using Nomarski optics (Yamamoto and Chujo, 1978). Under these conditions, both Purkinje cells and smaller neurons were visible. Using an extracellular microelectrode, spikes could be recorded by positioning the electrode tip near a visualized cell.

Five months later, Takahashi also published a paper in which he visualized motoneurons in thin spinal cord slices (130–150 μm) using Nomarski optics (Takahashi, 1978). In this study, intracellular recordings were obtained from the identified motoneurons. In 1980, Llinás and Sugimori used Hoffman modulation microscopy to visualize cerebellar Purkinje cell somata and obtain intracellular recordings from identified neurons in very thin cerebellar slices (200 µm; Llinás and Sugimori, 1980). These early attempts at visualizing neurons in brain slices were breakthroughs, but they suffered from the inability to clearly resolve neurons in slices thicker than 100 to 200 µm. Thin sections result in large numbers of unhealthy neurons due to damage at the slice surface. Sections this thin typically sever many of the dendrites, even in the deepest cells.

Brian MacVicar was the first to publish a manuscript describing the benefits of infrared (IR) light and a video camera for visualizing neurons in brain slices (MacVicar, 1984). The longer wavelength of IR light results in less light scattering and therefore better imaging in thicker slices (Gibson, 1978; Keenan et al., 1988; MacVicar, 1984). In the late 1980s, Brown and Keenan were the first to combine video microscopy with differential interference contrast (DIC) microscopy and IR-filtered light to visualize neurons in brain slices (Brown and Keenan, 1987; Keenan et al., 1988). Using an inverted microscope, their studies of rat brain slices (amygdala and hippocampus) clearly demonstrated the optical benefits of IR-DIC video microscopy. Resolution was explored as a function of both the slice thickness (150–450 µm) and the age of the animal (postnatal day 7 to adult). In absolute terms, the resolution *decreased* with increasing slice thickness and age. However, the advantage of IR-DIC video microscopy, relative to previous types of imaging techniques, clearly *increased* with increasing slice thickness and age.

Two years later, Dodt and Zieglgänsberger (1990) used IR-DIC video microscopy to show numerous examples of healthy neocortical and hippocampal neurons in brain slices. Using an inverted microscope, they found that neurons could be visualized in 400-µm thick slices from adult rats at depths of up to 40 µm. Shortly thereafter, Bert Sakmann and colleagues used IR-DIC video microscopy to obtain the first examples of WCRs from visually preselected neurons (Stuart et al., 1993). Previous studies had used an inverted microscope configuration (lens below the slice) because the available objective lenses were not well suited for an upright configuration (lens above the slice). The manufacture of suitable water-immersion lenses caused us (and most others) to switch to an upright configuration. Water-immersion lenses that are designed for this purpose have a smaller "footprint" in the bath, allowing better access by the patch pipette; they have better thermal characteristics, causing less effect on the bath temperature; they have an appropriate working distance, usually 2 mm or longer; and the numerical aperture (NA; typically 0.75 or larger) is adequate for the required resolution.

The benefit of IR-DIC video microscopy for in vitro neurophysiological studies resulted in a tremendous increase in the number of laboratories using the technique since the early 1990s. We use an upright microscope (Zeiss Axioskop or Olympus BX51) equipped with a water-immersion objective (Zeiss 40×, Olympus 60×), infrared filtered light, DIC optics, a Hamamatsu C2400 video camera, and a video enhancement device (Beggs et al., 2000; Faulkner and Brown, 1999; McGann et al., 2001; Moyer and Brown, 1998; Moyer et al., 2002; Xiang and Brown, 1998; Xiang et al., 1994). In 400 μm-thick slices of perirhinal cortex, suitable images can be obtained at depths of 100 to 120 μm below the surface of the tissue using the Zeiss microscope and a 40× water-immersion lens (NA = 0.75; working distance = 3.6 mm). Using the newer Olympus microscope and Olympus 60× lens (NA = 0.90; working distance = 2.0 mm), suitable images can be obtained at depths of 120 to 150 μm in the same tissue. Some of the newer microscopes and water-immersion objective lenses are specifically designed to accommodate light in the IR spectrum.

## 2.2. Using IR-DIC Video Microscopy to Select Healthy Neurons

In any given slice there are both healthy neurons and unhealthy neurons. The unhealthy cells can either be unpatchable or they may die shortly after beginning a recording. This section describes how to distinguish between healthy and unhealthy neurons.

### 2.2.1. Characteristics of Dead or Unhealthy Neurons

Numerous "dead" neurons are always present within the first 25 μm below the surface of a brain slice (Fig. 3). They typically have a characteristic swollen appearance. In many cases, all that can be discerned is a large nucleus and a prominent nucleolus. Over time, these dead cells often become invisible. Figure 5A illustrates two examples of these round and transparent dead neurons that cannot be patched. They explode under the positive pressure of the elec-

Fig. 5. Infrared differential interference contrast (IR-DIC) video micrographs of unhealthy and healthy neurons in 400-μm-thick brain slices. (**A**) Several examples of dead neurons in layer V of perirhinal cortex (PR). The only structures visible are a round soma and a prominent nucleus, presumably due to breakdown of the somatic membrane. The nucleus of one neuron is outlined by a dashed circle, and two other cells are indicated by asterisks. (**B**) Several examples of unhealthy neurons in layer V of PR. In these cells, the nucleus is clearly visible through the somatic membrane (outlined by dashed circle in far left cell and indicated by an asterisk in far right cell). These cells would probably look like healthy, patchable neurons without the IR-DIC video enhancement. In fact, some of them can be patched but they usually die within 5 to 30 minutes (see dimple caused by patch pipette on soma of middle cell). (**C**) Examples of healthy, patchable pyramidal cells in layer V of perirhinal cortex. Note that the nucleus is not visible. At least in perirhinal cortex, only cells without a visible nucleus (like those in **C**) yield healthy, stable long-term whole-cell recordings (1 to 5 hours). Scale bar is 10 μm (A–C).

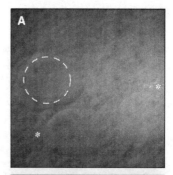

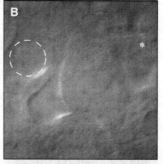

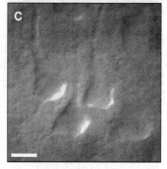

trode as it gets close to the cell. In addition to these dead cells, there are also cells that appear unhealthy because they have either a swollen, two-dimensional appearance or because they are shrunken. Unhealthy neurons differ from dead ones in that significant dendritic structures can still be seen. The somatic images of these neurons appear to be two-dimensional (seem to lack depth) when focusing up and down. Gigaohm seals can often be obtained on cells that appear two-dimensional or shrunken (see Section 3.4, below), but subsequent WCRs cannot be obtained, even using an enormous amount of suction.

Some cells that may appear healthy when using DIC video microscopy and white light may look obviously unhealthy when using IR-DIC video microscopy. For example, the cell body may be somewhat translucent, such that the nucleus is apparent (Fig. 5B). In these cells, high-quality WCRs can sometimes be obtained for as long as 10 to 30 minutes, after which there is a reduction in membrane potential, input resistance, and action potential amplitude. An important benefit to IR-DIC video microscopy is the ability to select healthy neurons that allow stable WCRs for several hours.

### 2.2.2. Characteristics of Healthy Neurons

Healthy neurons are easy to distinguish from dead or unhealthy neurons using IR-DIC video microscopy. First, the nucleus is not visible in healthy neurons. Instead, the cell membrane appears smooth and bright (Fig. 5C). There are no outward blebs or small round indentations in the somatic cell membrane. Second, healthy neurons have a more three-dimensional appearance while many unhealthy neurons have a two-dimensional appearance. This can be appreciated by focusing up and down through the neuron of interest. Generally, one initially encounters many examples of unhealthy cells before finding one that is healthy. Among neurons that are selected based on these visual criteria, about 80% or more yield stable somatic WCRs. The success rate assumes minimal difficulties from mechanical sources such as manipulator drift or debris in the electrode tip.

### 2.2.3. Considerations Regarding Cell Depth

We usually patch neurons that are deeper than 50 μm and preferably 70 to 140 μm below the tissue surface (Beggs et al., 2000; Faulkner and Brown, 1999; McGann et al., 2001; Moyer and Brown, 1998, 2002; Moyer et al., 2002; Xiang and Brown, 1998). As noted earlier (Section 2.1), some of the newer microscopes and objective lenses (which are better designed for IR-filtered light) enable imaging at depths up to ~150 μm from the tissue surface. In considering the depth of imaging, it is worth noting that many of our recordings are from perirhinal cortex, which is remarkably low in myelin (Brown and Furtak, 2006; Burwell, 2001; Zilles and Wree, 1995). It is possible that the low myelin improves the optical properties of perirhinal cortex. This conjecture could be tested by comparing the depth at which good images can be obtained from perirhinal cortex with comparable images from entorhinal cortex, which is adjacent to perirhinal cortex and where the myelin level is more typical of cortex.

## 3. Recording from Neurons in Brain Slices

In the first part of this section we discuss several issues that are pertinent to patch-clamp recordings, including the intracellular pipette solution, the use of an agar bridge, and liquid–liquid junction potentials. In the second part of this section we describe the actual procedures that are involved in obtaining whole-cell recordings in brain slices and show some examples of the results.

### 3.1. Selecting an Appropriate Patch-Pipette Solution

Since the contents of the patch pipette diffuse throughout the cell, it is important to consider adverse effects of the pipette solution on neuronal function, including run-down of membrane currents. A wide range of patch-pipette solutions is used, depending on the experimental design (for review see Kay, 1992). One of the major constituents of a pipette solution is usually a potassium salt, commonly in the form of KCl, K-methylsulfate, or K-gluconate, which we routinely use. K-methylsulfate is sometimes preferable because it has been reported to minimize run-down of calcium-activated potassium currents (Zhang et al., 1994). Some researchers in the field of aging use a K-methylsulfate–based pipette solution (e.g., Power et al., 2001; Wu et al., 2004). A KCl-based pipette solution results in a lower pipette resistance and it minimizes liquid–liquid junction potentials (because chloride is a major part of the ground wire and the aCSF; see Section 3.3, below). On the other hand, the resulting high intracellular chloride concentration can cause some GABA$_A$-mediated inhibitory synaptic currents to be depolarizing, which can be troublesome or confusing.

A number of other ions and chemicals are also usually added to the standard patch-pipette solutions. Our standard solution contains the following (in mM): 110 K-gluconate, 10 N-[2-Hydroxyethyl]piperazine-N′-[2-ethanesulfonic acid] (HEPES), 1.0 ethyleneglycotetraacetic acid (EGTA), 20 KCl, 2.0 MgCl$_2$, 2.0 Na$_2$ATP, 0.25 Na$_3$GTP, 10 phosphocreatine (di-tris), pH 7.3, 290 mOsm. Sometimes we also add 0.5% biocytin for intracellular labeling and identification of neuronal morphology. The pipette solutions are aliquoted into small microcentrifuge tubes and stored at −20°C. Each day, fresh-patch pipette solutions are thawed and filtered (using a 0.2-μm syringe filter) before use. For labeling cells, the tip of the patch pipette is filled with the standard patch solution and then the pipette is back-filled with the biocytin-containing solution. This procedure minimizes biocytin leakage into the tissue prior to obtaining a gigaohm seal.

Depending on the experiment, other compounds may also be added to the patch-pipette solution. For example, a cesium salt may be included to block certain potassium currents, or a charged local anesthetic (such as QX-314) might be added to block sodium action potentials (Frazier et al., 1970; Kelso et al., 1986). We include guanosine triphosphate (GTP) in the pipette solution to conserve G-protein–mediated responses (Trussell and Jackson, 1987). We also include an ATP regenerative system to prevent rundown of calcium channels (Forscher and Oxford, 1985) and to supply energy for other intracellular reactions (Kay, 1992). A small amount of EGTA is also included to help buffer intracellular calcium ions. Since 1,2-bis(o-amino phenoxy)ethane-N,N,N′,N′-tetra acetic acid (BAPTA) is a much faster calcium chelator than EGTA, BAPTA is preferred when rapid elimination of intracellular calcium buildup is required (Tsien, 1980). Our standard K-gluconate solution allows stable recordings for several hours without changes in the input resistance, resting membrane potential, or firing properties (Fig. 6), which can be of several types (Beggs et al., 2000; Faulkner and Brown, 1999; McGann et al., 2001; Moyer and Brown, 1998; Moyer et al., 2002), as described in Section 3.6.

## 3.2. Grounding, Shielding, and the Use of an Agar Salt Bridge

Proper grounding is essential for obtaining low noise patch-clamp recordings from neurons in any preparation including brain slices. To reduce extraneous noise, it is necessary to minimize ground loops as well as pick up from electromagnetic radiation (for example, from computers, monitors, fluorescent lights, and power lines). This requires careful grounding of all instruments through low-resistance cables and possibly the use of a Faraday cage. Excellent discussions of topics relevant to various noise sources and proper grounding can be found in the Axon Guide (Sherman-Gold, 1993) and elsewhere (Levis and Rae, 1992; Penner, 1995).

For recording, the bath solution is grounded using a sintered silver/silver chloride (Ag/AgCl) wire (for additional information on Ag/AgCl wires, see Alvarez-Leefmans, 1992; Geddes, 1972; Ives and Janz, 1961; Janz and Ives, 1968). The sintered Ag/AgCl wire is more durable than a bare Ag/AgCl wire, which also works well (we have used both). The wire is loosely wound into one or more loops and positioned at the end of the chamber next to the port from which aCSF exits. Every couple of weeks, we re-chloride the ground wire by placing it into a vial containing Clorox for 20 to 30 minutes. The recording wire (a silver wire that connects the shank of the pipette to the preamplifier) also requires periodic

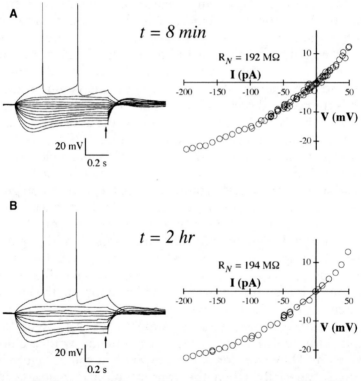

Fig. 6. Stable whole-cell recordings from a healthy perirhinal neuron. (**A**) Voltage traces and voltage-current *(V-I)* relation of a layer V pyramidal neuron in rat perirhinal cortex (PR). The data were collected shortly after establishment of the whole-cell configuration (resting membrane potential –78 mV). (**B**) Whole-cell recording from the same cell 127 minutes later. No changes in resting membrane potential, discharge pattern, action potential characteristics, spike threshold, or input resistance were seen over the 2-hour time period. Arrows in **A** and **B** indicate the point at which the *V-I* relation was measured.

re-chloriding. Re-chloriding the ground and recording wires can also be done through electrolysis, using a DC-power source such as a 1.5-V alkaline battery. The Ag/AgCl ground wire is connected at the positive pole, another silver wire at the negative pole, and current is passed while both electrodes are immersed in a chloride solution (e.g., 0.1M HCl). A switch can be used to reverse the current flow intermittently and a potentiometer in series with the power source adjusts the current flow. Alternatively, dipping the wire in Clorox is easier and seems to work just as well.

The Ag/AgCl ground wire is connected to a common ground block that is located on (but is electrically insulated from) the surface of the vibration isolation table. This copper block has 16 holes for connecting banana jacks from different sources. Grounds from all metallic surfaces (including the head stage, the microscope, the isolation table, the isolated stimulators, the specimen stage, and the manipulators) go through this common ground, which is connected directly to the signal ground of the patch-clamp amplifier. The Faraday cage can also be grounded through the copper block, but it is also common to ground it through either the instrument rack ground or the ground terminal of a grounded outlet. It is important to make sure that all metallic surfaces near the head stage are properly grounded through the common ground. Failure to ground in this manner will enable the surfaces to act as antennae that pick up line-frequency signals. Going through a common ground also eliminates ground loops. If additional noise persists, it is important to troubleshoot. Starting with the amplifier and oscilloscope, equipment is added one piece at a time, until the noise source is identified. Careful troubleshooting in the beginning will save valuable time in the long run.

As mentioned earlier, we use a sintered Ag/AgCl wire for our ground wire, which works well in conjunction with a chloride-based aCSF. However, for some purposes, a salt bridge may be preferred, especially when the concentration of chloride in the aCSF bath is changed during an experiment. Salt bridges are composed of a thin agar tube containing KCl throughout. The salt bridge is made by adding 2% to 3% agar to a solution of 3M KCl, boiling the mixture, sucking it up into a thin capillary or Teflon tube (~1 mm), and allowing it to cool. The tube is shaped, using gentle heat, before adding the agar mixture. One end of the agar tube is in contact with the aCSF bath solution and the other end is placed in a well containing a 3M KCl solution. When using a salt bridge, one end of the Ag/AgCl ground wire is placed in the 3M KCl solution and the other end is connected to the ground block. Thus the ground wire is *indirectly* in contact with the aCSF bath through the agar salt bridge. When using a salt bridge, care should be taken in cases where vastly different aCSF solutions are switched for extended periods of time (for further discussion see Neher, 1992).

### 3.3. Liquid–Liquid Junction Potentials and Correcting for Them

In electrophysiological studies, liquid–liquid junction potentials (hereafter referred to as liquid junction potentials) arise from differences in ionic composition and ionic mobilities at the junction between two solutions (e.g., the intracellular pipette solution and the extracellular bath solution). Left uncorrected, liquid junction potentials can contribute significantly to errors in membrane potential measurements. Such errors typically range from 1 or 2 mV (using KCl electrode solutions) to more than 10 mV (using K-gluconate or K-methylsulfate). With the use of charge carriers other than KCl, corrections for liquid junction potentials are usually important. While some authors do correct for liquid junction potentials, others not only fail to correct but also fail to mention the value of the liquid junction potentials, which can complicate comparisons among studies. The following sections briefly describe how to measure liquid junction potentials as well as how to calculate them using the generalized Henderson liquid junction potential equation.

### 3.3.1. Measuring Liquid Junction Potentials

Measuring the liquid junction potential between the bath and pipette solution is fairly straightforward. In the first step, the recording chamber is filled with the same solution that is used in the patch pipette. The filled patch pipette is placed into the bath, which is static (no flow), and the potential is measured under current-clamp conditions. This potential is then zeroed (canceled). With the pipette remaining in place, the bath is then quickly replaced with aCSF and the resulting voltage (the junction potential) is measured. Finally, the bath solution is replaced with the pipette solution to verify reversibility. When measuring the liquid junction potential, it is necessary to use a salt bridge if the chloride ion concentration in the patch pipette solution is much lower than the chloride ion concentration in the aCSF bathing solution. Erwin Neher has published a detailed description of how to measure liquid junction potentials along with a listing of approximate junction potentials for some commonly used pipette solutions (Neher, 1992).

In our studies, voltages are typically corrected for a measured liquid junction potential of about +13 mV between the bath and patch pipette solutions (when using standard aCSF and a K-gluconate-based pipette solution). Corrections are applied according to the equation $V_M = V_P - V_L$, where $V_M$ is the corrected membrane voltage, $V_P$ is the observed pipette voltage, and $V_L$ is the liquid–liquid junction potential (McGann et al., 2001; Moyer and Brown, 1998; Moyer et al., 2002). Our measured liquid junction potential of +13 mV is similar to that obtained based on theoretical calculations.

### 3.3.2. Calculating Liquid Junction Potentials

Calculating liquid junction potentials involves the use of the generalized Henderson liquid junction potential equation (Barry and Lynch, 1991):

$$E^S - E^P = (RT / F)S_F \ln\left\{ \sum_{i=1}^{N} z_i^2 u_i a_i^P \bigg/ \sum_{i=1}^{N} z_i^2 u_i a_i^S \right\} \tag{1}$$

where

$$S_F = \sum_{i=1}^{N} [(z_i u_i)(a_i^S - a_i^P)] \bigg/ \sum_{i=1}^{N} [(z_i^2 u_i)(a_i^S - a_i^P)] \tag{2}$$

and $E^S - E^P$ represents the potential ($E$) of the solution ($S$) relative to the pipette ($P$). The variables $u$, $a$, and $z$ represent, respectively, the mobility, activity, and valence of each of the ionic species ($i$). $R$, $T$, and $F$ represent, respectively, the gas, temperature (in Kelvin), and Faraday constants.

Theoretical calculation of liquid junction potentials has become more convenient with the development of JPCalc by Peter Barry (1994). This software is inexpensive and can be extended by the user to include the mobilities and valences of additional ions. We have used JPCalc to calculate the liquid junction potential in our experiments and found the value to be in excellent agreement with the empirical measurements (within ~1%). For a measured liquid junction potential +13.1 mV, the calculated value using JPCalc was +13.0 mV. Ordinarily, theoretical calculations of the liquid junction potential provide an accurate estimate of the liquid junction potential, but the value should be checked empirically, especially when high concentrations of divalent ions are used (Barry and Lynch, 1991). There are several excellent reviews of liquid junction potentials that also list the mobilities of some commonly used ions (Barry, 1994; Barry and Diamond, 1970; Barry and Lynch, 1991; Neher, 1992; Ng and Barry, 1995).

## 3.4. Obtaining Whole-Cell Recordings in Brain Slices

Whole-cell recordings (WCRs) from neurons in brain slices involve obtaining a tight seal between the patch pipette and cell membrane and then rupturing the seal. This is done by application of gentle suction (negative pressure, to rupture the membrane within the lumen of the patch pipette) to gain access to the inside of the cell. Although this section focuses on WCRs, other configurations (cell-attached, inside-out, outside-out, and perforated-patch) are also popular (for a brief review see Cahalan and Neher, 1992). Once a whole-cell configuration has been achieved, and the current- or voltage-clamp experiment is completed, one can proceed to additional histological or molecular analyses.

### 3.4.1. The Gigaohm Seal and Obtaining Access to the Cell

After waiting for 45 to 60 minutes for the slices to recover, one slice is removed from the holding chamber, placed into the recording chamber and secured by positioning a net on top of the slice. The securing net can be made by flattening a piece of silver or platinum wire and shaping it to fit in the recording chamber (commercial products are also available from Warner Instruments, Hamden, CT). Four or five single strands of nylon hosiery are glued across the wire using cyanoacrylate. When placed on the slice, the nylon hosiery serves to anchor the slice in several locations. After securing the slice, the brain region of interest is identified using a 2.5× objective and the microscope aligned for Köhler illumination. This is done by stopping down the luminous field diaphragm, removing the polarizer, and focusing the condenser (with lens in place) until a crisp outer ring is observed through the oculars. This procedure is then repeated using the 40× or 60× water-immersion objective lens. An infrared light filter is then placed directly onto the top of the luminous field diaphragm.

Once the neuron of interest is visualized, a patch pipette is pulled using a Flaming/Brown micropipette puller (Model P-97, Sutter Instruments, Novato, CA). Obtaining the right patch pipette tip size and geometry are both important for achieving a high success rate (for an excellent review, see Sakmann and Neher, 1995). The patch pipette is then fire polished and filled with an appropriate patch solution that has been filtered using a 0.2-μm syringe filter. Fire polishing the pipette tip is helpful for obtaining high-resistance seals (our seals are typically about 5 GΩ and we have had seals as high as 25 GΩ). After attaching the pipette to the head stage, positive pressure is applied to the pipette by blowing into a Tygon or Nalgene (Fisher Scientific, Pittsburgh, PA) tube that is directly connected with the pipette holder. A stopcock is used to maintain or release the positive pressure. The pipette is then lowered into the bath and the amplifier voltage is zeroed. In voltage-clamp mode, with the command potential set to 0 mV (with "tracking" turned on), a 5-mV hyperpolarizing voltage step is applied. Based on the measured current, the pipette resistance (2–5 MΩ for somatic and 5–10 MΩ for dendritic recordings) is calculated from Ohm's law.

As the pipette gets closer to the cell, we usually reapply positive pressure. This helps to clear the membrane of debris. The plume of fluid radiating out from the pipette tip increases visibility, which can be helpful in selecting an optimal location on the cell membrane to patch. The pipette is then lowered onto the cell until the positive pressure creates a small "dimple" (Fig. 7A). At this point the positive pressure is quickly removed (by turning the stopcock) and very gentle suction is applied by mouth to form a gigaohm seal (>1 GΩ or $10^9 \Omega$), which we estimate by measuring the current in response to a 25-mV hyperpolarizing voltage step. It is important to remove the positive pressure as soon as a small but distinct dimple is visible. Otherwise, the positive pressure will create a very large dimple, which may damage the cell membrane (Fig. 7B).

After obtaining a stable gigaohm seal, the fast and slow capacitive transients are removed or minimized using dials on the amplifier. The capacitance can be minimized by keeping the bath solution level very low and by coating the pipette with Sylgard (Dow Corning, Midland, MI) or some other insulating compound. The

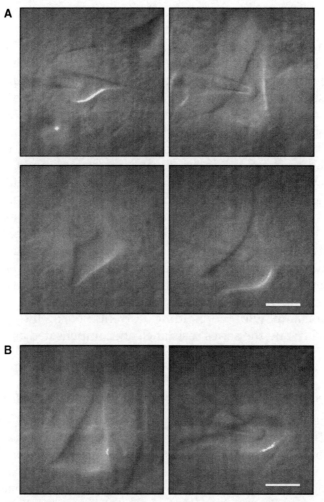

Fig. 7. Examples of the dimple created by positive pressure through the whole-cell patch pipette. (**A**) IR-DIC video micrographs of typical dimple sizes. (**B**) IR-DIC video micrographs of dimples that are too large. In visually healthy neurons, this results either from excessive positive pressure or failure to quickly release the positive pressure as soon as a dimple is observed. In these cases, obtaining a stable whole-cell recording is unlikely. Scale bar for **A** and **B** is 10μm.

Sylgard should be applied to within 20μm of the tip and extend back so none of the pipette glass is in direct contact with the aCSF. The Sylgard can be cured using a heat gun. This should be done before fire-polishing the electrode tip (for information regarding tip geometry and patch membrane estimates, see Sakmann and Neher, 1995). Unless quartz electrodes are used, coating is extremely important for achieving the low noise necessary for recording single-channel currents in either the cell-attached, inside-out, or outside-out configurations. Detailed information regarding different noise sources and types of glass used in patch-clamp recording can be found in the literature (for review see Benndorf, 1995; Sherman-Gold, 1993).

After canceling the fast and slow capacitive current, we switch from the tracking mode to the voltage-clamp mode (with the command potential set to –70 mV, uncorrected for liquid junction potentials) before attempting to go into the whole-cell configuration. To break into the cell, extremely gentle suction is applied by mouth every couple of seconds until a sudden increase in the capacitive current is observed as the membrane capacitance of the cell is revealed. If the cell is healthy, little holding current should be required to keep the membrane potential at –70 mV. The resting potential can be quickly assessed by switching to I = 0 mode (current-clamp with no holding current). Next, the whole-cell capacitance and the series resistance are compensated as described below.

### 3.4.2. Series Resistance Compensation and Measurement

In the whole-cell configuration, access resistance ($R_a$) refers to the resistance to current flow through the pipette tip and into the cell cytoplasm, while series resistance ($R_s$) refers to the combination of $R_a$ and any other external resistances, such as that contributed by an agar salt bridge. With WCRs these external resistances are generally small enough to ignore, effectively making $R_a = R_s$. Typically, $R_s$ is two to five times larger than the pipette resistance. However, in some cases this ratio can be reduced to 2 to 3 with extremely gentle suctions applied to the pipette (e.g., for review see Marty and Neher, 1995). The presence of a large uncompensated capacitance can pose a problem in attempting to compensate for $R_s$. As noted earlier, coating the pipette with Sylgard and maintaining a low fluid level in the bath can reduce the capacitance associated with the pipette.

Uncompensated series resistance can lead to significant voltage-clamp errors. If the uncompensated resistance is 10 MΩ, we know from Ohm's law that this will cause a 20-mV voltage error in response to 2 nA of injected current. Although it is not possible to compensate fully for series resistance (because amplifiers use positive feedback and "ringing" occurs at 100% compensation), it should be possible to compensate for about 80% of the series resistance. In the preceding example, this amount of compensation would result in a 4-mV error. It is important to check the series resistance periodically throughout the experiment because it is not uncommon for the patch membrane to reseal partially or completely. It is usually possible to decrease $R_s$ to its original value with a few gentle suctions. Of course, if $R_s$ increases appreciably during an experiment, and repeated attempts to decrease it fail, the experiment should be terminated (for excellent reviews of series resistance and compensation see Armstrong and Gilly, 1992; Marty and Neher, 1995).

### 3.4.3. Current-Clamp Recordings

In the current-clamp configuration, current is injected into the cell and the membrane voltage is measured. Current-clamp recordings are typically used to examine voltage responses to intracellular current injections and synaptic stimulation. They are used to study a wide range of topics including but not limited to synaptic transmission, synaptic plasticity, neuronal connectivity, calcium dynamics, aging, and epilepsy (Barrionuevo and Brown, 1983; Brown et al., 1979; Dan and Poo, 2006; Deuchars et al., 1994; Kelso et al., 1986; Landfield and Pitler, 1984; Magee and Johnston, 1997, 2005; Markram et al., 1997b; Moyer et al., 1996, 2000; Scharfman, 1996; Thomson and Deuchars, 1997; Williams and Johnston, 1988). In interpreting recordings (in *either* current- or voltage-clamp mode), it is important to recognize that neurons are *not* isopotential during transient conductance changes and that recorded electrical signals may be much smaller and slower than those at the originating site.

Many researchers are now using WCRs rather than sharp microelectrodes for current-clamp recordings (Beggs et al., 2000; Jaffe and Brown, 1997; Johnston et al., 2000; McGann et al., 2001; Moyer and Brown, 1998; Power et al., 2002; Reyes et al., 1998; Stuart et al., 1997a; Tarczy-Hornoch et al., 1998; Wu et al., 2002; Yeckel et al., 1999). It should be realized, however, that the patch-clamp amplifier is a current follower (or current to voltage converter) rather than a voltage follower (as in traditional current-clamp amplifiers).

Current-clamp recordings from patch-clamp amplifiers can lead to distortions of fast signals such as action potentials (Magistretti et al., 1996, 1998).

Whole-cell and sharp-electrode recordings from similar cell types typically reveal similar resting membrane potentials, but the input resistance is generally much higher using WCRs (Staley et al., 1992). The large difference in input resistance is likely due to an artificial leak conductance around the site of the microelectrode impalement. This leak artifact is not present in WCRs due to the tight gigaohm seal formed between the pipette and cell membrane. Of course, a legitimate concern regarding WCRs is the diffusion of pipette contents into the cell, which may wash out important intracellular constituents. In our current-clamp WCR studies, we see little or no change in neuronal discharge pattern, spike height, or input resistance over several hours in healthy neurons (see Fig. 6).

On the other hand, washout effects can impair the induction of certain forms of synaptic plasticity, such as long-term synaptic plasticity, that are thought to play a role in learning and memory (see Brown et al., 2004; Dan and Poo, 2006; Frick et al., 2004; Malinow, 2003; Sah et al., 2003). When washout or run-down effects are suspected, the perforated patch technique can be employed. Instead of breaking into the cell, electrical access to the interior of the neuron is gained by including within the pipette solution certain ion-pore–forming compounds (such as nystatin, amphotericin, gramicidin, or β-escin) that diffuse into the membrane that is within the lumen of the pipette tip (for review see Marty and Neher, 1995; Sherman-Gold, 1993). The process of forming these ion-permeable pores can require 2 to 30 minutes, and the access resistance is typically higher than that obtained using WCRs.

### 3.4.4. Voltage-Clamp Recordings

In a voltage-clamp recording, the transmembrane voltage is controlled and the resulting transmembrane current is measured (Johnston and Wu, 1995; Sherman-Gold, 1993; Sigworth, 1995; Sigworth et al., 1995). The application of a voltage clamp does not eliminate the "space-clamp" problem (the non-isopotentiality of the neuron), which was introduced earlier in the discussion of current-clamp analysis (see Section 3.4.3). When a voltage-clamp is applied to one part of a neuron, it cannot be assumed that the entire cell is being maintained at the intended holding potential. For most neurons, the most reasonable assumption is just the opposite. Transient voltage excursions (or "escape" from voltage-clamp) are the norm rather than the exception in membrane that is distal to the somatic or dendritic site at which the voltage clamp is applied. In studies of cortical neurons, the important issue concerns the size of the space-clamp error and its consequences. Since many or most dendrites can produce regenerative sodium or calcium spikes, small errors can be amplified.

Although space-clamp errors may have little practical or theoretical significance, knowing with certainty whether this is the case is a nontrivial problem (e.g., Armstrong and Gilly, 1992; Carnevale et al., 1997; Mainen and Sejnowski, 1996; Mainen et al., 1996; Spruston et al., 1993; White et al., 1995). Nevertheless, there are several reasons for sometimes preferring voltage-clamp recordings to current-clamp recordings. First, there are circumstances in which some control of voltage may be preferable to no control whatsoever. Second, depending on the purpose of the experiment, certain exacerbating conditions, such as dendritic spikes, can be curbed pharmacologically by blocking calcium or sodium channels. Third, by reducing capacitive membrane currents, voltage-clamp recordings may afford superior temporal resolution. This is often the case in analyzing synaptic responses, where voltage-clamp recordings may reveal complexity in the underlying sequences of conductance changes (Claiborne et al., 1993; Xiang and Brown, 1998). Finally, the outcome of an experiment may furnish useful information about the success of the attempt to control the membrane potential (Kelso et al., 1986).

An alternative approach to the space-clamp problem is to simplify the neuron. For example, the dendritic tree can be pruned by acutely dissociating neurons from a brain slice using enzymes and mechanical agitation (Huguenard and Alger, 1986; Kay and Wong, 1987; Mitterdorfer and Bean, 2002; Moyer et al., 1994; Thibault and Landfield, 1996). A different tactic is to use cultured neurons that have little or no dendritic arbor (Chikwendu and McBain, 1996; Yuan et al., 2005). A more extreme solution is to eliminate the dendrites altogether by using the "nucleated patch" configuration (Bekkers, 2000; Jung et al., 2001; Korngreen and Sakmann, 2000; Sather et al., 1992). Of course, these methods are obviously unsuitable if the original research questions concern, for example, the physiology of higher-order dendrites. In this case, it may be necessary to record simultaneously from multiple parts of a neuron using either electrophysiological (Section 3.4.5) or optical methods (Section 3.7).

## 3.4.5. Simultaneous Recordings with Multiple Electrodes

Recording from multiple neurons or multiples sites on the same neuron is a powerful way to study communication both between and within neurons. The development and application of IR-DIC video microscopy to brain slices has facilitated these efforts, especially those involving simultaneous patching on two or more parts of the same cell (Chen and Johnston, 2005; Golding et al., 2001; Stuart et al., 1997a). Care must be taken, however, to minimize disruption of one recording while obtaining another. We find that it is helpful first to lower both electrodes' locations near the targets of interest. This way, little movement of the electrode is required to form the gigaohm seal. This is especially helpful if the cells of interest are close to each other or if recordings are made from both the soma and proximal dendrite.

Dual recordings have enabled elegant studies of the origin of sodium spikes at the soma and their back-propagation into the dendrites (Magee and Johnston, 1997; Stuart and Sakmann, 1994; Stuart et al., 1997b). Dendritic spikes can cause calcium ion entry into dendritic spines (Jaffe and Brown, 1997; Jaffe et al., 1994; Yuste and Denk, 1995), which can control activity-dependent synaptic plasticity (Magee and Johnston, 1997; Markram et al., 1997b). Dual somatic and dendritic recordings have also been used to study propagation of synaptic signals from the dendrites to the soma and to investigate differences in the density of various types of ion channels as a function of distance from the soma (Hoffman et al., 1997; Magee and Johnston, 1995b).

Simultaneous recordings from separate neurons in the same brain slice can be critical for exploring basic properties of synaptic transmission. Paired recordings from two neurons have been successfully obtained blind using sharp microelectrodes (Deuchars et al., 1994; Thomson and West, 1993). Recently, simultaneous patch-clamp recordings from multiple, visually identified neurons have also been performed in brain slices (Markram et al., 1997a). Visualization is of obvious benefit for preselecting different classes of neurons in specific locations or layers. By recording from two or more neurons, it is possible to determine whether the connections are excitatory or inhibitory and to evaluate frequency-dependent characteristics of the postsynaptic responses as a function of the presynaptic firing rate (Galarreta and Hestrin, 1998; Gibson et al., 1999; Reyes et al., 1998).

A major problem with recording from pairs of neurons in cortical brain slices is the low probability of finding two cells that are synaptically connected (Bianchi and Wong, 1994; Deuchars and Thomson, 1996; Gupta et al., 2000; Michelson and Wong, 1994; Reyes et al., 1998). Recently, Raphael Yuste's group has attempted to overcome this problem using optical techniques. By bulk-loading the cells with a fluorescent calcium indicator, they found that it was possible to identify postsynaptic neurons by stimulating one presynaptic cell and observing those cells whose somatic calcium response is time-locked to the action potentials of the presynaptic neuron (Peterlin et al., 2000). One drawback to this technique is that it does not conveniently reveal inhibitory connections. Another is that all of the cells are loaded with a calcium buffer, which could alter the neurophysiology in important ways. Nevertheless, this technique is useful for rapidly determining candidate postsynaptic neurons as well as identifying microcircuits within a brain region of interest.

### 3.5. Hardware and Software for Patch-Clamp Studies

There are several excellent reviews that describe how to set up a patch-clamp rig (e.g., see Heinemann, 1995; Levis and Rae, 1992). We currently use an upright microscope (Zeiss Axioskop or Olympus BX51) equipped with a variety of lens objectives, including a 2.5× air lens, a 40× water-immersion (0.75 NA) lens, and 60× water-immersion (0.9 NA) lens. A Hamamatsu C2400 video camera equipped with a digital enhancement device is used for visualizing the cell body and dendrites in slices that are transilluminated with infrared-filtered light (see Section 3.4.1). Video images are either captured onto a PC using BioRad's COMOS software in video capture mode or captured onto a Macintosh using NIH Image and a frame grabber card. Some of our slice rigs are also equipped for confocal laser scanning microscopy or two-photon laser scanning microscopy. In these cases, the microscope is fixed to maintain alignment with the scan head, and the stage is movable.

In the standard rigs, the microscope is positioned on top of an XY translation device so that the scope is movable and the stage is fixed. With either microscope configuration, the specimen stage has a circular groove cut out to allow insertion of a recording chamber and platform (Warner Instruments). A dual-channel heater controller (Warner Instruments) can be used to maintain the bath temperature automatically during the experiment. We have used both hydraulic (MX630, SD Instruments, Grants Pass, OR) and motorized (MP-225, Sutter Instruments) micromanipulators for holding the amplifier head stage and advancing the patch pipette. We have also used several different patch-clamp amplifiers, including the Axopatch 1D, the Axopatch 200, and the EPC-7.

For acquisition and analysis of patch-clamp data, a variety of hardware and software packages are available from different companies such as Axon Instruments (Molecular Devices, Sunnyvale, CA) or HEKA (HEKA Instruments, Southboro, MA). We use IgorPro (WaveMetrics, Lake Oswego, OR) running on both Macintosh computers and PCs for both on-line data acquisition and off-line analysis. Electrophysiological data are digitized using an Instrutech ITC-16 converter (Instrutech Corp., Great Neck, NY). The ITC-16 uses 16-bit analog/digital (A/D) and digital/analog (D/A) converters for its analog inputs and outputs with voltage ranges of +10.23 to −10.24 volts. Software drivers for allowing IgorPro to acquire data through the ITC-16 are available through Instrutech. The data are low-pass filtered and acquired at sampling rates greater than five times the filtering frequency to minimize distortions due to aliasing (Heinemann, 1995). After completion of an experiment, the data are stored on the hard disk and then transferred to zip disks, optical disks, or CDs for long-term storage. Electrophysiological data can also be simultaneously digitized at 44 kHz using a Neuro-Corder (Neuro Data Instruments, New York, NY) and stored on VCR tape as an additional backup.

Several devices are used to program different paradigms for electrophysiological experiments. One such device that we have been using for years is the Master-8 programmable pulse generator (AMPI, Jerusalem, Israel). It is easy to use and allows storage and rapid retrieval of eight different programs. More recently, we have programmed our IgorPro acquisition software so that it can run predefined experimental protocols, including the ability to vary both current injection and voltage step amplitudes. In this way, the current- or voltage-clamp data are collected in real time directly through the IgorPro interface and stored on the computer.

## 3.6. Use of Patch-Clamp Techniques to Study Perirhinal Cortex

This section describes the use of whole-cell recordings to characterize the cellular neurophysiology of rat perirhinal (PR) cortex, which differs in several interesting ways from other cortical regions. In a series of studies, we used IR-DIC video microscopy to select *visibly healthy* neurons of various morphological types throughout the different layers of PR (see Fig. 4). In many cases, the pipettes were filled with biocytin (Section 3.1) so that the cells could be anatomically reconstructed following the recordings. Thus far, published findings have included 217 reconstructions (Beggs et al., 2000; Faulkner and Brown, 1999; McGann et al., 2001; Moyer and Brown, 1998; Moyer et al., 2002). By comparing the video images with the reconstructions, we were able to validate the impressions one has about morphological cell types in the living state. The neurophysiological properties were evaluated as a function of the morphological cell type and cortical layer.

Rat PR neurons exhibit at least five different firing patterns (Fig. 8). In response to a just-threshold depolarizing current step, *regular spiking* (RS) cells fire the first action potential (AP) near the onset of the current pulse (within <100 ms) and they exhibit variable amounts of spike frequency adaptation (accommodation; Fig. 8A). With stronger depolarization, APs are recruited at progressively longer latencies. In perirhinal cortex, RS cells commonly exhibit considerable spike frequency accommodation, meaning that the interspike interval progressively increases during a train of APs. Sometimes RS cells stop firing entirely in spite of continued depolarization. Pyramidal neurons commonly have a RS firing pattern (see Figs. 8A, 9C, and 12), but the morphology of RS neurons is variable (Faulkner and Brown, 1999). This is the most common firing pattern in most regions of cortical. *Burst spiking* (BS) cells fire an initial burst of 2 to 3 APs shortly after onset of a depolarizing current step (with <100 ms; Fig. 8B). Stronger depolarization may elicit a second burst of APs or a spike train that exhibits frequency accommodation. BS cells are most common in layer V, where they tend to be pyramidal neurons that have a thick apical dendrite that shows relatively little taper as it extends toward the pial surface (Fig. 9A).

In response to a just-threshold current step, *late spiking* (LS) cells fire the first AP near the end of the current step. The spike latency

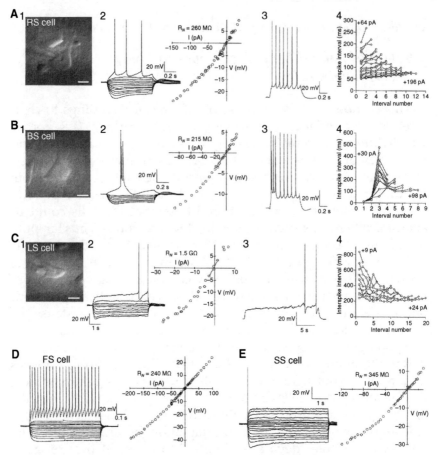

Fig. 8. Varieties of cell firing patterns in perirhinal cortex. (**A**) An IR-DIC video micrograph and physiology of a regular spiking (RS) pyramidal neuron in layer V. This neuron had a resting membrane potential ($E_R$) of −76 mV. RS cells comprise ~50% of the pyramidal cells in layer II/III, ~76% of the pyramidal cells in layer V, and ~1% of the principal cells in layer VI. (**B**) An IR-DIC image and physiology of a burst spiking (BS) pyramidal neuron in layer V ($E_R$ = −75 mV). BS cells comprise ~9% of the pyramidal neurons in layer V and only in cells from rats of age 14 days and older. (**C**) An IR-DIC image and physiology of a late spiking (LS) nonpyramidal neuron in layer VI ($E_R$ = −82 mV). Notice the unusually long latency to first spike in panel 3 (delay of 16 seconds in response to a 20 second, 8 pA current injection). LS cells comprise ~50% of the pyramidal cells in layer II/II, ~14% of pyramidal cells in layer V, and ~90% of the principal cells in layer VI. (**D**) Physiology of a fast spiking (FS) nonpyramidal neuron in layer I ($E_R$ = −78 mV). FS cells are found in all layers of perirhinal cortex. Small, round cells (soma ≤10 μm) with diminutive dendrites invariably have a FS firing pattern (Faulkner and Brown, 1999; McGann et al., 2001). (**E**) Physiology of a single spiking (SS) nonpyramidal neuron in layer VI ($E_R$ = −82 mV). SS cells comprise ~7% of the principal neurons in layer VI.

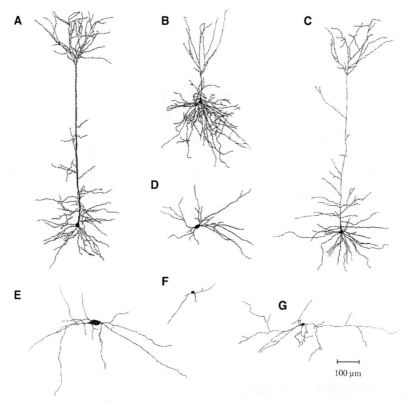

Fig. 9. Morphology of neurons in rat perirhinal cortex. Illustrated are various examples from each of the five different cell classes. (**A**) Burst spiking (BS) spiny pyramidal neuron in layer V. Notice the thick apical dendrite, which is typical of BS pyramidal cells. (**B**) Late spiking (LS) spiny pyramidal neuron in layer II/III. (**C**) Regular spiking (RS) spiny pyramidal neuron in layer V. Notice the relatively thin apical dendrite. (**D,E**) LS spiny and aspiny nonpyramidal neurons located in layer VI. (**F**) Fast spiking aspiny nonpyramidal neuron in layer I. (**G**) Single spiking aspiny nonpyramidal neuron in layer VI. All neurons are oriented with the pial surface at the top and the external capsule at the bottom.

can be several seconds (Figs. 8C and 11). With larger current steps, additional spikes are recruited at progressively shorter latencies. The subthreshold voltage response consists of a rapid depolarization followed by a slower ramp up to the spike threshold. LS cells tend to exhibit less extreme spike accommodation, and they sometimes exhibit what we have termed "anti-accommodation," meaning

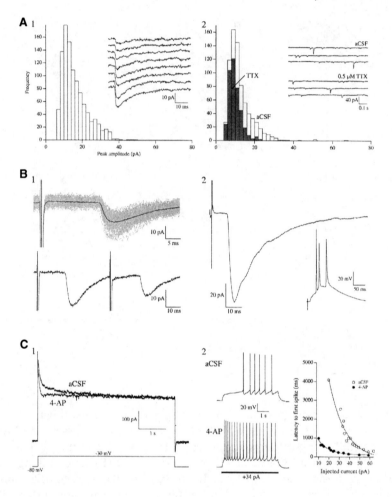

that the interspike interval actually decreases across the spike train.

Although LS neurons have been reported in other brain regions, they appear to be uniquely prevalent in layer VI of PR, where one study found that about 90% of the larger neurons had a LS firing pattern (McGann et al., 2001). Preliminary voltage-clamp data suggest that the LS firing pattern results from the presence of a slowly inactivating potassium current (Fig. 10C). When LS neurons are depolarized, activation of this potassium current initially resists depolarization to the spike threshold. As the potassium conductance slowly inactivates, the cell gradually approaches the spike

Fig. 10. Synaptic and other currents recorded in rat perirhinal neurons. (**A1**) Spontaneous postsynaptic currents (PSCs) recorded from a layer V pyramidal neuron in a brain slice prepared from an aged rat. The PSCs were collected for 6 minutes at a holding potential of –78 mV. A total of 1,032 individual PSCs were analyzed (mean amplitude of 15.9 pA). Spontaneous PSCs from this and other layer V pyramidal neurons have a positively skewed amplitude frequency distribution. Current sweeps are individual examples of spontaneous PSCs. (**A2**) Effects of tetrodotoxin (TTX) on spontaneous PSCs. Spontaneous PSCs were collected from a layer V pyramidal neuron in a brain slice prepared from a 3-week-old rat. Notice that the amplitude frequency distribution was positively skewed before and after bath application of 0.5 μM TTX (holding potential – 87 mV). TTX decreased both the frequency (aCSF 3.1 Hz; TTX 1.7 Hz) and amplitude (aCSF 12.5 ± 0.2 pA; TTX 9.9 ± 0.2 pA) of the spontaneous PSCs (presumably by eliminating multi-quantal release). (**B1**) Stimulation of layer I elicits long-latency currents in layer V pyramidal neurons. The neuron was held at –92 mV. Top trace is an overlay of 67 consecutive evoked currents and their average (solid current trace). Bottom trace illustrates paired pulse depression (average of 14 consecutive evoked PSCs in the same neuron). (**B2**) Evoked synaptic current in a layer V burst spiking neuron held at –74 mV. Current trace is an average of 58 consecutive PSCs evoked by layer I stimulation. Under current-clamp conditions, the same stimulation elicited a burst of 3 action potentials. (**C1**) Voltage-clamp analyses of late spiking in a perirhinal neuron from layer VI. Preliminary data indicate that bath application of 30 μM 4-aminopyridine (4-AP) blocks a slowly inactivating potassium current. (**C2**) Under current-clamp conditions, the same concentration of 4-AP blocks delayed firing, as shown in the leftward and downward compression of the plot of firing delay as a function of current-step amplitude (D-I plot). D-I plots of LS neurons bathed in aCSF containing 4-AP are similar to those of RS neurons bathed in standard aCSF.

←————————————————————————————————————

threshold. LS is blocked by a low concentration (30 μM) of 4-aminopyridine (4-AP, see Fig. 10C1), which has also been shown to block late spiking in PR neurons (Fig. 10C2) as well as striatal neurons (Nisenbaum et al., 1994). LS neurons have been suggested, based on theoretical considerations, to participate in temporal encoding (McGann and Brown, 2000; Padlubnaya et al., 2006; Tieu et al., 1999).

*Fast spiking* (FS) cells (Fig. 8D) always fire with a short latency (<10 ms), regardless of the amplitude of the suprathreshold current step. They can fire at very high frequencies (>100 Hz), and they show little or no spike frequency adaptation. FS cells commonly have a very small cell body and diminutive dendritic arbor (Fig. 9F; see also Faulkner and Brown, 1999; McGann et al., 2001). *Single spiking* (SS) cells fire only a single AP in response to suprathreshold current steps (Fig. 8E). We have not reconstructed a sufficient number of SS neurons to know whether they are associated with any particular morphology. The morphology of one SS neuron is shown in Fig. 9G. The relative frequencies of these different firing patterns vary in different layers of PR (see legend to Fig. 8).

We have also been using the patch-clamp technique to study synaptic neurophysiology in PR (Fig. 10). Recordings of spontaneous excitatory postsynaptic currents (EPSCs) are shown in Fig. 10A (inset). As shown in the histogram, spontaneous EPSCs tend to be small (<40 pA; mean ~15 pA) in PR relative to those in the adjacent lateral nucleus of the amygdala (Faulkner and Brown, 1999; Moyer et al., 2002). Interestingly, the evoked EPSCs (Fig. 10B) have an unusually long onset latency (>5 ms). The long onset latency appears to reflect an unusually slow AP conduction velocity, which was estimated to be ~0.2 m/s (Moyer et al., 2002). This estimate was based on the slope of the regression of inter-electrode distance against EPSC onset latency. The slow conduction velocity is not surprising because PR is remarkably low in myelin (Brown and Furtak, 2006; Zilles and Wree, 1995), as mentioned above. The value of the conduction velocity is comparable to that reported for other unmyelinated fibers (see Berg-Johnsen and Langmoen, 1992). Given the large numbers of LS neurons in PR and relatively low levels of myelin, we tend to think of PR as "slow" cortex (Brown and Furtak, 2006) (Figs. 11 and 12).

### 3.7. Alternative Methods for Recording Current or Voltage

For some purposes, the classical method of using "sharp" microelectrodes may be preferred, especially when dialysis of the recorded neuron is an issue. Both current- and voltage-clamp microelectrode recordings are possible using a discontinuous or switch clamp (Johnston and Brown, 1984) that alternates at several thousand cycles per second between measuring voltage and passing current. When properly adjusted, the switch clamp eliminates the

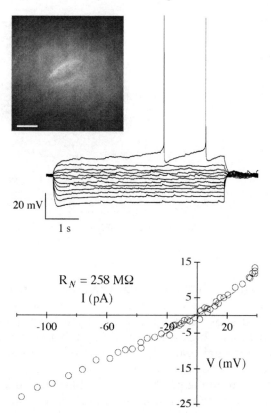

Fig. 11. Visually guided whole-cell recordings from a perirhinal neuron in a brain slice prepared from an adult rat. An IR-DIC video micrograph and neurophysiology of a layer VI nonpyramidal late spiking neuron in a 3-month-old rat. The neuron was located 85 µm below the slice surface and would not be visible without the benefit of IR-DIC video microscopy. This cell had a resting membrane potential of –78 mV and remained stable for 3 hours.

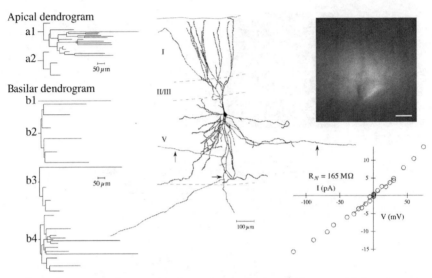

Fig. 12. Morphology and physiology of a layer V pyramidal neuron in perirhinal cortex of an aged rat. An IR-DIC video microscopy permitted visualization and subsequent placement of the patch pipette onto the soma of this neuron, which was located 76 μm below the surface of this 400-μm-thick horizontal brain slice. In aged tissue, a neuron this deep below the slice surface is not visible without IR-DIC video microscopy. Inclusion of biocytin in the patch pipette enabled subsequent analysis of neuronal morphology. Shown are the apical and basilar dendrograms, obtained using a Neurolucida system (MicroBrightfield, Colchester, VT), along with a camera lucida serial reconstruction. The Neurolucida system enables three-dimensional reconstructions that can be rotated to visualize better the organization of the dendritic and axonal processes. The Neurolucida system is also useful for quantifying dendritic length, spine density, branching patterns, and soma size. The morphological data can be exported for compartmental modeling in programs such as NEURON (Carnevale and Hines, 2006; Hines and Carnevale, 1997, 2001). The axon (indicated by arrows) from this cell descended and ramified within layers V and VI en route to the external capsule. This regular spiking neuron had a resting membrane potential of –76 mV and remained stable for 40 minutes. Scale bar within the IR-DIC video micrograph is 10 μm.

voltage that develops across the microelectrode resistance when current is passed. The resistance of a suitable microelectrode is usually at least an order of magnitude higher than that of a suitable patch pipette. An important limitation of this method is that the microelectrode/amplifier combination can be slow and noisy relative to comparable whole-cell recordings. However, for many purposes this does not present any serious problems (Kelso et al., 1986; Thomson and Destexhe, 1999; Thomson and West, 2003; Thomson et al., 2002).

Another important alternative method for measuring transmembrane potential makes use of optical recordings from neurons containing voltage-sensitive dyes (see Zochowski et al., 2000). A major advantage of this approach is that it enables simultaneous recordings of transmembrane potential from many more parts of a neuron than would be practicable using either patch pipettes or sharp microelectrodes (Djurisic and Zecevic, 2005; Djurisic et al., 2004; Ebner and Chen, 1995; Tominaga et al., 2002; Zecevic and Antic, 1998; Zochowski et al., 2000). Optical recordings of transmembrane potential have been productively used in studies of both invertebrate and vertebrate neurons or circuits (Bozza et al., 2004; Bureau et al., 2004; Higashi et al., 1999; Wachowiak and Cohen, 2001; Zhu and Zhu, 2004; Zochowski et al., 2000). When used in conjunction with patch-clamp recordings, optical techniques can provide invaluable data on information flow through individual neurons or neuronal circuits (Djurisic and Zecevic, 2005; Djurisic et al., 2004; Helmchen et al., 1999; Kerr et al., 2005; Knott et al., 2006; Mainen et al., 1999; Miesenbock and Kevrekidis, 2005; Stosiek et al., 2003).

## 4. Whole-Cell Recordings In Vivo

Although the focus of this chapter is on brain slices, it is important to note that whole-cell recording methods are now also being used to study neurophysiology in vivo. First performed by David Ferster's group (Ferster and Jagadeesh, 1992), in vivo whole-cell patch-clamp techniques are now being used in several laboratories (Cang and Isaacson, 2003; Lee et al., 2006; Margrie et al., 2002, 2003; Pei et al., 1991; Zhu and Connors, 1999; Zhu and Zhu, 2004). Most of this work has been done on rat neocortex because of its accessibility. The use of patch-clamp techniques for in vivo recordings presents many of the same challenges inherent in applying it to brain slices: pulling a suitable pipette, maintaining positive pressure while lowering the pipette, forming a GΩ seal, breaking into a cell, and compensating for capacitance and series resistance (see Section 3, above). Additional challenges include maintaining an adequate state of anesthesia/pain control, in accordance with protocol review boards, and minimizing cardiopulmonary-produced tissue movement. Cell selection is often a problem because most of the recordings are done "blind." However, the combinations of genetics (e.g., green fluorescence protein (GFP) expression) and two-photon microscopy have been used to select specific neurons for patch-clamping (referred to as two-photon targeted patching, see Margrie et al., 2002).

In spite of the additional level of technical difficulty, there are obvious merits in applying patch-clamp techniques in vivo. A few of these include the ability to record from neurons without damaging circuit connections or normal blood supply, preservation of a more normal hormonal/neuromodulatory "milieu," and the ability to apply natural sensory stimulation (such as visual stimulation or vibrissae stimulation). More recently, whole-cell recordings have been used in awake rats under restraint (Margrie et al., 2002) and even freely moving rats (Lee et al., 2006). Application of patch-clamp techniques to freely moving animals will enable scientists to study neuronal physiology during various natural behaviors, including learning and memory tasks. There is no debate over the ultimate need to know how the information obtained from brain slice studies applies to the intact and functioning nervous system.

## Acknowledgments

The authors thank Pawel Boguszewski and Vicky Leung for critical comments and helpful discussion.

## References

Aghajanian, G. K. and Rasmussen, K. (1989) Intracellular studies in the facial nucleus illustrating a simple new method for obtaining viable motoneurons in adult rat brain slices. *Synapse* **3**, 331–338.

Aitken, P. G., Breese, G. R., Dudek, F. E., et al. (1995) Preparative methods for brain slices: a discussion. *J. Neurosci. Meth.* **59**, 139–149.

Alger, B. E. and Teyler, T. J. (1976) Long-term and short-term plasticity in the CA1, CA3, and dentate regions of the rat hippocampal slice. *Brain Res.* **110**, 463–480.

Alvarez-Leefmans, F. J. (1992) Extracellular reference electrodes, in *Practical Electrophysiological Methods: A guide for In Vitro Studies in Vertebrate Neurobiology*, (Kettenmann, H. and Grantyn, R., eds.), Wiley-Liss, New York, pp. 171–182.

Armstrong, C. M. and Gilly, W. F. (1992) Access resistance and space clamp problems associated with whole-cell patch clamping. *Meth. Enzymol.* **207**, 101–122.

Avshalumov, M. V. and Rice, M. E. (2002) NMDA receptor activation mediates hydrogen peroxide-induced pathophysiology in rat hippocampal slices. *J. Neurophysiol.* **87**, 2896–2903.

Barrionuevo, G. and Brown, T. H. (1983) Associative long-term potentiation in hippocampal slices. *Proc. Natl. Acad. Sci. (USA)* **80**, 7347–7351.

Barry, P. H. (1994) JPCalc, a software package for calculating liquid junction potential corrections in patch-clamp, intracellular, epithelial and bilayer measurements and for correcting junction potential measurements. *J. Neurosci. Meth.* **51**, 107–116.

Barry, P. H. and Diamond, J. M. (1970) Junction potentials, electrode standard potentials, and other problems in interpreting electrical properties of membranes. *J. Membrane Biol.* **3**, 93–122.

Barry, P. H. and Lynch, J. W. (1991) Liquid junction potentials and small cell effects in patch-clamp analysis. *J. Membrane Biol.* **121**, 101–117.

Beggs, J. M., Moyer, J. R., Jr., McGann, J. P., and Brown, T. H. (2000) Prolonged synaptic integration in perirhinal cortical neurons. *J. Neurophysiol.* **83**, 3294–3298.

Bekkers, J. M. (2000) Properties of voltage-gated potassium currents in nucleated patches from large layer 5 cortical pyramidal neurons of the rat. *J. Physiol.* **525**, 593–609.

Benndorf, K. (1995) Low-noise recording, in *Single-Channel Recording*, 2nd ed. (Sakmann, B. and Neher, E., eds.), Plenum Press, New York, pp. 129–145.

Berg-Johnsen, J. and Langmoen, I. A. (1992) Temperature sensitivity of thin unmyelinated fibers in rat hippocampal cortex. *Brain Res.* **576**, 319–321.

Bianchi, R. and Wong, R. K. (1994) Carbachol-induced synchronized rhythmic bursts in CA3 neurons of guinea pig hippocampus in vitro. *J. Neurophysiol.* **72**, 131–138.

Blanton, M. G., LoTurco, J. J., and Kriegstein, A. R. (1989) Whole cell recordings from neurons in slices of reptilian and mammalian cerebral cortex. *J. Neurosci. Meth.* **30**, 203–210.

Borst, J. G. G., Helmchen, F., and Sakmann, B. (1995) Pre- and postsynaptic whole-cell recordings in the medial nucleus of the trapezoid body of the rat. *J. Physiol.* **489**, 825–840.

Bozza, T., McGann, J. P., Mombaerts, P., and Wachowiak, M. (2004) In vivo imaging of neuronal activity by targeted expression of a genetically encoded probe in the mouse. *Neuron* **42**, 9–21.

Brahma, B., Forman, R. E., Stewart, E. E., Nicholson, C., and Rice, M. E. (2000) Ascorbate inhibits edema in brain slices. *J. Neurochem.* **74**, 1263–1270.

Brown, T. H. and Furtak, S. C. (2006) Low myelin staining in rat perirhinal cortex and parts of the amygdala. *Soc. Neurosci. Abstr.* **32**, Program No. 638.17.

Brown, T. H. and Keenan, C. L. (1987) Visualization of hippocampal synapses in brain slices using video microscopy. *Soc. Neurosci. Abstr.* **13**, 1515.

Brown, T. H., Lindquist, D. H., and Furtak, S. C. (2004) Hebbian Synapses, in *Encyclopedia of Neuroscience,* 3rd ed. (Adelman, G. and Smith, B. H., eds.), Elsevier Science, New York.

Brown, T. H., Wong, R. K. S., and Prince, D. A. (1979) Spontaneous miniature synaptic potentials in hippocampal neurons. *Brain Res.* **177,** 194–199.

Bureau, I., Shepherd, G. M., and Svoboda, K. (2004) Precise development of functional and anatomical columns in the neocortex. *Neuron* **42,** 789–801.

Burwell, R. D. (2001) Borders and cytoarchitecture of the perirhinal and postrhinal cortices in the rat. *J. Comp. Neurol.* **437,** 17–41.

Cahalan, M. and Neher, E. (1992) Patch clamp techniques: an overview. *Meth. Enzymol.* **207,** 3–66.

Cang, J. and Isaacson, J. S. (2003) In vivo whole-cell recording of odor-evoked synaptic transmission in the rat olfactory bulb. *J. Neurosci.* **23,** 4108–4116.

Carnevale, N. T. and Hines, M. L. (2006) *The Neuron Book,* Cambridge University Press, Cambridge, UK.

Carnevale, N. T., Tsai, K. Y., Claiborne, B. J., and Brown, T. H. (1997) Comparative electrotonic analysis of three classes of rat hippocampal neurons. *J. Neurophysiol.* **78,** 703–720.

Chen, X. and Johnston, D. (2005) Constitutively active G-protein-gated inwardly rectifying K+ channels in dendrites of hippocampal CA1 pyramidal neurons. *J. Neurosci.* **25,** 3787–3792.

Chikwendu, A. and McBain, C. J. (1996) Two temporally overlapping "delayed-rectifiers" determine the voltage-dependent potassium current phenotype in cultured hippocampal interneurons. *J. Neurophysiol.* **76,** 1477–1490.

Choi, D. W. (1992) Excitotoxic cell death. *J. Neurobiol.* **23,** 1261–1276.

Choi, D. W. (1994) Calcium and excitotoxic neuronal injury. *Ann. N. Y. Acad. Sci.* **747,** 162–171.

Choi, D. W. (1995) Calcium: still center-stage in hypoxic-ischemic neuronal death. *TINS* **18,** 58–60.

Christie, B. R., Eliot, L. S., Ito, K.-I., Miyakawa, H., and Johnston, D. (1995) Different $Ca^{2+}$ channels in soma and dendrites of hippocampal pyramidal neurons mediate spike-induced $Ca^{2+}$ influx. *J. Neurophysiol.* **73,** 2553–2557.

Claiborne, B. J., Xiang, Z., and Brown, T. H. (1993) Hippocampal circuitry complicates analysis of long-term potentiation in mossy fiber synapses. *Hippocampus* **3,** 115–122.

Dan, Y. and Poo, M. M. (2006) Spike timing-dependent plasticity: from synapse to perception. *Physiol. Rev.* **86,** 1033–1048.

Davson, H., Welch, K., and Segal, M. B. (1987) *The Physiology and Pathophysiology of the Cerebrospinal Fluid,* Churchill Livingstone, New York, 15–33.

Desagher, S., Glowinski, J., and Prémont, J. (1997) Pyruvate protects neurons against hydrogen peroxide-induced toxicity. *J. Neurosci.* **17,** 9060–9067.

Deuchars, J. and Thomson, A. M. (1996) CA1 pyramid-pyramid connections in rat hippocampus in vitro: dual intracellular recordings with biocytin filling. *Neuroscience* **74,** 1009–1018.

Deuchars, J., West, D. C., and Thomson, A. M. (1994) Relationships between morphology and physiology of pyramid-pyramid single axon connections in rat neocortex in vitro. *J. Physiol.* **478,** 423–435.

Djurisic, M., Antic, S., Chen, W. R., and Zecevic, D. (2004) Voltage imaging from dendrites of mitral cells: EPSP attenuation and spike trigger zones. *J. Neurosci.* **24**, 6703–6714.

Djurisic, M. and Zecevic, D. (2005) Imaging of spiking and subthreshold activity of mitral cells with voltage-sensitive dyes. *Ann. N.Y. Acad. Sci.* **1048**, 92–102.

Dodt, H.-U. and Zieglgänsberger, W. (1990) Visualizing unstained neurons in living brain slices by infrared DIC-videomicroscopy. *Brain Res.* **537**, 333–336.

Duffy, C. J. and Teyler, T. J. (1975) A simple tissue slicer. *Physiol. Behav.* **14**, 525–526.

Ebner, T. J. and Chen, G. (1995) Use of voltage-sensitive dyes and optical recordings in the central nervous system. *Prog. Neurobiol.* **46**, 463–506.

Edwards, F. A. and Konnerth, A. (1992) Patch-clamping cells in sliced tissue preparations. *Meth. Enzymol.* **207**, 208–222.

Edwards, F. A., Konnerth, A., Sakmann, B., and Takahashi, T. (1989) A thin slice preparation for patch clamp recordings from neurones of the mammalian central nervous system. *Pflügers Arch.* **414**, 600–612.

Faulkner, B. and Brown, T. H. (1999) Morphology and physiology of neurons in the rat perirhinal-lateral amygdala area. *J. Comp. Neurol.* **411**, 613–642.

Feig, S. and Lipton, P. (1990) N-methyl-D-aspartate receptor activation and $Ca^{2+}$ account for poor pyramidal cell structure in hippocampal slices. *J. Neurochem.* **55**, 473–483.

Ferster, D. and Jagadeesh, B. (1992) EPSP-IPSP interactions in cat visual cortex studied with in vivo whole-cell patch recording. *J. Neurosci.* **12**, 1262–1274.

Forscher, P. and Oxford, G. S. (1985) Modulation of calcium channels by norepinephrine in internally dialyzed avian sensory neurons. *J. Gen. Physiol.* **85**, 743–763.

Frazier, D. T., Narahashi, T., and Yamada, M. (1970) The site of action and active form of local anesthetics. *J. Pharmacol. Exp. Ther.* **171**, 45–51.

Frick, A., Magee, J., and Johnston, D. (2004) LTP is accompanied by an enhanced local excitability of pyramidal neuron dendrites. *Nat. Neurosci.* **7**, 126–135.

Galarreta, M. and Hestrin, S. (1998) Frequency-dependent synaptic depression and the balance of excitation and inhibition in the neocortex. *Nat. Neurosci.* **1**, 587–594.

Geddes, L. A. (1972) *Electrodes and the Measurement of Bioelectric Events*, John Wiley & Sons, New York.

Gibson, H. L. (1978) *Photography by Infrared*, John Wiley & Sons, New York, p. 545.

Gibson, J. R., Beierlein, M., and Connors, B. W. (1999) Two networks of electrically coupled inhibitory neurons in neocortex. *Nature* **402**, 75–79.

Golding, N. L., Kath, W. L., and Spruston, N. (2001) Dichotomy of action-potential backpropagation in CA1 pyramidal neuron dendrites. *J. Neurophysiol.* **86**, 2998–3010.

Golding, N. L., Mickus, T. J., Katz, Y., Kath, W. L., and Spruston, N. (2005) Factors mediating powerful voltage attenuation along CA1 pyramidal neuron dendrites. *J. Physiol.* **568**, 69–82.

Gulledge, A. T. and Stuart, G. J. (2005) Cholinergic inhibition of neocortical pyramidal neurons. *J. Neurosci.* **25**, 10308–10320.

Gupta, A., Wang, Y., and Markram, H. (2000) Organizing principles for a diversity of GABAergic interneurons and synapses in the neocortex. *Science* **287**, 273–278.

Heinemann, S. H. (1995) Guide to data acquisition and analysis, in *Single-Channel Recording*, 2nd ed. (Sakmann, B. and Neher, E., eds.), Plenum Press, New York, pp. 53–91.

Helmchen, F., Svoboda, K., Denk, W., and Tank, D. W. (1999) In vivo dendritic calcium dynamics in deep-layer cortical pyramidal neurons. *Nat. Neurosci.* **2**, 989–996.

Hestrin, S. and Armstrong, W. E. (1996) Morphology and physiology of cortical neurons in layer I. *J. Neurosci.* **16**, 5290–5300.

Higashi, S., Crair, M. C., Kurotani, T., Inokawa, H., and Toyama, K. (1999) Altered spatial patterns of functional thalamocortical connections in the barrel cortex after neonatal infraorbital nerve cut revealed by optical recording. *Neurosci.* **91**, 439–452.

Hille, B. (2001) *Ion Channels of Excitable Membranes*, 3rd ed., Sinauer Associates, Sunderland, MA.

Hines, M. L. and Carnevale, N. T. (1997) The NEURON simulation environment. *Neural Comput.* **9**, 1179–1209.

Hines, M. L. and Carnevale, N. T. (2001) NEURON: a tool for neuroscientists. *Neuroscientist* **7**, 123–135.

Hoffman, D. A. and Johnston, D. (1998) Downregulation of transient K$^+$ channels in dendrites of hippocampal CA1 pyramidal neurons by activation of PKA and PKC. *J. Neurosci.* **18**, 3521–3528.

Hoffman, D. A., Magee, J. C., Colbert, C. M., and Johnston, D. (1997) K+ channel regulation of signal propagation in dendrites of hippocampal pyramidal neurons. *Nature* **387**, 869–875.

Huguenard, J. R. and Alger, B. E. (1986) Whole-cell voltage-clamp study of the fading of GABA-activated currents in acutely dissociated hippocampal neurons. *J. Neurophysiol.* **56**, 1–18.

Ives, J. G. D. and Janz, G. J. (1961) *Reference Electrodes: Theory and Practice*, Academic Press, New York.

Jaffe, D. B. and Brown, T. H. (1997) Calcium dynamics in thorny excrescences of CA3 pyramidal neurons. *J. Neurophysiol.* **78**, 10–18.

Jaffe, D. B., Fisher, S. A., and Brown, T. H. (1994) Confocal laser scanning microscopy reveals voltage-gated calcium signals within hippocampal dendritic spines. *J. Neurobiol.* **25**, 220–233.

Janz, G. J. and Ives, J. G. D. (1968) Silver, silver chloride electrodes. *Ann. N.Y. Acad. Sci.* **148**, 210–221.

Johnston, D. and Brown, T. H. (1984) Biophysics and microphysiology of synaptic transmission in hippocampus, in *Brain Slices* (Dingledine, R., ed.), Plenum, New York, pp. 51–86.

Johnston, D., Hoffman, D. A., Magee, J. C., et al. (2000) Dendritic potassium channels in hippocampal pyramidal neurons. *J. Physiol.* **525**, 75–81.

Johnston, D. and Wu, S. M.-S. (1995) *Foundations of cellular neurophysiology*, MIT Press, Cambridge, MA.

Jung, H.-Y., Staff, N. P., and Spruston, N. (2001) Action potential bursting in subicular pyramidal neurons is driven by a calcium tail current. *J. Neurosci.* **21**, 3312–3321.

Kay, A. R. (1992) An intracellular medium formulary. *J. Neurosci. Meth.* **44**, 91–100.

Kay, A. R. and Wong, R. K. S. (1987) Calcium current activation kinetics in isolated pyramidal neurones of the CA1 region of the mature guinea-pig hippocampus. *J. Physiol.* **392**, 603–616.

Keenan, C. L., Chapman, P. F., Chang, V. C., and Brown, T. H. (1988) Videomicroscopy of acute brain slices from amygdala and hippocampus. *Brain Res. Bull.* **21**, 373–383.

Keller, J. N., Kindy, M. S., Holtsberg, F. W., et al. (1998) Mitochondrial manganese superoxide dismutase prevents neural apoptosis and reduces ischemic brain injury: suppression of peroxynitrite production, lipid peroxidation, and mitochondrial dysfunction. *J. Neurosci.* **18**, 687–697.

Kelso, S. R., Ganong, A. H., and Brown, T. H. (1986) Hebbian synapses in hippocampus. *Proc. Natl. Acad. Sci. (USA)* **83**, 5326–5330.

Kerr, J. N., Greenberg, D., and Helmchen, F. (2005) Imaging input and output of neocortical networks in vivo. *Proc. Natl. Acad. Sci. (USA)* **102**, 14063–14068.

Knott, G. W., Holtmaat, A., Wilbrecht, L., Welker, E., and Svoboda, K. (2006) Spine growth precedes synapse formation in the adult neocortex in vivo. *Nat. Neurosci.* **9**, 1117–1124.

Korngreen, A. and Sakmann, B. (2000) Voltage-gated K+ channels in layer 5 neocortical pyramidal neurons from young rats: subtypes and gradients. *J. Physiol.* **525**, 621–639.

Kovachich, G. B. and Mishra, O. P. (1980) Lipid peroxidation in rat brain cortical slices as measured by the thiobarbituric acid test. *J. Neurochem.* **35**, 1449–1452.

Kovachich, G. B. and Mishra, O. P. (1983) The effects of ascorbic acid on malonaldehyde formation, $K^+$, $Na^+$ and water content of brain slices. *Exp. Brain Res.* **50**, 62–68.

Landfield, P. W. and Pitler, T. A. (1984) Prolonged $Ca^{2+}$-dependent afterhyperpolarizations in hippocampal neurons of aged rats. *Science* **226**, 1089–1092.

Lee, A. K., Manns, I. D., Sakmann, B., and Brecht, M. (2006) Whole-cell recordings in freely moving rats. *Neuron* **51**, 399–407.

Levis, R. A. and Rae, J. L. (1992) Constructing a patch clamp setup. *Meth. Enzymol.* **207**, 14–66.

Li, C.-L. and McIlwain, H. (1957) Maintenance of resting membrane potentials in slices of mammalian cerebral cortex and other tissues in vitro. *J. Physiol.* **139**, 178–190.

Lipton, P., Aitken, P. G., Dudek, F. E., et al. (1995) Making the best of brain slices: comparing preparative methods. *J. Neurosci. Meth.* **59**, 151–156.

Llinás, R. and Sugimori, M. (1980) Electrophysiological properties of *in vitro* Purkinje cell somata in mammalian cerebellar slices. *J. Physiol. (Lond.)* **305**, 171–195.

MacVicar, B. A. (1984) Infrared video microscopy to visualize neurons in the in vitro brain slice preparation. *J. Neurosci. Meth.* **12**, 133–149.

Magee, J. C., Avery, R. B., Christie, B. R., and Johnston, D. (1996) Dihydropyridine-sensitive, voltage-gated $Ca^{2+}$ channels contribute to the resting intracellular $Ca^{2+}$ concentration of hippocampal CA1 pyramidal neurons. *J. Neurophysiol.* **76**, 3460–3470.

Magee, J. C. and Johnston, D. (1995a) Characterization of single voltage-gated $Na^+$ and $Ca^{2+}$ channels in apical dendrites of rat CA1 pyramidal neurons. *J. Physiol.* **487** (pt 1), 67–90.

Magee, J. C. and Johnston, D. (1995b) Synaptic activation of voltage-gated channels in the dendrites of hippocampal pyramidal neurons. *Science* **268**, 301–304.

Magee, J. C. and Johnston, D. (1997) A synaptically controlled, associative signal for Hebbian plasticity in hippocampal neurons. *Science* **275**, 209–213.

Magee, J. C. and Johnston, D. (2005) Plasticity of dendritic function. *Cur. Opin. Neurobiol.* **15**, 334–342.

Magistretti, J., Mantegazza, M., de Curtis, M., and Wanke, E. (1998) Modalities of distortion of physiological voltage signals by patch-clamp amplifiers: a modeling study. *Biophys. J.* **74**, 831–842.

Magistretti, J., Mantegazza, M., Guatteo, E., and Wanke, E. (1996) Action potentials recorded with patch-clamp amplifiers: are they genuine? *TINS* **19**, 530–534.

Mainen, Z. F., Carnevale, N. T., Zador, A. M., Claiborne, B. J., and Brown, T. H. (1996) Electrotonic architecture of hippocampal CA1 pyramidal neurons based on three-dimensional reconstructions. *J. Neurophysiol.* **76**, 1904–1923.

Mainen, Z. F., Maletic-Savatic, M., Shi, S. H., Hayashi, Y., Malinow, R., and Svoboda, K. (1999) Two-photon imaging in living brain slices. *Methods* **18**, 231–239.

Mainen, Z. F. and Sejnowski, T. J. (1996) Influence of dendritic structure on firing pattern in model neocortical neurons. *Nature* **382**, 363–366.

Malinow, R. (2003) AMPA receptor trafficking and long-term potentiation. *Philos. Trans. R. Soc. Lond. B. Biol. Sci.* **358**, 707–714.

Margrie, T. W., Brecht, M., and Sakmann, B. (2002) In vivo, low-resistance, whole-cell recordings from neurons in the anaesthetized and awake mammalian brain. *Pflügers Arch.* **444**, 491–498.

Margrie, T. W., Meyer, A. H., Caputi, A., et al. (2003) Targeted whole-cell recordings in the mammalian brain in vivo. *Neuron* **39**, 911–918.

Markram, H., Lubke, J., Frotscher, M., Roth, A., and Sakmann, B. (1997a) Physiology and anatomy of synaptic connections between thick tufted pyramidal neurones in the developing rat neocortex. *J. Physiol. (London)* **500**, 409–440.

Markram, H., Lübke, J., Frotscher, M., and Sakmann, B. (1997b) Regulation of synaptic efficacy by coincidence of postsynaptic APs and EPSPs. *Science* **275**, 213–215.

Marty, A. and Neher, E. (1995) Tight-seal whole-cell recording, in *Single-Channel Recording*, 2nd ed. (Sakmann, B. and Neher, E., eds.), Plenum, New York, pp. 31–52.

Mattson, M. P. (1998) Modification of ion homeostasis by lipid peroxidation: roles in neuronal degeneration and adaptive plasticity. *TINS* **21**, 53–57.

McGann, J. P. and Brown, T. H. (2000) Fear conditioning model predicts key temporal aspects of conditioned response production. *Psychobiology* **28**, 303–313.

McGann, J. P., Moyer, J. R., Jr., and Brown, T. H. (2001) Predominance of late-spiking neurons in layer VI of rat perirhinal cortex. *J. Neurosci.* **21**, 4969–4976.

McKernan, M. G. and Shinnick-Gallagher, P. (1997) Fear conditioning induces a lasting potentiation of synaptic currents in vitro. *Nature* **390**, 607–611.

Michelson, H. B. and Wong, R. K. (1994) Synchronization of inhibitory neurones in the guinea-pig hippocampus in vitro. *J. Physiol.* **477** (pt 1), 35–45.

Miesenbock, G. and Kevrekidis, I. G. (2005) Optical imaging and control of genetically designated neurons in functioning circuits. *Annu. Rev. Neurosci.* **28**, 533–563.

Mitterdorfer, J. and Bean, B. P. (2002) Potassium currents during the action potential of hippocampal CA3 neurons. *J. Neurosci.* **22**, 10106–10115.

Moyer, J. R., Jr. and Brown, T. H. (1998) Methods for whole-cell recording from visually preselected neurons of perirhinal cortex in brain slices from young and aging rats. *J. Neurosci. Meth.* **86**, 35–54.

Moyer, J. R., Jr. and Brown, T. H. (2002) Patch-clamp techniques applied to brain slices, in *Advanced Techniques for Patch-Clamp Analysis* (Walz, W., Boulton, A. A., and Baker, G. B., eds.), Humana Press, Totowa, NJ, pp. 135–194.

Moyer, J. R., Jr. and Disterhoft, J. F. (1994) Nimodipine decreases calcium action potentials in an age- and concentration-dependent manner. *Hippocampus* **4**, 11–18.

Moyer, J. R., Jr., Disterhoft, J. F., Black, J. P., and Yeh, J. Z. (1994) Dihydropyridine-sensitive calcium channels in acutely-dissociated hippocampal CA1 neurons. *Neurosci. Res. Comm.* **15**, 39–48.

Moyer, J. R., Jr., McNay, E. C., and Brown, T. H. (2002) Three classes of pyramidal neurons in layer V of rat perirhinal cortex. *Hippocampus* **12**, 218–234.

Moyer, J. R., Jr., Power, J. M., Thompson, L. T., and Disterhoft, J. F. (2000) Increased excitability of aged rabbit CA1 neurons after trace eyeblink conditioning. *J. Neurosci.* **20**, 5476–5482.

Moyer, J. R., Jr., Thompson, L. T., Black, J. P., and Disterhoft, J. F. (1992) Nimodipine increases excitability of rabbit CA1 pyramidal neurons in an age- and concentration-dependent manner. *J. Neurophysiol.* **68**, 2100–2109.

Moyer, J. R., Jr., Thompson, L. T., and Disterhoft, J. F. (1996) Trace eyeblink conditioning increases CA1 excitability in a transient and learning-specific manner. *J. Neurosci.* **16**, 5536–5546.

Narahashi, T. (1974) Chemicals as tools in the study of excitable membranes. *Physiol. Rev.* **54**, 813–889.

Narahashi, T., Moore, J. W., and Scott, W. R. (1964) Tetrodotoxin blockage of sodium conductance increase in lobster giant axons. *J. Gen. Physiol.* **47**, 965–974.

Neher, E. (1992) Correction for liquid junction potentials in patch clamp experiments. *Meth. Enzymol.* **207**, 123–131.

Newman, G. C., Qi, H., Hospod, F. E., and Grundmann, K. (1992) Preservation of hippocampal brain slices with in vivo or in vitro hypothermia. *Brain Res.* **575**, 159–163.

Ng, B. and Barry, P. H. (1995) The measurement of ionic conductivities and mobilities of certain less common organic ions needed for junction potential corrections in electrophysiology. *J. Neurosci. Meth.* **56**, 37–41.

Nisenbaum, E. S., Xu, Z. C., and Wilson, C. J. (1994) Contribution of a slowly inactivating potassium current to the transition to firing of neostriatal spiny projection neurons. *J. Neurophysiol.* **71**, 1174–1189.

Okada, Y., Tanimoto, M., and Yoneda, K. (1988) The protective effect of hypothermia on reversibility in the neuronal function of the hippocampal slice during long lasting anoxia. *Neurosci. Lett.* **84**, 277–282.

Padlubnaya, D. B., Parekh, N. H., and Brown, T. H. (2006) Neurophysiological theory of kamin blocking in fear conditioning. *Behav. Neurosci.* **120**, 337–352.

Paxinos, G. and Watson, C. (1998) *The Rat Brain in Stereotaxic Coordinates*, 3rd ed, Academic Press, San Diego.

Pei, X., Volgushev, M., Vidyasagar, T. R., and Creutzfeldt, O. D. (1991) Whole cell recording and conductance measurements in cat visual cortex in-vivo. *Neuroreport* **2**, 485–488.

Penner, R. (1995) A practical guide to patch clamping, in *Single-Channel Recording*, 2nd ed. (Sakmann, B. and Neher, E., eds.), Plenum Press, New York, pp. 3–30.

Peterlin, Z. A., Kozloski, J., Mao, B.-Q., Tsiola, A., and Yuste, R. (2000) Optical probing of neuronal circuits with calcium indicators. *Proc. Natl. Acad. Sci. (USA)* **97**, 3619–3624.

Power, J. M., Oh, M. M., and Disterhoft, J. F. (2001) Metrifonate decreases $sI_{AHP}$ in CA1 pyramidal neurons in vitro. *J. Neurophysiol.* **85**, 319–322.

Power, J. M., Wu, W. W., Sametsky, E., Oh, M. M., and Disterhoft, J. F. (2002) Age-related enhancement of the slow outward calcium-activated potassium current in hippocampal CA1 pyramidal neurons in vitro. *J. Neurosci.* **22**, 7234–7243.

Reyes, A., Lujan, R., Rozov, A., Burnashev, N., Somogyi, P., and Sakmann, B. (1998) Target-cell-specific facilitation and depression in neocortical circuits. *Nat. Neurosci.* **1**, 279–284.

Rice, M. E. (1999) Use of ascorbate in the preparation and maintenance of brain slices. *Methods* **18**, 144–149.

Rice, M. E. (2000) Ascorbate regulation and its neuroprotective role in the brain. *TINS* **23**, 209–216.

Rice, M. E., Pérez-Pinzón, M. A., and Lee, E. J. K. (1994) Ascorbic acid, but not glutathione, is taken up by brain slices and preserves cell morphology. *J. Neurophysiol.* **71**, 1591–1596.

Rothman, S. M. (1985) The neurotoxicity of excitatory amino acids is produced by passive chloride influx. *J. Neurosci.* **5**, 1483–1489.

Saeed, D., Goetzman, B. W., and Gospe, S. M., Jr. (1993) Brain injury and protective effects of hypothermia using triphenyltetrazolium chloride in neonatal rat. *Pediatric Neurology* **9**, 263–267.

Sah, P., Faber, E. S., Lopez De Armentia, M., and Power, J. (2003) The amygdaloid complex: anatomy and physiology. *Physiol. Rev.* **83**, 803–834.

Sakmann, B. and Neher, E. (1995) Geometric parameters of pipettes and membrane patches, in *Single-Channel Recording*, 2nd ed. (Sakmann, B. and Neher, E., eds.), Plenum Press, New York, pp. 637–650.

Sakmann, B. and Stuart, G. (1995) Patch-pipette recordings from the soma, dendrites, and axon of neurons in brain slices, in *Single-Channel Recording*, 2nd ed. (Sakmann, B. and Neher, E., eds.), Plenum Press, New York, pp. 199–211.

Sather, W., Dieudonne, S., MacDonald, J. F., and Ascher, P. (1992) Activation and desensitization of N-methyl-D-aspartate receptors in nucleated outside-out patches from mouse neurones. *J. Physiol.* **450**, 643–672.

Scharfman, H. E. (1996) Hyperexcitability of entorhinal cortex and hippocampus after application of aminooxyacetic acid (AOAA) to layer III of the rat medial entorhinal cortex in vitro. *J. Neurophysiol.* **76**, 2986–3001.

Schurr, A., Payne, R. S., Miller, J. J., and Rigor, B. M. (1997a) Brain lactate is an obligatory aerobic energy substrate for functional recovery after hypoxia: further in vitro validation. *J. Neurochem.* **69**, 423–426.

Schurr, A., Payne, R. S., Miller, J. J., and Rigor, B. M. (1997b) Brain lactate, not glucose, fuels the recovery of synaptic function from hypoxia upon reoxygenation: an in vitro study. *Brain Res.* **744**, 105–111.

Schwartzkroin, P. A. (1975) Characteristics of CA1 neurons recorded intracellularly in the hippocampal in vitro slice preparation. *Brain Res.* **85**, 423–436.

Schwartzkroin, P. A. and Andersen, P. (1975) Glutamic acid sensitivity of dendrites in hippocampal slices in vitro, in *Advances in Neurology* (Kreutzberg, G. W., ed.), Raven Press, New York, pp. 45–51.

Sharp, P. E. and La Regina, M. C. (eds.) (1998) The Laboratory Rat. *The Laboratory Animal Pocket Reference Series*, (Suckow, M. A., ed.), CRC Press, New York, p. 13.

Sherman-Gold, R. (ed.) (1993) The Axon Guide. Axon Instruments, Foster City, CA.

Sigworth, F. J. (1995) Design of the EPC-9, a computer-controlled patch-clamp amplifier. 1. Hardware. *J. Neurosci. Meth.* **56**, 195–202.

Sigworth, F. J., Affolter, H., and Neher, E. (1995) Design of the EPC-9, a computer-controlled patch-clamp amplifier. 2. Software. *J. Neurosci. Meth.* **56**, 203–215.

Skrede, K. K. and Westgaard, R. H. (1971) The transverse hippocampal slice: a well-defined cortical structure maintained in vitro. *Brain Res.* **35**, 589–593.

Spruston, N., Jaffe, D., Williams, S. H., and Johnston, D. (1993) Voltage- and space-clamp errors associated with the measurement of electrotonically remote synaptic events. *J. Neurophysiol.* **70**, 781–802.

Staley, K. J., Otis, T. S., and Mody, I. (1992) Membrane properties of dentate gyrus granule cells: comparison of sharp microelectrode and whole-cell recordings. *J. Neurophysiol.* **67**, 1346–1358.

Stosiek, C., Garaschuk, O., Holthoff, K., and Konnerth, A. (2003) In vivo two-photon calcium imaging of neuronal networks. *Proc. Natl. Acad. Sci. (USA)* **100**, 7319–7324.

Stuart, G. and Sakmann, B. (1995) Amplification of EPSPs by axosomatic sodium channels in neocortical pyramidal neurons. *Neuron* **15**, 1065–1076.

Stuart, G., Schiller, J., and Sakmann, B. (1997a) Action potential initiation and propagation in rat neocortical pyramidal neurons. *J. Physiol.* **505**, 617–632.

Stuart, G., Spruston, N., Sakmann, B., and Hausser, M. (1997b) Action potential initiation and backpropagation in neurons of the mammalian CNS. *TINS* **20**, 125–131.

Stuart, G. J., Dodt, H.-U., and Sakmann, B. (1993) Patch clamp recordings from the soma and dendrites of neurons in brain slices using infrared video microscopy. *Pflügers Arch.* **423**, 511–518.

Stuart, G. J. and Sakmann, B. (1994) Active propagation of somatic action potentials into neocortical pyramidal cell dendrites. *Nature* **367**, 69–72.

Takahashi, T. (1978) Intracellular recording from visually identified motoneurons in rat spinal cord slices. *Proc. R. Soc. Lond. (B)* **202**, 417–421.

Tarczy-Hornoch, K., Martin, K. A. C., Jack, J. J. B., and Stratford, K. J. (1998) Synaptic interactions between smooth and spiny neurones in layer 4 of cat visual cortex in vitro. *J. Physiol. (Lond.)* **508**, 351–363.

Thibault, O. and Landfield, P. W. (1996) Increase in single L-type calcium channels in hippocampal neurons during aging. *Science* **272**, 1017–1020.

Thompson, L. T., Moyer, J. R., Jr., and Disterhoft, J. F. (1996) Transient changes in excitability of rabbit CA3 neurons with a time-course appropriate to support memory consolidation. *J. Neurophysiol.* **76**, 1836–1849.

Thomson, A. M. and Destexhe, A. (1999) Dual intracellular recordings and computational models of slow inhibitory postsynaptic potentials in rat neocortical and hippocampal slices. *Neuroscience* **92**, 1193–1215.

Thomson, A. M. and Deuchars, J. (1997) Synaptic interactions in neocortical local circuits: dual intracellular recordings *in vitro*. *Cereb. Cortex* **7**, 510–522.

Thomson, A. M. and West, D. C. (1993) Fluctuations in pyramid–pyramid excitatory postsynaptic potentials modified by presynaptic firing pattern and postsynaptic membrane potential using paired intracellular recordings in rat neocortex. *Neuroscience* **54**, 329–346.

Thomson, A. M. and West, D. C. (2003) Presynaptic frequency filtering in the gamma frequency band; dual intracellular recordings in slices of adult rat and cat neocortex. *Cereb. Cortex* **13**, 136–143.

Thomson, A. M., West, D. C., Wang, Y., and Bannister, A. P. (2002) Synaptic connections and small circuits involving excitatory and inhibitory neurons in layers 2–5 of adult rat and cat neocortex: triple intracellular recordings and biocytin labelling in vitro. *Cereb. Cortex* **12**, 936–953.

Tieu, K. H., Keidel, A. L., McGann, J. P., Faulkner, B., and Brown, T. H. (1999) Perirhinal-amygdala circuit-level computational model of temporal encoding in fear conditioning. *Psychobiology* **27**, 1–25.

Tominaga, T., Tominaga, Y., and Ichikawa, M. (2002) Optical imaging of long-lasting depolarization on burst stimulation in area CA1 of rat hippocampal slices. *J. Neurophysiol.* **88**, 1523–1532.

Trussell, L. O. and Jackson, M. B. (1987) Dependence of an adenosine-activated potassium current on a GTP-binding protein in mammalian central neurons. *J. Neurosci.* **7**, 3306–3316.

Tsien, R. Y. (1980) New calcium indicators and buffers with high selectivity against magnesium and protons: design, synthesis, and properties of prototype structures. *Biochemistry* **19**, 2396–2404.

Wachowiak, M. and Cohen, L. B. (2001) Representation of odorants by receptor neuron input to the mouse olfactory bulb. *Neuron* **32**, 723–35.

Watson, G. B., Lopez, O. T., Charles, V. D., and Lanthorn, T. H. (1994) Assessment of long-term effects of transient anoxia on metabolic activity of rat hippocampal slices using triphenyltetrazolium chloride. *J. Neurosci. Meth.* **53**, 203–208.

White, J. A., Sekar, N. S., and Kay, A. R. (1995) Errors in persistent inward currents generated by space-clamp errors: a modeling study. *J. Neurophysiol.* **73**, 2369–2377.

Williams, S. and Johnston, D. (1988) Muscarinic depression of long-term potentiation in CA3 hippocampal neurons. *Science* **242**, 84–87.

Wu, W. W., Chan, C. S., and Disterhoft, J. F. (2004) Slow afterhyperpolarization governs the development of NMDA receptor-dependent afterdepolarization in CA1 pyramidal neurons during synaptic stimulation. *J. Neurophysiol.* **92**, 2346–2356.

Wu, W. W., Oh, M. M., and Disterhoft, J. F. (2002) Age-related biophysical alterations of hippocampal pyramidal neurons: implications for learning and memory. *Ageing Res. Rev.* **1**, 181–207.

Xiang, Z. and Brown, T. H. (1998) Complex synaptic current waveforms evoked in hippocampal pyramidal neurons by extracellular stimulation of dentate gyrus. *J. Neurophysiol.* **79**, 2475–2484.

Xiang, Z., Greenwood, A. C., Kairiss, E. W., and Brown, T. H. (1994) Quantal mechanism of long-term potentiation in hippocampal mossy-fiber synapses. *J. Neurophysiol.* **71**, 2552–2556.

Yamamoto, C. (1975) Recording of electrical activity from microscopically identified neurons of the mammalian brain. *Experientia* **31**, 309–311.

Yamamoto, C. and Chujo, T. (1978) Visualization of central neurons and recording of action potentials. *Exp. Brain Res.* **31**, 299–301.

Yamamoto, C. and McIlwain, H. (1966) Electrical activities in thin sections from the mammalian brain maintained in chemically-defined media in vitro. *J. Neurochem.* **13**, 1333–1343.

Yeckel, M. F., Kapur, A., and Johnston, D. (1999) Multiple forms of LTP in hippocampal CA3 neurons use a common postsynaptic mechanism. *Nat. Neurosci.* **2**, 625–633.

Yuan, W., Burkhalter, A., and Nerbonne, J. M. (2005) Functional role of the fast transient outward K+ current IA in pyramidal neurons in (rat) primary visual cortex. *J. Neurosci.* **25**, 9185–9194.

Yuste, R. and Denk, W. (1995) Dendritic spines as basic functional units of neuronal integration. *Nature* **375**, 682–684.

Zecevic, D. and Antic, S. (1998) Fast optical measurement of membrane potential changes at multiple sites on an individual nerve cell. *Histochem. J.* **30**, 197–216.

Zhang, L., Weiner, J. L., Valiante, T. A., et al. (1994) Whole-cell recording of the $Ca^{2+}$-dependent slow afterhyperpolarization in hippocampal neurones: effects of internally applied anions. *Pflügers Arch.* **426**, 247–253.

Zhu, J. J. and Connors, B. W. (1999) Intrinsic firing patterns and whisker-evoked synaptic responses of neurons in the rat barrel cortex. *J. Neurophysiol.* **81**, 1171–1183.

Zhu, Y. and Zhu, J. J. (2004) Rapid arrival and integration of ascending sensory information in layer 1 nonpyramidal neurons and tuft dendrites of layer 5 pyramidal neurons of the neocortex. *J. Neurosci.* **24**, 1272–1279.

Zilles, K. and Wree, A. (1995) Cortex: Areal and laminar structure, in *The Rat Nervous System*, 2nd ed. (Paxinos, G., ed.), Academic Press, San Diego, pp. 649–685.

Zochowski, M., Wachowiak, M., Falk, C. X., et al. (2000) Imaging membrane potential with voltage-sensitive dyes. *Biol. Bull.* **198**, 1–21.

# 7

# In Vivo Patch-Clamp Technique

*Hidemasa Furue, Toshihiko Katafuchi,
and Megumu Yoshimura*

## 1. Introduction

The whole-cell patch-clamp recording technique (Marty and Neher, 1995) is nowadays a standard method for studying electrophysiological properties of the cellular membranes and synaptic inputs. This technique has been applied mainly to in vitro preparations such as culture cells, dissociated cells, and brain slices, contributing greatly to our understanding of ionic mechanisms of channels/receptors, and synaptic transmission in the neuronal circuits. However, the physiological significance of the neuronal and synaptic activities observed in such in vitro preparations remains to be clarified.

Classically, studies in in vivo neuronal and synaptic activities have been performed by intracellular recordings with sharp microelectrodes. In these studies, it is, however, quite difficult to obtain stable recordings and to analyze synaptic activities in detail, since the impalement of neurons with microelectrodes can cause significant damage, and the high resistance makes it difficult to analyze the synaptic responses under voltage-clamp conditions. Recently, the in vivo whole-cell patch-clamp recording technique has been developed to overcome these problems. The first successful voltage recordings from in vivo mammalian neurons using a patch pipette were reported in 1991 (Pei et al., 1991). Thereafter the in vivo whole-cell voltage recordings were obtained from various areas in mammals including the spinal cord (Furue et al., 1999; Graham et al., 2004a,b; Kato et al., 2006; Kawamata et al., 2006; Light and Willcockson, 1999), thalamus (Brecht and Sakmann, 2002b; Margrie et al., 2002), cerebellum (Chadderton et al., 2004; Loewenstein

From: *Neuromethods, Vol. 38: Patch-Clamp Analysis: Advanced Techniques, Second Edition*
Edited by: W. Walz @ Humana Press Inc., Totowa, NJ

et al., 2005), olfactory bulb (Margrie et al., 2001), visual cortex (Anderson et al., 2000a–c, 2001; Boudreau and Ferster, 2005; Carandini and Ferster, 1997; Chung and Ferster, 1998; Ferster and Jagadeesh, 1992; Jagadeesh et al., 1992, 1993; Pei et al., 1991, 1994; Priebe and Ferster, 2005, 2006; Priebe et al., 2004; Volgushev et al., 1993, 1995, 1996), auditory cortex (Deweese and Zador, 2004; Kaur et al., 2004; Machens et al., 2004; Metherate and Ashe, 1993, 1994; Wehr and Zador, 2003, 2005), somatosensory cortex (Chung et al., 2002; Dittgen et al., 2004; Margrie et al., 2003; Moore and Nelson, 1998; Nelson et al., 1994), barrel cortex (Brecht and Sakmann, 2002a; Brecht et al., 2003; Larkum and Zhu, 2002; Manns et al., 2004; Petersen et al., 2003a,b, 2004; Waters et al., 2003; Zhu and Connors, 1999; Zhu and Zhu, 2004; Zhu et al., 2004), and motor cortex (Brecht et al., 2004). Furthermore, the successful in vivo recordings of excitatory postsynaptic currents (EPSCs) and inhibitory postsynaptic currents (IPSCs) have been performed under voltage-clamp conditions from the spinal cord (Furue et al., 1999; Kawamata et al., 2006; Koga et al., 2005, 2006; Narikawa et al., 2000; Sonohata et al., 2004), cerebellum (Chadderton et al., 2004; Loewenstein et al., 2005), and auditory cortex (Wehr and Zador, 2005). The in vivo whole-cell recording technique has become an extremely powerful tool for studying synaptic details of natural physiological responses. This chapter describes methods for in vivo whole-cell recording technique from spinal dorsal horn and cortex neurons and their synaptic responses to natural sensory stimulation in the rodent.

## 2. Technique for In Vivo Preparations

An essential aspect of in vivo patch-clamp recordings is to maintain the preparations in good condition for the longest time possible. It is important to reduce the bleeding during surgery, to complete the surgery as quickly as possible, and to keep vital signs within a physiological range under a stable anesthesia condition. Surgery for the in vivo preparation is usually completed within 1 or 2 hours, from the time of administration of anesthetics to the start of recording. Recordings can be obtained from the healthy in vivo preparation for more than 12 hours, and stable recordings from single neurons can be acquired for up to 3 hours. Usually two to eight cells can be thoroughly investigated in a day. The following subsections discuss the induction and maintenance of anesthesia, monitoring vital signs from the preparations, and surgical procedures for exposing the surface of the spinal cord and cortex.

## 2.1. Anesthesia

Anesthesia for in vivo experiments is a prerequisite for the relief of pain caused by surgical procedures and physiological examinations. Anesthesia also immobilizes the preparations and minimizes the risk of injuries caused by animal movements (Shibutani, 2000). However, the effects of the anesthetics on the nervous system, tissues, and organs need to be taken into consideration when analyzing the neuronal response in the anesthetized animals (Hara and Harris, 2002; Rudolph and Antkowiak, 2004). Urethane, α-chloralose, barbiturates, ketamine, and their combinations are commonly used as general anesthetics for in vivo whole-cell recordings like those for in vivo extracellular unit recordings or intracellular recordings. Urethane, α-chloralose, or their combination is better to use in the in vivo experiments because the anesthetic effect lasts longer than barbiturates and ketamine. The anesthetic action induced by the first injection of urethane (or α-chloralose) lasts for more than 8 hours. If the long-acting anesthetics are used, intravenous cannulation for supplemental administration of anesthetics may not be needed. It is known that the addition of α-chloralose to urethane suppresses the urethane-induced hyperglycemia. On the other hand, the anesthesia by barbiturates or ketamine lasts less than 1 hour. In our experiments, mice and rats are deeply anesthetized with either urethane or pentobarbital. They are administered intraperitoneally at 1.2 to $1.5\,g\cdot kg^{-1}$ and $40\,mg\cdot kg^{-1}$, respectively. If withdrawal reflexes in response to a noxious stimulus such as pinching the skin are observed during the experiment, urethane is supplementary injected at $0.4\,g\cdot kg^{-1}$ intraperitoneally to maintain areflexia. In the pentobarbital anesthesia, pentobarbital should be continuously supplemented at $4\,mg\,kg^{-1}\cdot h^{-1}$ into the femoral veins. When the first injection of the anesthetics fails to induce enough anesthesia, supplementary administration should be performed carefully so as not to cause an overdose of the anesthetics in the in vivo experiments.

## 2.2. Monitoring Vital Signs and Artificial Ventilation

Under the anesthesia, the left or right carotid artery is cannulated for blood pressure monitoring. The catheter's tip lies just inside the

aortic arch, which is filled with a heparin-saline prior to catheter-
ization. The opposite end of the catheter is connected to a trans-
ducer to monitor the mean arterial systolic blood pressure
(80–140 mm Hg). If the blood pressure falls below 80 mm Hg, further
recordings from the preparations are not made, because low blood
pressure is likely to alter neuronal function in the central and
peripheral nervous systems. After a tracheotomy, a tracheal cannula
is inserted. Animals are artificially ventilated with oxygen-enriched
room air. End tidal $pCO_2$ is monitored throughout the experiment
and maintained at around 3.8% by adjusting the ventilation rate or
tidal volume. Mice and rats are ventilated at a stroke volume of 200
to 400 µL at 150 to 250 breaths/min and at a stroke volume of 2 to
3 mL at 100 to 150 breaths/min, respectively. The reason for the use
of artificial ventilation with the in vivo patch-clamp recordings is
threefold. First, artificial ventilation is essential for maintaining the
normal $pO_2$, $pCO_2$, and acid-base balance of the blood. Second,
ventilation at a small tidal volume at a high respiration rate in
combination with pneumothorax reduces the respiratory-induced
movement of the spinal cord and brain, thereby maintaining the
formation of whole-cell patch and stable recordings. Third, muscle
relaxants can be used under artificial ventilation. When neuronal
nuclei or afferent fibers are stimulated by electrodes, the stimulation sometimes causes a motor activation that also disturbs stable
recordings. In this case, the administration of muscle relaxants
under artificial ventilation is essential for keeping stable recordings. Rectal temperature is also monitored and maintained constant
at 37° to 38°C by means of a circulating hot water blanket beneath
the abdomen. All vital signs should be kept within the normal
physiological range.

### 2.3. Laminectomy and Setting the Spine in a Stereotaxic Apparatus

A laminectomy should be carefully performed in order to avoid
unexpected accidents such as spinal injuries and bleeding caused
by sharp operating tools. The following discussion describes the
laminectomy procedure at the thoracolumbar level, where neurons
that respond to hindlimb stimulation can be obtained. At least three
spinal segments of vertebral arches are removed to allow enough
working space for the recordings. A posterior midline skin incision

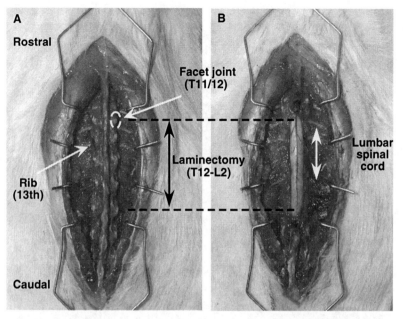

Fig. 1. The lumbar spinal cord exposed by a thoracolumbar laminectomy. (**A**) Thoracolumbar vertebrae exposed under urethane anesthesia.
The 13th rib is a guide for identification of the spinal levels. The laminectomy is performed from the facet joints at the thoracic level (T) of 11 and
12. (**B**) The lumbar enlargement was exposed by the laminectomy.

is performed from the thoracolumbar spinal level. Lidocaine (0.5%)
with 1:100,000 epinephrine is intramuscularly injected into both
sides of the spine in order to decrease bleeding during the surgery.
The thoracolumbar vertebrae are exposed as shown in Fig. 1A.
After the facet joints of T11-T12 are incised, a laminectomy is performed by cutting the pedicles bilaterally from the level of T12 to
L2 using a fine scissors to expose the lumbar enlargement of the
spinal cord at L3-L5 (Fig. 1B). Although micro-drills are also available for this laminectomy, it takes more time to remove the three
spinal levels of vertebral arches. After the surface of the dura matter
is washed with saline, the dura matter is opened. Then the animal
is placed in a stereotaxic apparatus and the lumbar spine is fixed
by the clamp arms (Fig. 2).

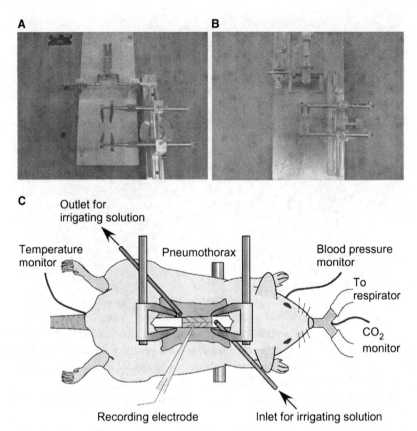

Fig. 2. Stereotaxic apparatus for the spinal cord and schematic illustration of in vivo preparation fixed by the arms. Stereotaxic apparatus (Model ST-SF and STS-SF, Narishige, Japan, USA, UK) for rats (**A**) and mice (**B**). (**C**) Thoracolumbar vertebrae of the in vivo preparation fixed by the two arms after the laminectomy. The animal is intubated and artificially ventilated through a tracheal tube. End-tidal $CO_2$ partial pressure is maintained at around 3.8%. A bilateral pneumothorax is made to reduce spinal cord movement by respiration. Blood pressure is measured through a cannulated carotid artery. Rectal temperature is kept at 37° to 38°C by a circulating hot water blanket beneath the rat abdomen. The surface of the spinal cord is irrigated with Krebs solution.

## 2.4. Setting the Head in a Stereotaxic Apparatus and Craniotomy for Recording from Somatosensory Cortex

An anesthetized rat is placed in a stereotaxic head frame as shown in Fig. 3. The incisor bar is adjusted until the height of

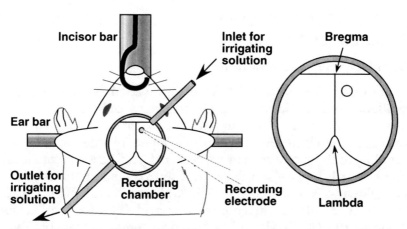

Fig. 3. Schematic illustration of in vivo preparation for recording from cortex neurons. The head of the animal is set in a stereotaxic head frame. A craniotomy is performed to make a small hole opened above the somatosensory cortex according to the stereotaxic coordinates to insert a patch pipette into the cortex. The surface of the cortex is perfused with Krebs solution.

lambda and bregma are almost equal. A midline incision into the scalp's skin is made with a scalpel from between the eyes to the level of the ears, and the edges are retracted. The junctions of the sagittal and transverse sutures (bregma) can be seen now through the bone. After clearing the underlying connective tissue, a plastic cylinder (1.5 cm diameter, 0.5 mm height) is mounted onto the center of the dorsal surface of the cranium and glued on using epoxy resins or dental cement to serve a bath where the surface of the cortex is irrigated with Krebs solution. Under visual guidance through a binocular microscope with 8× to 40× magnifications, a craniotomy is carefully performed using a dental drill. A hole is opened above the right or left somatosensory cortex according to the stereotaxic coordinates. Following removal of the bone, the surface of the dura matter is washed with saline and cut with a sharp needle or fine forceps to allow access for the patch pipette. The dura needs to be opened just before the electrode penetrations to avoid blood clotting. If required, brain pulsations are minimized by making a bilateral pneumothorax and suspending the animal on the thoracic vertebra.

## 3. Whole-Cell Patch-Clamp Recording from Spinal Cord and Cortex Neurons In Vivo

### 3.1. Electrodes and Solutions

The electrodes are pulled from thin-walled borosilicate glass capillaries (outer diameter 1.5 mm) using a puller. They have a resistance of 8 to 15 MΩ. Compared to electrodes designed for in vitro use, the electrodes have a long taper in order to minimize dimpling to the spinal cord or cortical surface. The K-gluconate and cesium patch-pipette solutions are used. The solutions consist of the following compositions (mM): K-gluconate 135, KCl 5, $CaCl_2$ 0.5, $MgCl_2$ 2, ethyleneglycotetraacetic acid (EGTA) 5, adenosine triphosphate (ATP)-Mg 5, N-2-Hydroxyethyl peperazine-N'-2-ethane sulfonic acid (HEPES)-KOH 5, pH 7.2; and $Cs_2SO_4$ 110, tetraethylammonium (TEA)-Cl 5, $CaCl_2$ 0.5, $MgCl_2$ 2, EGTA 5, ATP-Mg 5, HEPES-CsOH 5, pH 7.2. The cesium solution is used in the voltage-clamp mode especially for recording IPSCs, because cesium and TEA stabilize recordings of IPSCs at a holding potential of 0 mV, due to the inhibition of voltage-gated potassium channels. The surface of the spinal cord and cortex is irrigated with 95% $O_2$–5% $CO_2$–equilibrated Krebs solution containing the following composition (in mM): NaCl 117, KCl 3.6, $CaCl_2$ 2.5, $MgCl_2$ 1.2, $NaH_2PO_4$ 1.2, glucose 11, and $NaHCO_3$ 25. The reversal potentials for EPSCs and IPSCs are equal to equilibrium potentials of cations (~0 mV) and chloride ion (~−70 mV), respectively. Therefore, only EPSCs can be recorded since IPSCs are negligible under voltage-clamp conditions at a holding potential of −70 mV. In contrast, IPSCs can be isolated at a holding potential of 0 mV.

### 3.2. Blind Patch-Clamp Recording

The blind patch-clamp recording technique to make a whole-cell configuration from both spinal cord and cortex neurons is similar to the one used in slice preparations (Yang et al., 2001; Yoshimura and Nishi, 1993). When we record from spinal dorsal horn neurons, the dura is cut and reflected under a binocular microscope with 8× to 40× magnifications. Since the dorsal roots enter the spinal cord just above the recording sites, the dorsal root is lifted using a glass hook so that a recording electrode can be advanced into the dorsal horn from the surface of the spinal cord. The superficial dorsal gray matter lateral to the dorsal root entry zone under the Lissauer's tract is discernible because of its translucence. The pia-arachnoid membrane is cut to make a window to penetrate the patch electrode into the spinal cord. Before recording from cortical neurons, the dura is cut and removed. The pia-arachnoid membrane is not necessarily removed in young animals. The surface of the exposed spinal cord and cortex is irrigated with Krebs solution. Using a binocular microscope, the patch pipette is positioned in the irrigating Krebs solution near the surface of the spinal cord or cortex. In the voltage-clamp mode at a holding potential of 0 mV, the evoked current in response to a square voltage step (5 mV, 10 ms, 50 Hz) is continuously monitored and used as an indicator of pipette resistance. Using a syringe connected to the patch pipette, a continuous high pressure (100–200 mm Hg) is applied to the electrode to prevent plugging the electrode tip with tissue debris. Then the patch electrode is advanced into the spinal dorsal horn or cortex from the surface with a micromanipulator. When the tip of the electrode reaches the surface, a DC shift is monitored. At a depth of more than 50 μm from the surface of the spinal cord or cortex, the pressure in the electrode is slightly reduced while the electrode is passed at a steady speed (2–5 μm s⁻¹). A small decrease in the current amplitude (80–90% of initial amplitude) indicates that the tip of the pipette is approaching a cell. At that time the pressure is released and then a slight negative pressure is applied through the attached syringe until a high-resistance seal is established.

After the holding potential is changed to −70 mV, which is near the resting membrane potentials, the patch membrane is ruptured by applying an additional negative pressure to obtain the whole-cell recording configuration. A transient membrane capacitive current appears in response to a square voltage step. Recordings from the neurons are recognized by either of the following two criteria: spontaneous synaptic responses (EPSCs at the holding potential of −70 mV and IPSCs at the holding potential of 0 mV) or action (sodium) currents being initiated by depolarizing pulses. It is difficult to establish a stable recording on occasion. The most common reason might be the poor condition of the seal between the membrane and pipette. It is important to reduce the spinal cord or brain pulsations by making a bilateral pneumothorax with a small amount of ventilation, as well as suspending the animal on the thoracic vertebra as mentioned above. Space-clamp errors under voltage-clamp conditions, especially in large neurons with extensive dendritic trees, should be taken into consideration in order to analyze the neuronal activities and synaptic responses. Capacitive and series resistance are compensated with a patch-clamp amplifier, and these are measured during the experiments. Seals often form on glial cells or on unidentified debris. Once the recording pipette fails to attain a neuron, the tip might be clogged because of a negative pressure being applied. Thus it is better to use a new electrode for the next recording.

### 3.3. Protocols for Natural Sensory Stimulation

Mechanical, thermal, and chemical stimuli can be applied to the skin of the ipsilateral hind limb as shown in Fig. 4. For mechanical stimulation, pinching of the skin folds with toothed forceps and brushing the skin are used as noxious and innocuous mechanical stimuli, respectively. For thermal stimulation, noxious (50°–60°C) and innocuous (38°–40°C) thermal stimuli are applied from a base temperature of about 32°C using a feedback-controlled thermal stimulator with radiant heat lamps or Peltier thermal electrical devices set on the skin. Skin temperature is measured with a thermometer set on the skin. Noxious cold (0°–10°C) stimuli are applied by placing Peltier thermal electrical devices. It takes several seconds to reach the target temperatures from the base temperature. For chemical stimulation, a small amount of chemicals is injected through a short-length 25-gauge needle into the skin.

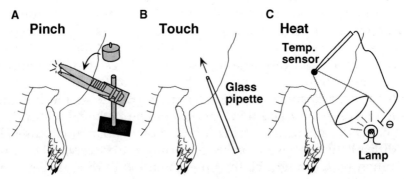

Fig. 4. Protocols for natural cutaneous stimulation. (**A**) Mechanical noxious stimulation applied to the hindlimb by pinching the skin with toothed forceps. The forceps is fixed by a stand. Different weights are applied to vary the pressure of the pinch stimulation. (**B**) Innocuous mechanical stimulation given by an air puff or brushing the skin. (**C**) Thermal stimulation is applied by a feedback-controlled thermal stimulation with a radiant heat lamp.

## 3.4. Recording Synaptic Responses from Spinal Cord Neurons

The success rates for forming high seal resistances of more than 5 GΩ and for achieving the whole-cell configuration from substantia gelatinosa (SG) neurons (lamina II of the spinal cord) of the spinal cord are about 50% and 20% of all trials, respectively. However, once a recording was successful, the resting membrane and firing properties of the mouse and rat SG neurons in the spinal dorsal horn were similar to those of whole-cell patch-clamp recordings from slice preparations. The resting membrane potentials and the input membrane resistances of SG neurons determined by applying a voltage step (5–10 mV) under voltage-clamp conditions were about −65 mV and 600 MΩ, respectively (Furue et al., 1999). Although the input membrane resistance was higher than those recorded with microelectrodes in spinal cord slices (Baba et al., 1994; Yoshimura and Jessell, 1989), the difference might be due to the better seal condition between the electrode and membrane with patch electrodes than with microelectrodes. As shown in Fig. 5, SG neurons recorded from the lumbar spinal level exhibits spontaneous EPSP under current-clamp conditions. When pinch stimulation

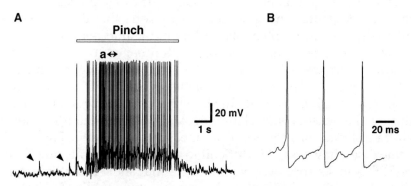

Fig. 5. Excitatory postsynaptic potentials (EPSPs) and action potentials elicited in spinal substantia gelatinosa neurons of mice in vivo in response to pinch stimulation. (**A**) Membrane potential recorded under a zero current-clamp condition. Pinch stimulation applied to the ipsilateral hindlimb evokes EPSPs and results in generation of action potentials during the stimulation. Resting membrane potential is −72 mV. Arrowheads, EPSPs. (**B**) The action potentials marked by "a" in **A** is shown in an expanded time scale.

is applied to the ipsilateral hindlimb, the frequency of EPSPs increases and the stimulation results in generation of repetitive action potentials with a significant overshoot. Stimulating the contralateral hindlimb does not elicit any synaptic responses or firing.

The signal-to-noise ratio of synaptic responses obtained under voltage-clamp conditions in vivo is high enough to analyze synaptic responses in detail with an accurate estimation and almost the same as that of synaptic responses obtained with slice patch-clamp recording techniques. Miniature and spontaneous EPSCs can be isolated at a holding potential of −70 mV under voltage-clamp conditions as shown in Fig. 6. Under voltage-clamp conditions at a holding potential of −70 mV, SG neurons exhibit spontaneous EPSCs (Fig. 6A). The average amplitude and frequency of miniature EPSCs under the presence of tetrodotoxin (TTX) are ~20 pA and ~5 Hz, respectively. These values are not distinct from those obtained from SG neurons—in vitro slice preparations (Kohno et al., 1999). However, the EPSCs with large amplitude (EPSCs indicated by arrowheads in Fig. 6A) are frequently observed in vivo but not in the slice preparations. These EPSCs might be evoked by firing of interneurons or primary afferent neurons innervating the recorded neurons, since the large amplitude of EPSCs are inhibited by TTX (Fig. 6B). The frequency of spontaneous EPSCs with large amplitude tends to be higher when spinal injuries have occurred near the recording site during surgery, suggesting that this increase is likely due to injury potentials evoked in the interneurons.

Under voltage-clamp mode, pinch stimulation to the skin of the hindlimb evokes a barrage of EPSCs at a holding potential of −70 mV (Fig. 7). The frequency of the evoked-EPSCs is different from cell to cell. Even when the baseline level (current) is changed during the sensory stimulation-evoked responses, the frequency and amplitude of EPSCs can be measured as shown Fig. 7C. In some cases, however, it is difficult to obtain the accurate amplitude of EPSCs in sensory-evoked responses that show a multiple summation resulting from high-frequency bursting of EPSCs. Then the total area of the responses is measured as a synaptic strength (see the analysis of IPSCs below; Fig. 8B,C). The rise time, decay time, and half width of the synaptic responses can be also estimated in order to examine the kinetics of the synaptic responses. Using the cesium-containing pipette solution, inhibitory synaptic responses are recorded. When the holding potential is changed from −70 mV to 0 mV, transient inward currents are elicited by the activation of voltage-dependent sodium channels and then immediately inactivated within several seconds. Then IPSCs are isolated as shown in Fig. 8A. When touch stimulation is applied to the skin of the hindlimb, the stimulation elicits a barrage of IPSCs. The synaptic strength can be measured as the total area of the responses (Fig. 8B,C).

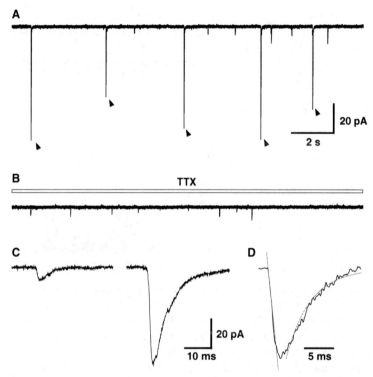

Fig. 6. Spontaneous EPSCs in spinal substantia gelatinosa neurons of mice recorded under voltage-clamp conditions. (**A**) Spontaneous excitatory postsynaptic currents (EPSCs). Arrowheads show EPSCs with large amplitude. (**B**) Miniature EPSCs recorded under the presence of 1 μM tetrodotoxin (TTX) applied to the surface of the spinal cord. EPSCs with large amplitude are suppressed by the TTX application. (**C**) A miniature EPSC (the left trace) and an EPSC with a large amplitude (the right trace) shown in an expanded time scale. (**D**) Averaged traces of 12 miniature EPSCs. The rise-time and decay-time constants for the EPSCs are 0.72 and 4.8 ms, respectively. The rise time is estimated by measuring the 20% to 80% time of the rising phase of the EPSC. The time constant is determined by using a single exponential fit to the decay phase of the EPSC. Holding potential, –70 mV.

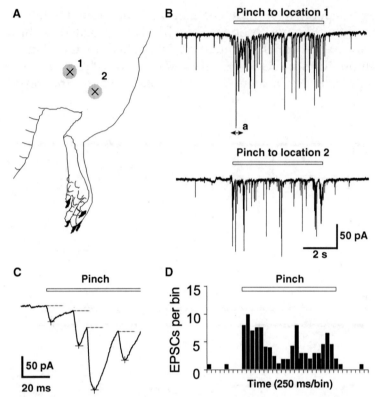

Fig. 7. The EPSCs in spinal substantia gelatinosa neurons of mice evoked by cutaneous pinch stimulation. (**A**) Locations of pinch stimuli applied to the ipsilateral hindlimb. (**B**) The EPSCs elicited by the stimulus applied to each of the locations. The stimulus evokes a barrage of EPSCs. The response to the stimuli applied to location 1 is larger than that to the location 2 stimuli. (**C**) The EPSCs marked by "a" in **B** is shown in an expanded time scale. (**D**) Time histogram of the pinch-evoked EPSCs with large amplitude (>20 pA) in the upper trace of **B** (one trial). Bin, 250 ms. Holding potential, –70 mV.

## 3.5. Recording of Synaptic Responses from Cortex Neurons

The success rates for achieving the tight seal and the whole-cell configuration from sensory cortex neurons are about 10% and below 5%, respectively. They are lower than those from SG neurons of the spinal cord, probably due to the difference in the densities of neurons; spinal SG neurons are more densely packed than cortex neurons. However, stable recordings from cortex neurons can be maintained for up to 7 hours, once the whole-cell mode is established. The EPSCs and IPSCs are recorded at holding potentials of −70 and 0 mV, respectively. As shown in Fig. 9, somatosensory

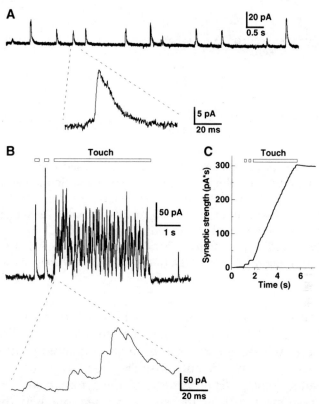

Fig. 8. Inhibitory postsynaptic currents (IPSCs) in spinal substantia gelatinosa neurons of mice evoked by cutaneous touch stimulation. (**A**) Spontaneous IPSCs recorded under the voltage-clamp condition at a holding potential of 0 mV. An IPSC is shown in an expanded time scale in the lower left trace. (**B**) The IPSCs evoked by touch stimulation applied to the ipsilateral hind limb. The IPSCs are shown in an expanded time scale in the lower trace. (**C**) Synaptic strength of the touch-evoked IPSCs.

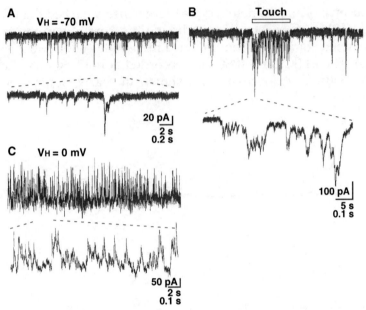

Fig. 9. Synaptic currents recorded from somatosensory cortex neurons of rats in vivo under voltage-clamp conditions. (**A**) Spontaneous EPSCs. The lower trace is shown in an expanded time scale. (**B**) The EPSCs evoked by touch stimulation applied to the contralateral whiskers. The lower trace is shown in an expanded time scale. EPSCs in **A** and **B** are recorded at a holding potential of −70 mV. (**C**) Spontaneous IPSCs recorded at a holding potential of 0 mV. (Dohi et al., 2003, with permission.)

cortex neurons exhibit spontaneous EPSCs (Fig. 9A) and IPSCs (Fig. 9C), and EPSCs evoked by touch stimulation applied to the contralateral whiskers show multiple summations resulting from high bursting of synaptic currents. The frequency and amplitude of spontaneous synaptic currents, and the synaptic strength of the evoked responses can be measured as mentioned in the above section.

## 3.6. Drug Application to the Surface of the Spinal Cord and Cortex

Drugs are administered by intravenous, intraperitoneal, intramuscular, intradermal, and subcutaneous injections in whole animals (Waynforth and Flecknell, 1992). In addition, drugs can be rapidly applied to neurons from the surface of the spinal cord and cortex. Drugs are added to Krebs solution, which is irrigated on the surface of the spinal cord or cortex. When $20\,\mu M$ 6-cyano-7-nitro-quinoxaline-2,3-dione (CNQX) is applied to an SG neuron that is located at a depth of $125\,\mu m$ from the surface of the spinal cord, the miniature EPSCs are dose-dependently and reversibly inhibited (Fig. 10). The half-maximum inhibitory concentration is consistent with that obtained from in slice experiments. Since there was no clear correlation between the amplitude or onset to peak of nor-

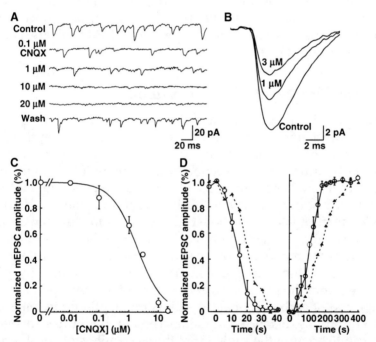

Fig. 10. Drug application from the surface of the rat spinal cord. (**A**) Miniature EPSCs recorded in the absence or presence of various concentrations of CNQX from the same substantia gelatinosa neurons. (**B**) Averaged traces of six miniature EPSCs in control and the actions of CNQX. (**C**) Amplitude of miniature EPSCs under the action of CNQX, relative to those in the control, which is plotted against the CNQX concentration. (**D**) Time courses of the amplitude of miniature EPSCs recorded from neurons at depths of 125 (solid line) and 250 (dotted line) $\mu m$ from the surface of the spinal cord following application (left) and recovery (right) of CNQX. Vertical bars indicate standard error of the mean (SEM).

adrenaline application and the depth of SG neurons that were located within a range from 50 to 150 μm, it is suggested that drugs perfused at the surface of the spinal cord acted equally on SG neurons as those administered by bath application in slice preparations (Sonohata et al., 2004). When recorded from neurons that are located at a depth of 250 μm from the surface, it takes a long time to inhibit the amplitude of the miniature EPSCs following the administration of the same concentration of CNQX. The effects of CNQX to abolish the EPSCs in these neurons are rapid, as is recovery after washout. These findings indicate that the effect is due to a direct action in the spinal cord and is not achieved via blood circulation. In neurons located at a depth of more than 500 μm, however, it is difficult to apply drugs from the surface, since higher concentrations of drugs and longer times (>30 min) are needed until the drugs start to act on the neurons.

### 3.7. Identification of Recorded Neurons

The recorded neurons are identified as being in the lamina of the spinal dorsal horn and cortex based on either the depth of the neurons from the surface of the spinal cord or their morphological features by intrasomatic injection of neurobiotin (0.2% in the electrode solution). When the patch electrodes are penetrated into the spinal cord or cortex, they are advanced at a constant angle using a micromanipulator. A whole-cell configuration is then formed with a cell at a regular depth measured from the point of contact with the cell to the surface of the spinal cord or cortex. This distance should be within the target lamina that is identified by slices obtained from the spinal cord and cortex of the same-aged animals. At the end of each neuronal recording, the depth of the cell is estimated from the distance that the micromanipulator has advanced, taking into account the angle. After the electrophysiological recordings are terminated, the rats are perfused transcardially with 4% paraformaldehyde in 0.1 M phosphate-buffered saline (PBs) (pH 7.4). The spinal cord or cortex is postfixed overnight in the same fixative at 4°C and then cut into 100-μm slices. The tissues are incubated for 12 hours with streptavidin-peroxidase conjugate and washed with Tris buffer. The peroxidase activity was revealed with diaminobenzidine (0.3 mg/mL) in the presence of hydrogen peroxide. Slices are mounted and morphological features of the injected neurons are examined under a light microscope (Fig. 11).

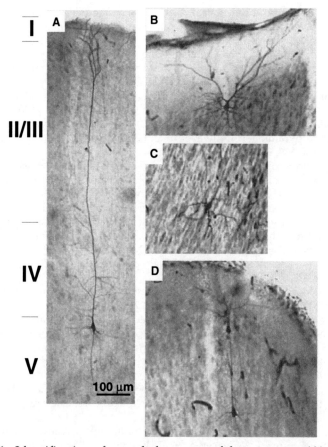

Fig. 11. Identification of recorded neurons of the rat cortex. (**A**) A pyramidal neuron is recorded from lamina V. (**B–D**) Neurons recorded from superficial laminae. After in vivo patch-clamp recordings are terminated, recorded neurons are stained with an intrasomatic injection of neurobiotin (see text). (Dohi et al., 2003, with permission.)

## References

Anderson, J. S., Carandini, M., and Ferster, D. (2000a) Orientation tuning of input conductance, excitation, and inhibition in cat primary visual cortex. *J. Neurophysiol.* **84**, 909–926.

Anderson, J. S., Lampl, I., Gillespie, D. C., and Ferster, D. (2000b) The contribution of noise to contrast invariance of orientation tuning in cat visual cortex. *Science* **290**, 1968–1972.

Anderson, J. S., Lampl, I., Gillespie, D. C., and Ferster, D. (2001) Membrane potential and conductance changes underlying length tuning of cells in cat primary visual cortex. *J. Neurosci.* **21**, 2104–2112.

Anderson, J., Lampl, I., Reichova, I., Carandini, M., and Ferster, D. (2000c) Stimulus dependence of two-state fluctuations of membrane potential in cat visual cortex. *Nat. Neurosci.* **3**, 617–621.

Baba, H., Yoshimura, M., Nishi, S., and Shimoji, K. (1994) Synaptic responses of substantia gelatinosa neurones to dorsal column stimulation in rat spinal cord *in vitro*. *J. Physiol.* **478**, 87–99.

Boudreau, C. E. and Ferster, D. (2005) Short-term depression in thalamocortical synapses of cat primary visual cortex. *J. Neurosci.* **25**, 7179–7190.

Brecht, M., Roth, A., and Sakmann, B. (2003) Dynamic receptive fields of reconstructed pyramidal cells in layers 3 and 2 of rat somatosensory barrel cortex. *J. Physiol.* **553**, 243–265.

Brecht, M. and Sakmann, B. (2002a) Dynamic representation of whisker deflection by synaptic potentials in spiny stellate and pyramidal cells in the barrels and septa of layer 4 rat somatosensory cortex. *J. Physiol.* **543**, 49–70.

Brecht, M. and Sakmann, B. (2002b) Whisker maps of neuronal subclasses of the rat ventral posterior medial thalamus, identified by whole-cell voltage recording and morphological reconstruction. *J. Physiol.* **538**, 495–515.

Brecht, M., Schneider, M., Sakmann, B., and Margrie, T. W. (2004) Whisker movements evoked by stimulation of single pyramidal cells in rat motor cortex. *Nature* **427**, 704–710.

Carandini, M. and Ferster, D. (1997) A tonic hyperpolarization underlying contrast adaptation in cat visual cortex. *Science* **276**, 949–952.

Chadderton, P., Margrie, T. W., and Hausser, M. (2004) Integration of quanta in cerebellar granule cells during sensory processing. *Nature* **428**, 856–860.

Chung, S. and Ferster, D. (1998) Strength and orientation tuning of the thalamic input to simple cells revealed by electrically evoked cortical suppression. *Neuron* **20**, 1177–1189.

Chung, S., Li, X., and Nelson, S. B. (2002) Short-term depression at thalamocortical synapses contributes to rapid adaptation of cortical sensory responses *in vivo*. *Neuron* **34**, 437–446.

Deweese, M. R. and Zador, A. M. (2004) Shared and private variability in the auditory cortex. *J. Neurophysiol.* **92**, 1840–1855.

Dittgen, T., Nimmerjahn, A., Komai, S., et al. (2004) Lentivirus-based genetic manipulations of cortical neurons and their optical and electrophysiological monitoring *in vivo*. *Proc. Natl. Acad. Sci. U S A* **101**, 18206–18211.

Dohi, A., Mizuno, M., Furue, H., and Yoshimura, M. (2003) [Method of in-vivo patch clamp recording from somatosensory cortex in rats.] (Japanese) *Nippon Seirigaku Zasshi* **65**, 322–329.

Ferster, D. and Jagadeesh, B. (1992) EPSP-IPSP interactions in cat visual cortex studied with *in vivo* whole-cell patch recording. *J. Neurosci.* **12**, 1262–1274.

Furue, H., Narikawa, K., Kumamoto, E., and Yoshimura, M. (1999) Responsiveness of rat substantia gelatinosa neurones to mechanical but not thermal stimuli revealed by *in vivo* patch-clamp recording. *J. Physiol.* **521**, 529–535.

Graham, B. A., Brichta, A. M., and Callister, R. J. (2004a) An *in vivo* mouse spinal cord preparation for patch-clamp analysis of nociceptive processing. *J. Neurosci. Methods.* **136**, 221–228.

Graham, B. A., Brichta, A. M., and Callister, R. J. (2004b) *In vivo* responses of mouse superficial dorsal horn neurones to both current injection and peripheral cutaneous stimulation. *J. Physiol.* **561**, 749–763.

Hara, K. and Harris, R. A. (2002) The anesthetic mechanism of urethane: the effects on neurotransmitter-gated ion channels. *Anesth Analg.* **94**, 313–318.

Jagadeesh, B., Gray, C. M., and Ferster, D. (1992) Visually evoked oscillations of membrane potential in cells of cat visual cortex. *Science* **257**, 552–554.

Jagadeesh, B., Wheat, H. S., and Ferster, D. (1993) Linearity of summation of synaptic potentials underlying direction selectivity in simple cells of the cat visual cortex. *Science* **262**, 1901–1904.

Kato, G., Yasaka, T., Katafuchi, T., et al. (2006) Direct GABAergic and glycinergic inhibition of the substantia gelatinosa from the rostral ventromedial medulla revealed by in vivo patch-clamp analysis in rats. *J. Neurosci.* **26**, 1787–1794.

Kaur, S., Lazar, R., and Metherate, R. (2004) Intracortical pathways determine breadth of subthreshold frequency receptive fields in primary auditory cortex. *J. Neurophysiol.* **91**, 2551–2567.

Kawamata, M., Furue, H., Kozuka, Y., Narimatsu, E., Yoshimura, M., and Namiki, A. (2006) Changes in properties of substantia gelatinosa neurons after surgical incision in the rat: in vivo patch-clamp analysis. *Anesthesiology* **104**, 432–440.

Koga, K., Furue, H., Rashid, M. H., Takaki, A., Katafuchi, T., and Yoshimura, M. (2005) Selective activation of primary afferent fibers evaluated by sine-wave electrical stimulation. *Mol. Pain* **1**, 13.

Kohno, T., Kumamoto, E., Higashi, H., Shimoji, K., and Yoshimura, M. (1999) Actions of opioids on excitatory and inhibitory transmission in substantia gelatinosa of adult rat spinal cord. *J. Physiol.* **518**, 803–813.

Larkum, M. E. and Zhu, J. J. (2002) Signaling of layer 1 and whisker-evoked $Ca^{2+}$ and $Na^+$ action potentials in distal and terminal dendrites of rat neocortical pyramidal neurons *in vitro* and *in vivo*. *J. Neurosci.* **22**, 6991–7005.

Light, A. R. and Willcockson, H. H. (1999) Spinal laminae I-II neurons in rat recorded in vivo in whole cell, tight seal configuration: properties and opioid responses. *J. Neurophysiol.* **82**, 3316–3326.

Loewenstein, Y., Mahon, S., Chadderton, P., et al. (2005) Bistability of cerebellar Purkinje cells modulated by sensory stimulation. *Nat. Neurosci.* **8**, 202–211.

Machens, C. K., Wehr, M. S., and Zador, A. M. (2004) Linearity of cortical receptive fields measured with natural sounds. *J. Neurosci.* **24**, 1089–1100.

Manns, I. D., Sakmann, B., and Brecht, M. (2004) Sub- and suprathreshold receptive field properties of pyramidal neurones in layers 5A and 5B of rat somatosensory barrel cortex. *J. Physiol.* **556**, 601–622.

Margrie, T. W., Brecht, M., and Sakmann, B. (2002) In vivo, low-resistance, whole-cell recordings from neurons in the anaesthetized and awake mammalian brain. *Pflugers Arch.* **444**, 491–498.

Margrie, T. W., Meyer, A. H., Caputi, A., et al. (2003) Targeted whole-cell recordings in the mammalian brain in vivo. *Neuron* **39**, 911–918.

Margrie, T. W., Sakmann, B., and Urban, N. N. (2001) Action potential propagation in mitral cell lateral dendrites is decremental and controls recurrent and lateral inhibition in the mammalian olfactory bulb. *Proc. Natl. Acad. Sci. USA* **98**, 319–324.

Marty, A. and Neher, E. (1995) Tight-seal whole-cell recording, in *Single-Channel Recording*, 2nd ed. (Sakmann, B. and Neher, E., eds.), Plenum Press, New York, pp. 31–52.

Metherate, R. and Ashe, J. H. (1993) Ionic flux contributions to neocortical slow waves and nucleus basalis-mediated activation: whole-cell recordings *in vivo*. *J. Neurosci.* **13**, 5312–5323.

Metherate, R. and Ashe, J. H. (1994) Facilitation of an NMDA receptor-mediated EPSP by paired-pulse stimulation in rat neocortex via depression of GABAergic IPSPs. *J. Physiol.* **481**, 331–348.

Moore, C. I. and Nelson, S. B. (1998) Spatio-temporal subthreshold receptive fields in the vibrissa representation of rat primary somatosensory cortex. *J. Neurophysiol.* **80**, 2882–2892.

Narikawa, K., Furue, H., Kumamoto, E., and Yoshimura, M. (2000) *In vivo* patch-clamp analysis of IPSCs evoked in rat substantia gelatinosa neurons by cutaneous mechanical stimulation. *J. Neurophysiol.* **84**, 2171–2174.

Nelson, S., Toth, L., Sheth, B., and Sur, M. (1994) Orientation selectivity of cortical neurons during intracellular blockade of inhibition. *Science* **265**, 774–777.

Pei, X., Vidyasagar, T. R., Volgushev, M., and Creutzfeldt, O. D. (1994) Receptive field analysis and orientation selectivity of postsynaptic potentials of simple cells in cat visual cortex. *J. Neurosci.* **11**, 7130–7140.

Pei, X., Volgushev, M., Vidyasagar, T. R., and Creutzfeldt, O. D. (1991) Whole cell recording and conductance measurements in cat visual cortex in-vivo. *Neuroreport* **2**, 485–488.

Petersen, C. C., Brecht, M., Hahn, T. T., and Sakmann, B. (2004) Synaptic changes in layer 2/3 underlying map plasticity of developing barrel cortex. *Science* **304**, 739–742.

Petersen, C. C., Grinvald, A., and Sakmann, B. (2003b) Spatiotemporal dynamics of sensory responses in layer 2/3 of rat barrel cortex measured in vivo by voltage-sensitive dye imaging combined with whole-cell voltage recordings and neuron reconstructions. *J. Neurosci.* **23**, 1298–1309.

Petersen, C. C., Hahn, T. T., Mehta, M., Grinvald, A., and Sakmann, B. (2003a) Interaction of sensory responses with spontaneous depolarization in layer 2/3 barrel cortex. *Proc. Natl. Acad. Sci. U S A* **100**, 13638–13643.

Priebe, N. J. and Ferster, D. (2005) Direction selectivity of excitation and inhibition in simple cells of the cat primary visual cortex. *Neuron* **45**, 133–145.

Priebe, N. J. and Ferster, D. (2006) Mechanisms underlying cross-orientation suppression in cat visual cortex. *Nat. Neurosci.* **9**, 552–561.

Priebe, N. J., Mechler, F., Carandini, M., and Ferster, D. (2004) The contribution of spike threshold to the dichotomy of cortical simple and complex cells. *Nat. Neurosci.* **7**, 1113–1122.

Rudolph, U. and Antkowiak, B. (2004) Molecular and neuronal substrates for general anaesthetics. *Nat. Rev. Neurosci.* **5**, 709–720.

Shibutani, M. (2000) Anesthesia, artificial ventilation and perfusion fixation, in *The Laboratory Rat* (Krinke, G. J., ed.), Academic Press, San Diego, pp. 511–521.

Sonohata, M., Furue, H., Katafuchi, T., et al. (2004) Actions of noradrenaline on substantia gelatinosa neurones in the rat spinal cord revealed by in vivo patch recording. *J. Physiol.* **555**, 515–526.

Volgushev, M., Pei, X., Vidyasagar, T. R., and Creutzfeldt, O. D. (1993) Excitation and inhibition in orientation selectivity of cat visual cortex neurons revealed by whole-cell recordings in vivo. *Vis. Neurosci.* **10**, 1151–1155.

Volgushev, M., Vidyasagar, T. R., and Pei, X. (1995) Dynamics of the orientation tuning of postsynaptic potentials in the cat visual cortex. *Vis. Neurosci.* **12**, 621–628.

Volgushev, M., Vidyasagar, T. R., and Pei, X. (1996) A linear model fails to predict orientation selectivity of cells in the cat visual cortex. *J. Physiol.* **496**, 597–606.

Waters, J., Larkum, M., Sakmann, B., and Helmchen, F. (2003) Supralinear Ca²⁺ influx into dendritic tufts of layer 2/3 neocortical pyramidal neurons in vitro and *in vivo. J. Neurosci.* **23**, 8558–8567.

Waynforth, H. B. and Flecknell, P. A. (1992) Administration of substances, in *Experimental and Surgical Technique in the Rat*, 2nd ed. Academic Press, San Diego, pp. 1–61.

Wehr, M. and Zador, A. M. (2003) Balanced inhibition underlies tuning and sharpens spike timing in auditory cortex. *Nature* **426**, 442–446.

Wehr, M. and Zador, A. M. (2005) Synaptic mechanisms of forward suppression in rat auditory cortex. *Neuron* **47**, 437–445.

Yang, K., Li, Y., Kumamoto, E., Furue, H., and Yoshimura M. (2001) Voltage-clamp recordings of postsynaptic currents in substantia gelatinosa neurons in vitro and its applications to assess synaptic transmission. *Brain Res. Brain Res. Protoc.* **7**, 235–240.

Yoshimura, M. and Jessell, T. M. (1989) Membrane properties of rat substantia gelatinosa neurons in vitro. *J. Neurophysiol.* **62**, 109–118.

Yoshimura, M. and Nishi, S. (1993) Blind patch-clamp recordings from substantia gelatinosa neurons in adult rat spinal cord slices: pharmacological properties of synaptic currents. *Neuroscience* **53**, 519–526.

Zhu, J. J. and Connors, B. W. (1999) Intrinsic firing patterns and whisker-evoked synaptic responses of neurons in the rat barrel cortex. *J. Neurophysiol.* **81**, 1171–1183.

Zhu, Y., Stornetta, R. L., and Zhu, J. J. (2004) Chandelier cells control excessive cortical excitation: characteristics of whisker-evoked synaptic responses of layer 2/3 nonpyramidal and pyramidal neurons. *J. Neurosci.* **24**, 5101–5108.

Zhu, Y. and Zhu, J. J. (2004) Rapid arrival and integration of ascending sensory information in layer 1 nonpyramidal neurons and tuft dendrites of layer 5 pyramidal neurons of the neocortex. *J. Neurosci.* **24**, 1272–1279.

# 8

# Perforated Patch-Clamp Techniques

## Constantine Sarantopoulos

## 1. Introduction

The patch-clamp technique, introduced by Erwin Neher and Bert Sakmann (Neher and Sakmann, 1976) in the mid-1970s facilitated recordings of currents through single-ion channels in living cells, advancing the ability to study the membrane function of excitable cells, such as neurons. Ionic currents through single membrane channels were recorded after sealing a micropipette tip onto a clean membrane patch. In the early 1980s, the technique evolved to allow recording of currents conveyed through membrane ionic channels in the whole cell: tight seals with high resistance in the G$\Omega$ range (gigaseals) were possible after drawing up a small patch of the membrane into the pipette tip by suction generated by application of gentle negative pressure (Hamill et al., 1981). Subsequent rupture of the patch with negative pressure or a voltage pulse allowed establishment of low-resistance electrical and physical access to the interior of the cell, thus allowing the recording of the entire population of ionic currents on the cellular membrane. This constitutes the basis of the whole-cell patch-clamp technique.

Within the following years, the whole-cell patch-clamp technique became a standard tool for electrophysiological recordings of ionic currents through membrane channels on neurons and other excitable cells, as well as for studying the effects of drugs on these currents (Hamill et al., 1981; Liem et al., 1995). Nevertheless, two significant drawbacks with regard to providing stable and reliable current recordings over time (Korn and Horn, 1989) limit the value of the technique, and these drawbacks are discussed in the following subsections.

From: *Neuromethods, Vol. 38: Patch-Clamp Analysis: Advanced Techniques, Second Edition*
Edited by: W. Walz @ Humana Press Inc., Totowa, NJ

## 1.1. Washout of Intracellular Constituents and Current Rundown

First, an irreversible loss of constituents from the interior of the cell develops over time after the rupture of the membrane patch, as a result of the dialysis with the internal solution in the recording pipette. The volume of the solution in the pipette is disproportionately larger compared to the volume of the intracellular compartment, so, after equilibration of the two compartments, dialysis and egress of the contents of the intracellular space is unavoidable (Fig. 1). This washout seems to be the result of diffusion of molecules of 100 to 500 molecular weight from the cytoplasm to the pipette (Horn and Marty, 1988). This loss impairs significantly the function of the ion channels in the membrane, which depend highly on intracellular factors, such as adenosine triphosphate (ATP) and phosphorylating molecules, and manifests as a time-dependent loss of the current through them (Armstrong and Eckert, 1987; Becq, 1996; Chad et al., 1987; Doroshenko et al., 1982; Forscher and Oxford, 1985; Horn and Korn, 1992; Hoshi, 1995; Li et al., 1992; Tang and Hoshi, 1999). Although many ionic currents undergo rundown as a result of the underlying dialysis, washout, and impairment of the channel function after whole-cell patch clamping, the progressive decrement of the current ($I_{Ca}$) through voltage-gated calcium channels (VGCC) is the most characteristic example. Sustained $I_{Ca}$ through functional VGCC requires channel phosphorylation, while during standard whole-cell patch-clamp, energy substrates, and substances involved in the phosphorylation process washout (Chad et al., 1987; Korn and Horn, 1989).

Current rundown is thus a major limitation of the whole-cell technique: typically, it results in significant current loss over a time course of minutes. It is obvious that, in the context of pharmacological or physiological studies, rundown introduces an important confounding factor, because results need to be interpreted taking into account the diminishing channel function.

In addition to current rundown, dialysis may alter the channel gating properties (Bezanilla et al., 1986), voltage-dependence of activation (Fernandez et al., 1984), and inactivation of some $K^+$ channels (Ciorba et al., 1997; Kupper et al., 1995; Marom et al., 1993; Tang and Hoshi, 1999).

Basic mechanisms implicated in the rundown process include the following:

Fig. 1. Conventional whole-cell patch technique after membrane rupture. Electrical and physical access to the interior of the cell is obtained by forming a tight gigaseal patch around a portion of cell membrane drawn into the pipette tip by suction, and subsequent rupture of the membrane. The technique results in dialysis by the disproportionately larger internal solution contained in the pipette, dialysis and wash-out of intracellular components, and current rundown.

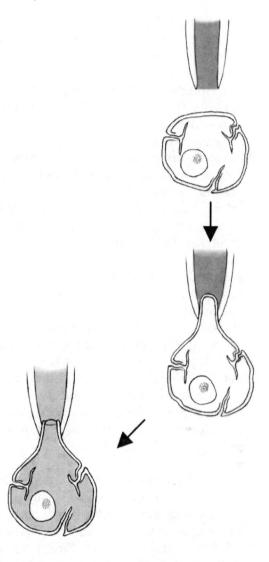

1. Dephosphorylation: a mechanism that is reversible by application of agents that promote cyclic adenosine monophosphate (cAMP)-dependent protein phosphorylation, such as the cAMP-dependent protein kinase A (PKA) (Armstrong and Eckert, 1987; Becq, 1996; Chad et al., 1987). Application of ATP together with PKA has been also shown to abolish rundown of the hyperpolarization-activated KAT1 channel in *Xenopus* oocytes, while alkaline phosphatase accelerates it, indicating the dependence

on channel dephosphorylation (Tang and Hoshi, 1999). Dephosphorylation also promotes rundown of VGCC in the whole-cell configuration, and is minimized by addition of agents such as 2 to 5 mM magnesium (Mg) ATP or an intracellular creatine phosphate/creatine phosphokinase nucleotide regeneration system (Forscher and Oxford, 1985), or the cAMP-dependent protein kinase, alone (Chad et al., 1987), or together with MgATP (Armstrong and Eckert, 1987; Tang and Hoshi, 1999).

2. Proteolysis: proteolytic degradation of channel protein. This is the second major mechanism that contributes to $I_{Ca}$ and other current rundown (Belles et al., 1988a,b). It can be inhibited by the calcium-dependent protease inhibitor leupeptin and is an irreversible process (Chad et al., 1987; Eckert et al., 1986). Proteolysis of channel protein by the membrane-associated proteinase calpaine is implicated in the irreversible rundown of $Ca^{2+}$ channels in *Helix* neurons (Chad and Eckert, 1986). Agents such as cytochalasin and desoxyribonuclease I, which depolymerize actin filaments, accelerate, whereas phalloidin (an actin filament stabilizer), phosphatidylinositol biphosphate, or $PIP_2$ (F-actin-serving protein inhibitors) have been shown to inhibit the rundown of the cardiac ATP-sensitive $K^+$ current (Furukawa et al., 1996; Hilgemann and Ball, 1996) and N-methyl-D-aspartate receptor current in cultured hippocampal neurons (Rosenmund and Westbrook, 1993), indicating that the channel rundown also depends on disruption of the actin cytoskeleton (Tang and Hoshi, 1999).

Both mechanisms are $Ca^{2+}$ dependent (Belles et al. 1988a,b; Forscher and Oxford, 1985). Calcium current rundown accelerates by increasing intracellular $[Ca^{2+}]$, and slows down by increasing ethyleneglycotetraacetic acid (EGTA) concentration (Belles et al., 1988; Byerly and Hagiwara, 1982; Forscher and Oxford, 1985; Horn and Korn, 1992; Korn and Horn, 1989).

## 1.2. Disruption of Intracellular Calcium Buffering Mechanisms

The disruption and rundown of energy-dependent, intracellular $Ca^{2+}$ buffering mechanisms is the second major drawback, due to dialysis, washout, and loss of ATP and diffusable $Ca^{2+}$ binding proteins (Korn and Horn, 1989; Korn and Weight, 1987). Energy-dependent buffering mechanisms, such as the plasma membrane $Na^+/Ca^{2+}$ exchange apparatus, run down due to the loss of ATP or other supporting intracellular factors. Furthermore, a ruptured patch results in significant diffusion of $Ca^{2+}$ out of the cell, or altered buffering by the introduction of exogenous $Ca^{2+}$ buffers, such as EGTA, in the internal pipette solution (Korn and Horn, 1989; Korn and Weight, 1987).

### *1.3. Prevention of Rundown*

Attempts to impede rundown include the addition of exogenous substances that can suppress proteolysis and support channel phosphorylation in the pipette solution. Various substances, such as ATP, EGTA, cAMP, cAMP-dependent protein kinase, creatine phosphate, creatine kinase, and protease inhibitors, added in the intracellular recording solution, have been shown to inhibit rundown (Armstrong and Eckert, 1987; Belles et al., 1988a,b; Chad and Eckert, 1986; Forscher and Oxford, 1985; Horn and Korn, 1992). Nevertheless this strategy is far from optimal; these exogenous substances may also produce significant alterations in a variety of intracellular molecules and signaling systems that may subsequently affect, in a variable and unpredictable way, the channels under investigation (Korn and Horn, 1989).

## 2. The Perforated Patch-Clamp Technique

The perforated patch-clamp technique is a modification of the whole-cell patch-clamp configuration, not dependent on the addition of exogenous active substances, that has significantly facilitated whole-cell studies of $I_{Ca}$ and other currents (Akaike and Harata, 1994; Horn and Marty, 1988; Rae et al., 1991; Sarantopoulos et al., 2004). The basic concept of the perforated patch technique is the establishment of access to the interior of the cell not by rupture, but through the formation of multiple tiny pores or channels in the membrane under the patch. These pores should ideally be impermeable to large molecules necessary for the proper function of the cells under investigation, but allow passage of certain ions. The perforated patch-clamp technique can thus significantly prevent washout and retard rundown, without loss of endogenous bioactive molecules from the intracellular compartment, and without the need for introduction of exogenous substances. It has also been suggested that perforated patch techniques minimize proteolysis, and maintain a relatively intact status in the endogenous $Ca^{2+}$ buffering mechanisms (Horn and Korn, 1992). Thus, electrophysiological recordings may yield a more realistic model of cellular events.

Practically, the perforated patch mode is achieved from the cell-attached, whole-cell configuration by adding various pore-forming compounds in the internal pipette solution. The molecules of these compounds insert into the membrane sealed in the pipette tip, and subsequently produce perforation, by forming multiple tiny channels that allow ions and small molecules to cross the patch. The size and biophysical properties of the channels are such that larger molecules and cell organelles remain trapped within the cell. Thus washout of essential molecules is avoided and current rundown is retarded (Fig. 2).

## 2.1. The Slow Whole-Cell Patch Configuration

Lindau and Fernandez (1986) were the first to develop a novel patch-clamp configuration in which the patch was not ruptured by suction but instead perforated so as to create multiple small pores, thus preventing the diffusion of large molecules out of rat perito-neal mast cells. Patches were permeabilized by exposing their extracellular side to ATP (400 μM), which was added into the pipette solution. Progressive permeabilization of the membrane was asso-ciated with the parallel emergence of a capacitive current transient in response to a voltage pulse, from which the membrane capaci-tance ($C_m$), conductance, and access resistance ($R_a$) could be derived. With this technique, $R_a$ values of 200 to 5000 MΩ were achieved (much higher than those in the standard whole-cell configuration), while membrane resistance (20–100 GΩ) and membrane capaci-tance values (5–10 pF) were in closer agreement with the corre-sponding values after the standard technique. Since the time constant (equal to $C_m \times R_a$) of the capacitive current transient was much slower after permeabilization (10 ms) compared to that after patch disruption (100 μs), Lindau and Fernandez named the tech-nique "slow whole-cell" as opposed to "fast whole-cell" after conventional patch disruption. The high-access resistance values (200–5000 MΩ), as well as the need for ATP receptors on the cell membrane (limiting the applicability of the technique to a limited type of cells) were significant shortcomings of the technique.

In this "slow whole-cell" configuration, mast cells still responded to secretagogue external stimulation, and the capacitance as well

Fig. 2. The perforated patch-clamp technique. The perforated patch mode is achieved from the cell-attached, whole-cell configuration by adding a pore-forming compound in the internal pipette solution. The molecules of the perforating agent diffuse and insert into the membrane sealed in the pipette tip, and subsequently produce perforation, by forming multiple tiny channels that allow only small ions and small molecules to cross the patch. The size and biophysical properties of the channels are such that larger molecules and cell organelles remain trapped within the cell. Thus washout of essential molecules is avoided and current rundown is retarded.

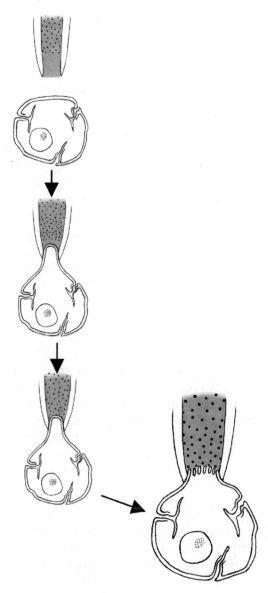

as the conductance of the cell membrane could be recorded during degranulation of the mast cells. Antigenic stimulation in these cells produced a transient increase in membrane conductance, followed by a threefold increase in cell capacitance indicative of exocytotic degranulation. In contrast, cells dialyzed with pipette solution in the standard whole-cell mode failed to respond to secretagogue stimulation, as a presumable consequence of dialysis with the pipette solution and subsequent washout of essential cytoplasmic components.

## 2.2. Perforated Patch Recording

The classical perforated patch technique was described by Horn and Marty (1988), who established electrical continuity to the interior of the cell by employing nystatin as an agent to perforate the cell membrane. Horn thus named the technique "perforated patch recording." The technique is based on the property of polyene antibiotic substances to form small pores through lipid-containing cell membranes. In addition to nystatin, amphotericin B is included in these antibiotics, the molecules of which insert into the cell membrane, polymerize, and generate channels that are permeable selectively to monovalent ions, but not to larger molecules or divalent cations such as $Ca^{2+}$ or $Mg^{2+}$. Permeabilization, by formation of these channels spanning the membrane, eventually reduces $R_a$ in a way sufficient to facilitate satisfactory current recordings, under stable conditions, with significantly retarded current rundown. One of the advantages of these agents is that they remain localized within the membrane under the patch; in addition, they neither pass into the cell (as to produce significant intracellular signaling alterations) nor diffuse laterally within the membrane beyond the margins of the patch under the pipette. Furthermore, the channels formed by the polyene antibiotics are not voltage-dependent, thus preventing uncontrolled variations of the access conductance to the cell when subjected to stimulation by voltage steps.

### 2.2.1. Polyene Antibiotics

Polyene antibiotics are produced by bacteria of the genus *Streptomyces*. Both nystatin and amphotericin B create channels in cell membranes, selectively permeable to small ions and molecules. This action was traditionally considered to be exclusive in membranes that contain sterol, and, in particular, ergosterol (Akaike and Harata, 1994), but more recent studies have shown the presence of channels in sterol-free membranes, as well (Cotero et al., 1998; Gruszecki, 2003; Venegas et al., 2003).

Both nystatin and amphotericin B contain a large polyhydroxylic lactone ring of 23 to 37 atoms with four to seven conjugated double bonds. They contain various hydrophilic groups on one side of the lactone ring, directed to the outside of the molecule. The other side of the ring is hydrophobic. This configuration renders an amphipathic molecule, ideal for the formation of aqueous pores traversing through lipid membranes. Each single pore is formed by eight to 10 monomeric polyene molecules arranged circumferentially, thus forming a barrel-shaped structure, whose hydrophilic interior allows the passage of small ions and small molecules (such as urea and glucose). The hydrophobic exterior and the hydrophilic interior of the barrel are established by appropriate orientation of the lactone ring, while the barrels are stabilized by sterol molecules (de Kruijff et al., 1974a,b; Finkelstein and Holz, 1973). The length of these pores has been estimated to be 2.8 nm, while the long dimension of the lactone ring is 2.2 nm.

Biophysical characteristics and permeability selectivity are different when polyene antibiotics are applied to one side, versus to both sides, of the membrane, reflective of two different types of channels formed with different biophysical properties. A single-barrel structure formed by the circumferential arrangement of the eight to 10 channel-forming monomers develops only after application of nystatin or amphotericin B to one side of the membrane. In contrast, two-sided application leads to a different type of channel resulting from two barrels connected end to end with hydrogen bonds (Kleinberg and Finkelstein, 1984). The permeability of one-sided–produced barrels is selective to univalent cations, while that of barrels after two-sided application is selective to univalent anions. In both cases, channels are impermeable to divalent ions (Akaike and Harata, 1994; Marty and Finkelstein, 1975). Nystatin and amphotericin B pores are also permeable to water and small hydrophilic nonelectrolytic molecules. Nystatin pores allow the passage of molecules with solute radius less than 0.4 nm, such as urea (0.18 nm) or glycerol (0.31 nm), but not glucose (0.42 nm) or sucrose (0.52 nm). Thus, the hydrophilic lumen of the channel has been estimated to have an effective radius of approximately 0.4 nm (Akaike and Harata, 1994; Hladky and Haydon, 1970; Holz and Finkelstein, 1970; Horn and Marty, 1988). The amphotericin B channel has been estimated to be larger in diameter, 0.8 nm (de Kruijff and Demel, 1974; de Kruijff et al., 1974a,b), thus allowing the passage of glucose and possibly sucrose (to which the nystatin pores are impermeable). Considering effective diameter, the pores of both nystatin and amphotericin B are also impermeable to large molecules such as glucose-6-phosphate dehydrogenase, nicotinamide adenine dinucleotide phosphate (NADP⁺), and glucose-6-phosphate. Prevention of rundown has been attributed to mechanical blockage of the passage of supporting intracellular compounds with a molecular weight higher than 200, or prevention of passage of the molecules, responsible for the washout, on the basis of the electrical charge.

Regarding ionic permeability after application to one side of the membrane, channels formed by either nystatin or amphotericin B are impermeable to multivalent ions, such as $Mg^{2+}$, $Ca^{2+}$, and $SO4^{2-}$, but allow passage of monovalent ions such as $Li^+$, $Na^+$, $K^+$, $Cs^+$, and $Cl^-$. The order of the size of permeable hydrated cations is proportional to their permeability rank: $Rb^+ > K^+ > Na^+ > Li^+$ (de Kruijff et al., 1974a,b). After one-sided membrane application, the permeability of channels to anions, such as $Cl^-$, is 10 times smaller than the permeability to cations (Marty and Finkelstein, 1975). Nevertheless, permeability to $Cl^-$ is not negligible during perforated patch recordings with polyene antibiotics.

The conductance of single polyene channels is of the order of only a few picoSiemens (pS), even at very high concentrations of the permeating ionophore (Ermishkin et al., 1976; Kasumov et al., 1979; Kleinberg and Finkelstein, 1984). Two-sided application results in channels of conductance of 2 pS for nystatin and 3.5 to 6 pS for amphotericin B in 2M KCl solution (Akaike and Harata, 1994; Ermishkin et al., 1976; Kasumov et al., 1979). Conductance of either nystatin or amphotericin B channels is not voltage-dependent over a wide voltage range, while no time-dependent changes have been reported. In contrast, some temperature dependence has been reported after one-sided application (Akaike and Harata, 1994). The lifetime of the open channels is in the order of minutes (Zeidler et al., 1995).

### 2.2.1.1. NYSTATIN AS A PERMEABILIZING AGENT IN PERFORATED PATCH RECORDINGS

Horn and Marty (1988) were the first to describe the nystatin-based perforated patch technique as a means to prevent rundown of muscarinic response to acetylcholine in rat lacrimal gland cells. Advancing the concept of the "slow whole-cell" patch technique of Lindau and Fernandez, they used nystatin as a pore-forming agent to decrease $R_a$ through a cell-attached patch, up to levels comparable to those in the standard whole-cell technique, thus obtaining high-quality voltage-clamp recordings.

Pipettes filled throughout with solution containing 20 to 100 μg/mL nystatin produced adequate perforation, but at the expense of reduced success rates of obtaining GΩ seal. In contrast, prefilling the tip of the pipette with nystatin-free solution, and then back-filling the remaining of the pipette shaft with nystatin-containing solution, improves the success rate; slow diffusion of nystatin molecules to the membrane interface over time avoids initial interference with the seal formation. However, this requires precise control of the filling distance of the nystatin-free solution within the pipette tip. Thus, the third technique, adopted by Horn and Marty, included obtaining seals in the absence of nystatin, and then introducing it (50–100 μg/mL in internal solution) through a fine polyethylene tube placed inside the patch pipette. The progress of permeabilization was assessed every 3 to 5 minutes by monitoring the $R_a$ of the cell, and recordings were initiated at an $R_a$ value of <50 MΩ. With nystatin, Horn and Marty (1988) and Korn and Horn (1989) reported $R_a$ values ranging between 18 and 50 MΩ, much lower than those reported previously by Lindau and Fernandez (100–5000 MΩ). Kurachi et al. (1990) reported even lower $R_a$ values (7–15 MΩ). With nystatin, low $R_a$ values are obtained within several minutes, and subsequently remain stable or slowly decrease for the next 1 to 2 hours. The technique facilitated stability in $Ca^+$-activated $K^+$ and $Cl^-$ current recordings, in response to muscarinic activation, for up to 1 hour, indicating that the underlying signaling cascade of inositol 1,4,5-triphosphate ($IP_3$) formation and $Ca^{2+}$ release from intracellular stores remained intact (Horn and Korn, 1992; Horn and Marty, 1988). In contrast, in standard whole-cell configuration, current rundown limited recording times to less than 5 minutes.

### 2.2.1.2. AMPHOTERICIN B AS A PERMEABILIZING AGENT IN PERFORATED PATCH RECORDINGS

Following the reports of successful use of nystatin, Rae et al. (1991) substituted amphotericin B for the former agent, thus achieving $R_a$ values even lower than those achieved by nystatin. In several cellular preparations Rae et al. obtained $R_a$ values of 3 to 10 MΩ, which are close to those after standard whole-cell configuration. They chose amphotericin B because of its larger pores and its single-channel conductance being twice as much as that of nystatin (Rae et al., 1991), translating apparently into much lower $R_a$ with the former ionophore. Furthermore, taking into consideration the concentration–conductance relationships of both agents, they assumed that amphotericin B might partition more fully at low concentrations compared to nystatin.

Rae et al. used a final concentration of amphotericin B 240 µg/mL, obtained by dilution into the pipette-filling solution aliquots of a 60 mg/mL stock solution in dimethylsulfoxide (DMSO). They used Sylgard-coated electrodes (Dow Corning, Midland, MI), the tips of which were filled (up to 300–500 µm) by brief (less than 1 second) immersion into an amphotericin B–free solution, and the remaining pipette shaft back-filled with the drug containing solution. Fresh pipette filling solutions were used (less than 1 hour old), since the drug potency cannot be maintained for more than 2 to 3 hours.

After an initial period of continuing decrease that lasted for about 30 minutes after the gigaseal formation, $R_a$ values remained stable for between 2 and 3 hours, without the drifts or upward creep that is observed in the standard whole-cell configuration. Other reported advantages of the technique include more effective whole-cell capacitance and series resistance compensation in permeabilized patch recordings, and the capacity to use pipettes from high-lead glasses that are easier to pull at the 580° to 630°C thermal range, and to fire-polish, than others. The lack of permeability of polyene antibiotic channels to multivalent ions that may be released from these glasses prevents any pertinent unwanted effects to the recorded currents.

### 2.2.1.3. Problems with Polyene Antibiotics in Perforated Patch Recordings

The use of polyene antibiotics as ionophores, though extensive, is not without certain problems, and has led to the development of techniques based on other agents. These problems include the following:

1. Instability of solutions: Polyenes are hydrophobic and require special techniques for solubilization in aqueous solutions. They are also sensitive to light and heat, and in solutions, they lose potency over time.
2. Interference with patch formation: The presence of polyene antibiotics at the orifice of the pipette tip may prevent gigaseal formation, so ideally these agents should be prevented from

reaching the interface of the patched membrane—by prefilling the tip of the pipettes with polyene-free solution—until after a gigaseal has been established.

3. Voltage and volume alterations secondary to chloride shifts: Permeability to univalent $Cl^-$ but not to larger anions leads to $Cl^-$ fluxes between the internal solution of the pipette and the intracellular compartment, in case the ratio of permeant/impermeant anions between the two compartments is not the same. The consequence of this $Cl^-$ redistribution between the interior of the cell and the pipette will be the development of a Donnan potential across the membrane, and alterations in the volume of the cell (due to water shifts). As a result of the Donnan potential, a voltage difference may develop between the interior of the pipette and the interior of the cell, proportional to the degree of the ionic redistribution. Voltage errors resulting from these Donnan forces can be as high as 10 mV. Rae et al. (1991) have also observed significant volume changes to either direction (shrinkage or swelling) in the cells they examined using perforating patches with amphotericin B, but both problems can be attenuated by choosing appropriate filling solutions. To prevent possible volume changes from osmotic discrepancies, Horn and Marty (1988) proposed the replacement of some of the $Cl^-$ in the pipette solution with a large nonpermeable anion such as $SO_4^{2-}$.

4. Failure: Perforation with polyene antibiotics also fails to produce low $R_a$ in some cells, in an unpredictable fashion, resulting in occasional rejection of cells on which the technique does not work.

### 2.2.2. The Gramicidin Perforated Patch Recording Configuration

Chloride redistribution is a limitation of polyene antibiotic channels (Akaike and Harata, 1994; Horn and Marty, 1988). Intracellular $Cl^-$ concentration ($[Cl^-]_i$) would tend to equilibrate with that in the pipette (Horn and Marty, 1988; Kleinberg and Finkelstein, 1984; Kyrozis and Reichling, 1995), introducing a source of bias at electrophysiological studies investigating phenomena wherein $[Cl^-]_i$ is an important determinant. This disruption of the physiological electrochemical gradient for $Cl^-$ may be detrimental for several signaling mechanisms studied with current or voltage-clamp techniques, such as current responses to γ-aminobutyric acid (GABA) ergic and glycinergic transmission in neurons. In addition, several other current responses, transport mechanisms, and transmitter release processes that depend on $[Cl]_i$ may be altered (Le Foll et al., 1998).

In an attempt to overcome this problem, gramicidin D was adopted next as a perforating agent (Ebihara et al., 1995; Kyrozis and Reichling, 1995). After partitioning in the cell membrane, gramicidin D results in formation of functional channels, which are selectively permeable to monovalent cations and small uncharged molecules, but not to $Cl^-$ (Hladky and Haydon, 1984), thus allowing recordings with the $[Cl^-]_i$ unaltered.

Commercially available gramicidin is a mixture of three similar antibiotics, named gramicidin A, B, and C. These are linear hydrophobic decapentapeptides in an alternating L-D sequence (Hladky and Haydon, 1984), obtained from the soil bacterial species *Bacillus brevis*, and called collectively gramicidin D. In lipid bilayer membranes, gramicidin D molecules are tied up as head-to-head dimmers and fold as right-handed helices, spanning the bilayer. The outer surface of the gramicidin dimmer is hydrophobic, thus interacting with the core of the lipid bilayer, while the hydrophilic lumen of the helix forms the channel that conveys the ions. Reversible formation of dimmers provides the basis of the opening and closing of the pore gating (Miloshevsky and Jordan, 2004).

Gramicidin D channels are voltage-insensitive, selectively permeable to univalent cations, with conductance with selectivity sequence $Rb^+ > K^+ > Cs^+ > Na^+ > Li^+$ (Andreoli et al., 1967; Mueller and Rudin, 1967; Tosteson et al., 1968). Divalent cations such as $Ca^{2+}$ do not pass, and the permeability to $Cl^-$ is negligible; thus the development of Donnan equilibrium between the pipette and cytosol is also negligible.

The gramicidin-perforated patch recording configuration technique was first described by Ebihara et al. (1995), who studied the effects of GABA on rat brain dissociated neurons and in slices, without disturbing the intracellular $[Cl^-]$. They achieved mean minimal $R_a$ of $16.1 \pm 1.2\,M\Omega$ within 40 minutes after establishing the $G\Omega$ seal, by using pipette-filling solutions containing $100\,\mu g/mL$ of gramicidin D. Intracellular chloride activities were not affected by different pipette solutions, indicating that $[Cl^-]_i$ was well maintained. Gramicidin channels were not permeable to $Cl^-$, even in the presence of large amounts of $[Cl^-]$ (150 mM) in the recording pipette solution. Intracellular chloride activities were maintained unperturbed in the holding potential range between −100 and −40 mV, indicating no voltage dependency within this range.

In a technique described on cultured dorsal horn neurons, Kyrozis and Reichling (1995) used unfiltered pipette-filling solutions containing 2 to 50 μg/mL gramicidin D, prepared from 2 mg/ml stocks in DMSO. As with the polyene antibiotics, prefilling the tip of the pipette with gramicidin-free solution increases the success rates of gigaseal formation, particularly when the higher concentrations are used (20–50 μg/mL). However, concentrations <5 μg/mL do not interfere with seal formation and do not require prefilling. Concentrations >20 μg/mL are complicated by frequent spontaneous rupturing of the membrane patch, something infrequent with concentrations <5 μg/mL. Perforation with 5 μg/mL of gramicidin D was consistently successful in yielding final $R_a$ values of 25 to 50 MΩ, slightly higher than the values with 100 μg/mL of amphotericin B (20–40 MΩ) used in comparison. Fresh solutions need to be made every 1–2 h, as the agent in solutions loses its pore-forming activity over time.

Perforated patches with gramicidin were stable, with no evidence of leakage beyond the borders of the membrane under patch, and no deterioration of the input resistance or the resting potential. Stability of [Cl⁻]ᵢ was confirmed by measuring the glycine or muscimol-induced voltage responses. The lack of interference of low concentrations of gramicidin with gigaseal formation in contrast to polyene antibiotics (Horn and Marty, 1988; Rae et al., 1991) is a significant advantage that may allow application of positive hydrostatic pressure in the pipette. This may be particularly helpful in patch recordings from tissue slices in order to prevent blockage of the tip by debris and processes from other cells, and even if it pushes the perforating agent to the tip, this may not interfere with seal formation.

## 2.2.3. β-Escin as Perforating Agent

Considerable failure rates, difficulties with handling polyene antibiotics and gramicidin in aqueous solutions due to their lipophilicity, loss of potency of prepared solution after 2 hours, long waiting times for perforation, interference with gigaseal formation requiring special attention to pipette geometry and filling technique, and possible cost issues have driven the search for other ionophore agents suitable for membrane permeabilization in perforated patch recordings (Akaike and Harata, 1994; Rae et al., 1991).

The saponin β-escin is an acceptable permeabilizing agent for perforated patch recordings in cardiac and neuronal cells. Saponins are glycosides composed of a steroid or triterpenoid nucleus with one or more carbohydrate branches. The original saponin compound has the property of partitioning with cholesterol molecules in the cell membrane and forming pores 8 nm in diameter (Bangham et al., 1962; Launikonis and Stephenson, 1999) without significantly affecting the structure and function of intracellular structures. β-escin is derived from the horse-chestnut plant (*Aesculus hippocastanum*). It has been used as a chemical "skinning" agent, in order to permeabilize smooth muscles for physiological and pharmacological studies, while preserving any receptor-coupled cellular signal transduction mechanisms (Akagi et al., 1999; Iizuka et al., 1997).

Fan and Palade (1998) first used β-escin as an ionophore in whole-cell perforated patch recordings in cardiac ventricular myocytes, in comparison with nystatin, amphotericin-B, and gramicidin. Success rates were highest with β-escin (65%), compared to 50% with nystatin, 33% with gramicidin, and 22% with amphotericin. They used β-escin concentrations of 30 to 50 µM in the pipette solution, either prepared daily from powder or diluted from a stock solution 50 mM in water, which could be stored for up to one week at $-20°C$ (a significant advantage secondary to hydrophilicity of β-escin). β-escin produced maximal permeabilization, as assessed by $R_a$ monitoring, at 10 to 35 minutes, and retarded rundown.

We also studied β-escin (50 µM) in perforated patch recordings from primary afferent rat neurons, in comparison with amphotericin B (240 µg/mL), and the standard, whole-cell technique (Sarantopoulos et al., 2004). β-escin, compared to amphotericin B, was easier to prepare and more stable in stock and internal pipette aqueous solutions, and did not interfere considerably with seal formation, which reflects the hydrophilic nature of β-escin. Rates of success were 59% with β-escin vs. 27% with amphotericin B. Time to minimal $R_a$ was much faster with β-escin (7 ± 9 minutes) compared to amphotericin (44 ± 14 minutes). This may also be due to hydrophilicity, allowing faster diffusion, or to a more potent mechanism of action by interacting with cholesterol in the lipid bilayer (Bangham et al., 1962). Minimal $R_a$ values were 6.8 ± 1.9 MΩ immediately after standard breakthrough, 7.9 ± 3.5 MΩ with amphotericin B, and 6.5 ± 1.6 MΩ after perforation with β-escin. $R_a$ after

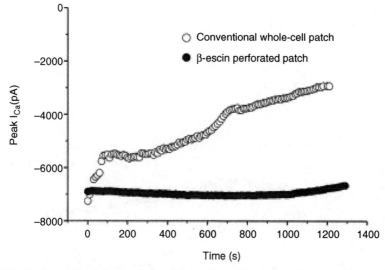

Fig. 3. β-escin perforated patch considerably delays $I_{Ca}$ rundown, compared to the standard technique. Graph shows time-course recordings of $I_{Ca}$ in a neuron perforated with β-escin in comparison to a neuron patched using the standard whole-cell technique. (Sarantopoulos et al., 2004, with permission from Elsevier.)

β-escin was significantly lower compared to amphotericin B, but not compared with the standard technique.

Perforation with β-escin significantly delayed the $I_{Ca}$ rundown, compared to either the standard whole-cell technique, or to the amphotericin B perforated patch (Figs. 3 and 4). The percentage loss of peak $I_{Ca}$ per minute, as an estimate of $I_{Ca}$ rundown, was significantly less after β-escin (0.05 ± 1.8%), vs. either amphotericin (1.8 ± 2%) or the standard patch (4.5 ± 2%) (Fig. 4). β-escin perforated patches were stable over 15 to 60 minutes and facilitated recordings of the effects of drugs, such as gabapentin or dexamethasone, on $I_{Ca}$, while the configuration of the current waveforms was well preserved (Fig. 4). Using $I_{Ca}$ reversal potential shifts, we estimated that β-escin pores were of a diameter large enough to allow free diffusion of $Ca^{2+}$ with an approximate internuclear bond length of 395 pm (Sutton, 1965). Concentrations of up to 50 μM in permeabilized skeletal myocytes resulted in passage of 17 to 18 kDa proteins (Konishi and Watanabe, 1995), while Fan and Palade (1998) showed the passage of fluorescent markers with a molecular weight up to 10 kDa through β-escin

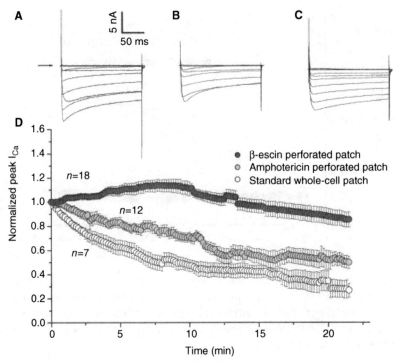

Fig. 4. Perforation with β-escin significantly delays the $I_{Ca}$ rundown, compared to either standard whole-cell technique, or amphotericin B perforated patch. Actual calcium current traces are obtained after standard whole-cell recordings (**A**), perforation with amphotericin (**B**), or β-escin (**C**). Traces were obtained by 200 ms square test pulses from a holding potential −90 to +30 mV in successive 10-mV increments. Normalized for capacitance and peak "starting" current $I_{Ca}$ traces, obtained from time-course recordings and averaged for each technique, show the course of the rundown over time for each patch technique (**D**). (Sarantopoulos et al., 2004, with permission from Elsevier.)

pores in patched cardiac myocytes. Nevertheless, despite the relatively large size of the pores, β-escin effectively delays $I_{Ca}$ rundown by providing a more effective diffusion barrier to macromolecules, which support channel function (Fan and Palade, 1998). Another explanation may be the better preservation of intracellular ATP concentrations by β-escin (Arnould et al., 1996), although this is controversial (Kerrick and Hoar, 1994; Konishi and Watanabe, 1995). Furthermore, despite the size of the pores, β-escin does not disrupt the cellular signaling apparatus (Akagi et al., 1999;

Iizuka et al., 1997). The capacity of cells to respond to a variety of ligands, second messengers, and bioactive chemicals after perforation with β-escin remains intact, thus providing an additional advantage of the technique (Akagi et al., 1999; Bremerich et al., 1997a,b; Iizuka et al., 1997). A few leaky neurons were a shortcoming of the technique. A leak of more than 10% of the peak $I_{Ca}$ was observed in 25% of neurons perforated with β-escin vs. 20% after amphotericin and 12% after standard technique.

# 3. Specific Protocols and Applications

## 3.1. Preparation of Solutions

### 3.1.1. Preparation of Solutions of Polyene Antibiotics

Nystatin and amphotericin B are lipophilic compounds with poor water solubility. For this reason, they require preparation of stock solutions in methanol or DMSO, but these solutions cannot be maintained more than a few weeks after freezing. Pipette-filling solutions, prepared by diluting aliquots of stock solutions into the internal pipette solutions, may also require special handling for complete solubilization, and should be prepared fresh every 1 to 2 hours because they lose potency after a few hours.

Stock aliquots and pipette filling solutions can be prepared following the following protocol (Axon Guide, 1993):

1. Place 3 to 6 mg of nystatin or amphotericin B powder into each of 40 to 50 microcentrifuge tubes. These stocks are stable and do not result in any loss of potency of the drug after prolonged storage in the freezer (−20°C).
2. On the day of the recording, prepare a fresh stock of solution by solubilizing the stocked powder by pipetting 100 µL of DMSO into the microcentrifuge tube, followed by subsequent vortex mixing for about 5 seconds. This provides a stock concentration of 0.03 to 0.06 mg/µL. Alternatively, stock solutions can be made up in microcentrifuge tubes by prediluting nystatin or amphotericin B as 3 to 6 mg per 100 µL stock solution in DMSO, keeping them in the freezer (−20°C) for up to a week, and thawing on the day of recordings (Sarantopoulos et al., 2004; Williams et al., 1999).
3. Aliquots from the freshly prepared stock solution can be added directly into the internal pipette solution by micropipetting; 4 µL/mL of the 0.03 mg/µL stock yields a final

concentration of 120 µg/mL, while 4 µL/mL of the 0.06 mg/µL stock solution results in 240 µg/mL. [With regard to amphotericin B, in the classical description of the technique, Rae et al. (1991) prepared stock solutions by adding 100 µL of DMSO to 6 mg of powder, and vortexing for 5 seconds. Subsequent addition by micropipetting of 4 µL/mL into the pipette solution yielded a final concentration of amphotericin B equal to 240 µg/mL.]

4. Initially, the added aliquot of the stock solution is triturated into the internal pipette-filling solution with the transfer *Pasteur* pipette. This usually results into a light-yellowish turbid solution. Maximum solubilization can be achieved by ultrasonication for 1 minute or vortex mixing for a few seconds. The saturating concentration of the polyene antibiotics in the final solution has been reported to be 120 µg/mL, and drug in excess may result in a turbid solution, precipitation, or adherence to the wall of the plastic tube or syringe. We avoid filtering these solutions, since some filters may retain or inactivate the perforating agent, although as per Rae et al. (1991) amphotericin B may be safely filtered through 0.2-µm Nalgene syringe filters No. 190–2020 (Nalgene® Labware, Rochester, NY), without affecting the efficacy of the ionophore.

5. The potency of either perforating agent in solution decreases within approximately 2 to 3 hours at room temperature, so preparation of fresh internal pipette-filling solutions every hour is strongly recommended. Solutions also need to be protected from light, in stocks as well as in the syringes used for filling the pipettes.

Alternatively, Akaike and Harata (1994) have reported the following protocol for preparing nystatin stock solutions in methanol (after Horn and Marty, 1988):

1. Stock solutions of nystatin can be prepared by dissolving nystatin in methanol at a concentration 10 mg/mL, usually 50 to 100 mg in 5 to 1Z0 mL, respectively.

2. Solubilization can be facilitated by (a) ultrasonication for 10 to 20 seconds, followed by (b) lowering the pH of the solution to less than 2.0 by adding 1 N HCl under constant stirring (this will result in changing the solution from turbid to clear), and (c) elevating pH back to 7.2 by 1 N KOH.

3. Divide stock solution into 1- to 1.5-mL aliquots, pipetted into microcentrifuge tubes.
4. Store in the freezer at –20°C for up to 4 weeks.
5. On the day of the recording, let the aliquot thaw, and subsequently vortex to resuspend in case a precipitate is visible at the bottom of the tube.
6. Dissolve the stock solution into the internal pipette solution by adding 15 to 40 µL/mL just before use. This will produce a final concentration of 150 to 400 µg/mL, respectively. The final methanol concentration in the pipette solution is estimated to be 1.5% to 4%. It has been reported that these concentrations have no effect on neurons (Akaike and Harata, 1994).
7. Because the saturating concentration of nystatin into the internal pipette solution is 120 µg/mL, higher concentrations may produce a turbid solution in the pipette, requiring brief vortex treatment prior to use.
8. Another approach is based on adding agents into the internal solution that enhance the solubility of nystatin, such as Pluronic F-127, a dispersing agent that may facilitate homogeneous dispersion of nystatin (Lucero and Pappone, 1990). Pluronic acid is daily dissolved at 37°C in DMSO (25 mg/mL), and, at concentrations of 0.05% into the pipette-filling solution, has been shown to shorten the perforation time, most likely by decreasing the time needed for membrane partitioning and channel formation (Axon Guide, 1993).
9. Yawo and Chuhma (1993) have added fluorescein in order to improve the success rate of the perforated patch with nystatin. Their stock solutions, containing 5 mg nystatin and 20 mg fluorescein sodium, dissolved in 1 mL methanol, may be kept at 4°C for several days. Immediately before use, 50 µL of the stock solution is dried with nitrogen gas stream in a polyethylene tube; 1 mL of the pipette solution is subsequently added into the tube and vortexed, yielding 250 µg/mL of nystatin and 1 mg/mL of fluorescein. Handling is carried out under yellow fluorescent light to prevent bleaching of fluorescein. The solution is then filtered through a 0.45-µm cellulose acetate filter, while tip prefilling is not necessary, and positive pressure may be applied at the pipette. The success rate of perforated patch recordings improved significantly in several preparations, including brain slices—something attributed to improved solubilization of nystatin in aqueous solution facilitated by the bipolar fluorescein molecules.

### 3.1.2. Preparation of Solutions for β-Escin

The hydrophilicity of β-escin confers a significant benefit over polyene antibiotics since it is readily soluble and stable in aqueous solutions. These can be maintained as stock solutions in the freezer, thawed on the day of the experiment, and reconstituted into the pipette-filling solutions much more easily than polyene antibiotics.

1. We prepare 25 mM stock solution by diluting 0.0275 mg of β-escin powder (Sigma E-1378, St. Louis, MO) into 1 mL of double-deionized water. This can be facilitated by brief vortex mixing.
2. The use of dark glass flasks or tubes, wrapped with aluminum foil or other cover to protect the solution from light is important, because β-escin is light-sensitive.
3. Stocks can be stored for up to 2 weeks at −20°C in a freezer, and thawed on the day of experiments.
4. Dilute 4 μL of this stock solution into 2 mL of the internal pipette solution. This provides a final concentration of 50 μM. (We usually use 3-mL filling syringes, wrapped with aluminum foil to protect from light). In case lower concentrations are needed, they can be accomplished by diluting the solution into 3 or 4 mL of pipette solution.
5. Mix up by vortexing for 1 minute.
6. Solution may turn slightly foamy (a characteristic saponin property), but foam microbubbles disappear after a few minutes.
7. In contrast to other agents, β-escin solutions remain stable for several hours after mixing in the filling syringe, at room temperature, with no evidence of precipitate formation, but protection from light is necessary. No filtering is necessary.

### 3.1.3. Preparation of Gramicidin D Solutions

1. Daily stocks are prepared by dissolving gramicidin D in DMSO at a concentration of 2 to 50 mg/mL (Kyrozis and Reichling, 1995).
2. Another approach is to prepare stock solutions by dissolving gramicidin in methanol, at 10 mg/mL (Ebihara, et al., 1995; Le Foll et al., 1998).
3. The final solution in the pipette should contain a concentration of gramicidin 2 to 100 μg/mL. Higher concentrations

(100 μg/mL) (Ebihara et al., 1995; Le Foll et al., 1998) may accelerate perforation speed (Ebihara et al., 1995), from 40 minutes (when 20 μg/mL are used) to 20 minutes with the former (D'Ambrosio, 2002). However, concentrations >20 μg/mL may result in frequent spontaneous rupturing of the membrane patch, something rare with concentrations of 5 μg/mL or less (Kyrozis and Reichling, 1995).

4. Kyrozis and Reichling (1995) conducted most recordings using final concentrations of 5 μg/mL prepared from stock solutions of 2 mg/mL DMSO. At these concentrations no special measures to enhance solubilization were needed, but for higher concentrations, after adding the gramicidin-containing stock solution into the pipette-filling solution, sonication for 15 seconds was necessary. Another advantage of the low concentrations (5 μg/mL or less) is the lack of interference with seal formation so that prefilling of the tip of pipettes with gramicidin-free solution is not necessary. In contrast, higher concentrations (>20 μg/mL) always necessitate tip prefilling. Investigators should be guided about which concentrations are more appropriate for their cellular preparations based on initial pilot experiments.

5. Prepare fresh gramicidin stock solution and pipette-filling solution every 1 to 2 hours, since gramicidin in solution has been reported to lose its activity fast (within a few hours) (Kyrozis and Reichling, 1995). Gramicidin is not light-sensitive.

## 3.2. Pulling and Filling Patch Pipettes

### 3.2.1. Pulling Suitable Pipettes

We fabricate patch pipettes from borosilicate glass capillaries, type 7052 (Garner Glass Company, Claremont, CA) (1.0 mm inner diameter, 1.5 mm outer diameter, 75 mm length), by pulling in a horizontal P-97 Flaming/Brown micropipette puller (Sutter Instruments, Novato, CA). Dimensions and pipette resistance vary depending on the type and size of the neuron under examination, but most commonly our pipettes have resistance ranging between 2 and 5 MΩ, with the solutions we use. In perforated patches, pipettes tapered in a short and steep fashion to a tip with a blunt geometrical shape confer a significant advantage over those with a long-tapered, straight, and sharp tip (Rae et al., 1991). The former

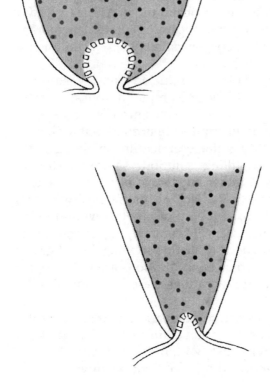

Fig. 5. Optimal tip geometry for perforated patch recordings. In perforated patches, pipettes tapered in a short and steep fashion to a tip with a blunt geometrical shape (top) confer a significant advantage over those with a long-tapered, straight and sharp tip (bottom), by accommodating a larger membrane portion, and exposure of a larger membrane surface area to the pore-forming effects of the perforating agents. Thus, the parallel opening of more pores decreases electrical resistance.

geometrical shape contributes significantly to low $R_a$, because it accommodates the insertion of a larger membrane portion by suction, and exposure of a larger membrane surface area to the pore-forming effects of the perforating agents. Thus, the parallel opening of more pores decreases electrical resistance (Fig. 5).

### 3.2.2. Considerations for Filling the Pipettes

Because most perforating agents may prevent successful GΩ seal formation, optimal filling of the pipettes is highly essential and subject to the following considerations.

Perforating agents establish electrical continuity with the interior of the cell by reducing $R_a$ to appropriate levels in a process that involves three major steps (Rae et al., 1991): (1) diffusion: the molecules of the ionophore move from the interface of the filling solution to the membrane surface; (2) partition: the ionophore molecules partition into the membrane; and (3) aggregation: the ionophore molecules combine with each other into multimers that form the channels. The time course of the perforating process (as indicated by the time needed to obtain the lowest possible $R_a$ at steady state) depends primarily on the diffusion kinetics (Rae et al., 1991). The time course of the perforation depends also on the process of evaporation of the solution from the tip of the pipette. This results in bulk flow of solution out of the pipette tip. This initially is the ionophore-free solution filling up the tip of the pipette, but this may enhance the onward movement of the ionophore-containing solution back-filling the shaft of the pipette. For electrodes filled to a distance of 500 μm, evaporation drives approximately 20 μm/10 sec (Rae et al., 1991), but this movement slows down after back-filling of the pipette (secondary to elimination of evaporation from the back surface of the ionophore-free solution) and ceases completely after the immersion of the tip of the back-filled pipette into the recording bath solution.

Nystatin, amphotericin B, and high concentrations of gramicidin D solutions at the tip of the patching pipette may impair GΩ seal formation. This can be prevented by prefilling the tip of the pipette, by brief immersion into an agent-free solution. Although β-escin is much safer in this context compared to polyene antibiotics (Fan and Palade, 1998; Sarantopoulos et al., 2004), we also apply the same technique with the former agent, too. The reason is that we have observed that concentrations of β-escin higher than 25 μM at the tip of the pipette may result in premature rupture of the membrane patch under the pipette tip, at the moment of application of negative pressure to establish the gigaseal. Ideally, we aim to fill the initial 500 μm of the pipette tip with a perforating agent-free internal solution, and then back-fill the pipette shaft with the perforating agent-containing solution (Fig. 6).

The height up to which the tip of the pipette is prefilled with the ionophore-free solution determines the time needed for the ionophore molecules to diffuse to the membrane interface and start partitioning. So the time needed for perforation depends highly on that filling distance. Suboptimal filling may result in faster diffusion and buildup of a high ionophore concentration at the tip, before actual gigaseal formation, introducing the risk of failure. Excessive filling may lead to delayed or incomplete diffusion with subsequent prolongation of the perforation time course of up to more than an hour.

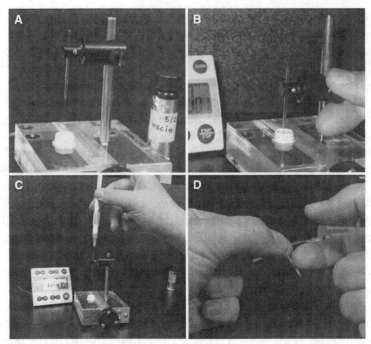

Fig. 6. Procedure for filling the pipette. A special device is used (**A**), with a moving arm in which the pipette is being held. We briefly immerse the tip of the pipette into the agent-free solution by dipping for 1 second (**B**), and subsequently back-fill the shaft (**C**) with the agent-containing solution. Solutions are protected from light by wrapping them in aluminum foil. Gentle tapping and rolling removes small bubbles of trapped air (**D**).

The filling height depends on factors that include the geometrical shape and dimensions of the tip, the type of the glass, and the time of the immersion. Rae et al. (1991) studied the length of filling as a function of the tip immersion time for several tip shapes, and reported similar general behavior for all pipette types tested: The filling distance increases with the tip diameter, and also increases exponentially with the immersion time. Brief immersion of pipette tips with a diameter of 1 to 2.5 μm, for 1 second or less, results in filling heights from 200 to 550 μm. Usually filling to about 500 μm provides enough time to establish a gigaseal before the agent reaches the tip by a combination of diffusion and evaporative bulk flow. In this case successful gigaseal formation is followed by relatively rapid partitioning and fast low $R_a$ establishment. With regard to the type of the glass, borosilicate glass requires shorter immersion times (1–5 seconds), while longer times (30–60 seconds) may be needed for lead-containing glass, which is more hydrophobic.

So filling needs to be optimized for each particular experimental setting by taking into consideration the type of the glass, the shape and dimensions of the pipette tip, the cell type, and the type and concentration of the perforating agent.

We briefly immerse the tip of the pipette, being held in the holding arm of a special pipette-filling device (Fig. 6A) for 1 second (Fig. 6B). We suggest that this step should be guided by control measurements under a microscope scale to confirm that filling of the tip after immersion has occurred up to no more than 500 μm. Based on these control measurements, as well as on initial observations to record the times needed for effective breakthrough, the immersion times needed for filling the pipette tips can be more precisely adjusted. The remainder of the pipette shaft is back-filled with a special pipette filler with the internal pipette solution containing the agent. We use 3-mL syringes, containing 2 mL of filling solution with the ionophore. We back-fill the pipette shafts via a nonmetallic syringe needle (Microfil MF28G, Microfil™, World Precision Instruments, Inc., Sarasota, FL) (Fig. 6C).

A special technique allows easy de-bubbling of small air bubbles trapped in between the two-solution interface, without mixing of the agent-containing solution with the agent-free solution at the tip. We hold the pipette in between the thumb and index finger of the left hand, rolling it gently by rotating movements, while we gently tap the pipette shaft with the index finger of the right hand a few times. The pipette is kept in a 45-degree (tip down) angle, which facilitates removal of the bubbles (Fig. 5D). Tapping does not result in mixing up of the two solutions and does not eliminate the free region (Marty and Neher, 1995).

Since polyene antibiotics are heat- and light-sensitive, their filling solutions should be protected from light and in a container with ice in between experiments. It has also been recommended to turn the light microscope off whenever it is not required during recordings. β-escin is light-sensitive, too, but gramicidin is not.

### 3.2.3. Cells Under Examination

We have found by experience that whenever acutely isolated neurons are examined, the degree of enzymatic digestion may be a significant factor affecting the success rate. Thus, we prefer to patch well-digested neurons, with a clean, smooth surface, free of any visible anomalies or undigested structures. Pilot experiments may guide the investigator using acutely dissociated cells, as to the optimum degree of enzymatic digestion, since underdigested cells may fail to perforate, while overdigested cells may get leaky.

### 3.3. Procedure for Patching

After appropriate filling and de-bubbling, we place the pipette on the pipette holder, without applying any positive pressure. This avoids premature diffusion of the agent-containing solution to the pipette tip. Then the pipette is dipped into the external solution in the recording chamber and guided toward the cell to be investigated. The approach toward the cell needs to be fast, before diffusion to the tip occurs; otherwise, the prefilling distance needs to be adjusted. The $G\Omega$ seal can be established by a transient, brief, and gentle application of negative pressure. This can be delivered by mouth suction via a connecting tube, or the plunger of a small syringe. We monitor the approach of the pipette toward the membrane of the cell on a video screen. A slight increase of the pipette resistance, when the tip of the pipette is superimposed on the membrane of the cell, indicates that the tip has touched the membrane. At that time, we apply minimal negative pressure to the interior of the pipette by gentle mouth suction, to establish the tight seal. This should ideally exceed $1\,G\Omega$. After a seal is obtained, holding potential is set at levels $-40$ to $-60\,mV$, and we wait for the perforating agent to act without applying any further negative pressure.

The progress of perforation can be monitored with repeated 10-mV hyperpolarizing steps from the holding potential, revealing the appearance of capacitive current transients with progressively increasing amplitudes, and progressively faster time constants (shorter decaying phase of the transient) of the falling phase, indicating that, as more and more channels open up and $R_a$ decreases, an increasing fraction of the cell capacity gets charged during the brief pulse. Usually $R_a$ decreases and the apparent capacitance transient increases over a time period, until stabilizing to a steady-state level. In most cells, pipettes pulled into a proper shape and filled with the proper solutions usually produce minimal $R_a$ that stabilizes by 20 to 40 minutes. After $R_a$ reaches the steady-state minimal value, it remains stable for 2 to 3 hours, and then begins to increase, presumably secondary to loss of activity of the perforating agent. Stability of $R_a$, in addition to retardation of rundown, is a major advantage of the technique.

After reaching this steady state, recordings may commence after appropriate compensation of membrane capacitance and series resistance. The perforated patch technique offers another significant advantage by allowing more effective compensation to take place.

Failure of perforation gets manifested as lack of appearance of the characteristic capacitive current transient, after a waiting period of a few minutes. Reasons include poor enzymatic digestion of the cells, in case they are acutely dissociated, inadequately low concentration of the perforating agent in the pipette, or excessive filling distance with perforating agent-free solution within the tip. In the latter case the molecules of the agent will take a longer time to diffuse through the prefilled tip up to the cell membrane. If this is the case, a longer waiting time may allow partitioning and effective, although delayed, perforation. Otherwise, the filling distance needs to be adjusted by shorter immersion times. Possible inactivation of the perforating agent is another reason for failure. This may occur as a result of using old stock solutions, or reconstituted filling solutions that remain for more than 1 or 2 hours at room temperature, or exposure to light. Premature rupture of the membrane prior to effective GΩ seal formation or failure to establish a GΩ seal may be a consequence of not prefilling the tip of the pipette with agent-free solution, or of a short filling distance in the tip. Appropriate adjustment may correct this problem.

# 4. Conventional and Special Applications of the Perforated Patch Technique

## 4.1. Conventional Applications

With the perforated patch technique, voltage-clamp recordings are feasible under stable conditions, without any significant rundown. The interior of the cell is maintained better, and significant alterations as a result of dialysis with the pipette solution are avoided, as a result of the retention of biologically pertinent substances within the intracellular compartment. The intracellular concentration of multivalent ions, such as $Ca^{2+}$, is also much better preserved. This advantage is clear in recordings of $I_{Ca}$, under stable conditions, without appreciable rundown for at least 60 minutes

(Akaike and Harata, 1994; Sarantopoulos et al., 2002, 2004). This can be more helpful in pharmacological studies, requiring sequential application of several modulators (Akaike and Harata, 1994). The perforated patch technique has also allowed reliable recordings of voltage-gated $Na^+$ currents, the activation and inactivation parameters of which shift to more negative potentials as a result of alterations in the junction potential between the pipette and interior of the cell in the standard whole-cell configuration. The perforated patch technique, in contrast, maintains stability in activation and inactivation parameters for up to 150 minutes (Akaike and Harata, 1994). Activation and inactivation kinetics of voltage-gated $K^+$ channels can be also affected by the standard, compared to the perforated, technique—something that has been attributed to washout of cytoplasmic second messengers by dialysis. Similarly, the activating effect of protein kinase C (PKC) on neuronal delayed outward rectifier $K^+$ channel is diminished by half in standard whole-cell, compared to perforated patch, recordings, which prevent washout of cytosolic PKC-I isoenzyme. The activity of other channels, such as inwardly rectifying $K^+$, is maintained in perforated patch recordings, as a result of maintenance of phosphatase activity that modulates them. Perforated patch recordings provide more stability in BK recordings, as well as $K_{ATP}$ channels. Nevertheless, β-escin should be avoided when $K_{ATP}$ channels are studied because of unpredictable effects in intracellular ATP concentrations. Other currents for which perforated patch recordings may be suitable include those elicited by ligand responses, such as adrenergic or muscarinic activation, glutamate via metabotropic receptors, and ionotropic receptors (Akaike and Harata, 1994). Perforated patch recordings also may be used in the investigation of intracellular signaling and second messenger pathways, such as those modulating synaptic plasticity, transport mechanisms across cytoplasmic membrane, such as $Na^+$-$Ca^{2+}$ exchange, or $Na^+$-$K^+$ transport. Because intracellular multivalent ions, such as $Ca^{2+}$, are not affected by the pipette-filling solution since the channels formed by the antibiotics are not permeated by these ions, the technique may be suitable for simultaneous whole-cell $I_{Ca}$ recordings with optical measurements.

### 4.2. Recordings from Perforated Vesicles

The outside-out patch is a configuration originating from the whole-cell attached mode, and results in recording from a membrane vesicle at the tip of the pipette. This is formed by slowly and gently pulling back the pipette from the membrane after gigaseal formation, so that a bleb-like projection of the membrane is being pulled out together with the withdrawing pipette. The ruptured ends of the membrane reseal, thus reforming as a vesicle on the tip of the pipette, the surface of which is the outside of the membrane. This configuration allows recordings from single-ion channels after direct application of agents to the extracellular side of the membrane (in the bath). The intracellular compartment is replaced with the pipette internal solution, but in the standard outside-out technique, after membrane rupture, dialysis may significantly wash out the signaling or second messenger molecules that physiologically modulate the channel function. In many cases this results in current rundown or loss of regulation of the channel, and is characteristically shown in GH3 pituitary tumor cells, where VGCC (L-type) and voltage-gated $K^+$ channels are modulated by intracellular $Ca^{2+}$ release, secondary to stimulation by thyrotropin-releasing hormone (TRH). This regulatory mechanism fails in standard outside-out patches, while VGCC activity runs down within 15 minutes. In contrast, by using the perforated vesicle technique both current and channel regulation remain intact, allowing detection of transmitter-induced inhibition of single calcium channels (Axon Guide, 1993; Levitan and Kramer, 1990) (Fig. 7).

In the perforated vesicle technique a perforating agent is included within the pipette internal solution to produce partitioning of the

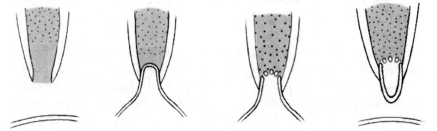

Fig. 7. The perforated vesicle technique. In the perforated vesicle technique, a gigaseal is formed around a portion of the cell membrane, and after permeabilization with a perforating agent, the pipette is pulled back slowly. This results in the formation of a vesicle in the tip, after the ends of the membrane reseal, the surface of which is the outside of the membrane. Dialysis is avoided and essential cytoplasmic factors and organelles are retained within the vesicle.

membrane of the vesicle facing the interior of the pipette, and formation of pores. Thus, membrane rupture and subsequent dialysis are avoided and essential cytoplasmic factors and organelles are retained within the vesicle.

Nystatin has been used as a perforating agent in this configuration, at a concentration of 100 µg/mL. This can be prepared by diluting 4 µL of nystatin stock solution in DMSO (25 mg/mL), into 1 mL of filtered pipette solution containing (in mM): 150 KCl, 5 MgCl$_2$, glucose 20, and 10 mM KHEPES (4-(2-hydroxyethyl) piperazine-1-ethanesulfonic potassium salt), at pH 7.1 (Axon Guide, 1993). Because nystatin pores are impermeable to nucleotides, proteins, and calcium buffers, they are not necessary in the pipette solution. Solubilization can be enhanced by vortexing and sonication for 30 seconds. Conventional patch pipettes can be used with resistance ranging between 2 and 4 MΩ when filled with the internal solutions such as the one above, but Sylgard coating may be necessary to reduce pipette capacitance, lower noise for single-channel recording, and permit accurate measurements. Pipette tips should be prefilled by brief immersion into a nystatin-free solution, and then back-filled with the nystatin-containing solution. Then, a seal is formed in the cell-attached mode, while, after compensating for pipette capacitance, perforation can be continuously monitored by capacitive current response to a 10 mV voltage step from −70 mV holding potential, from which $R_a$ can be deducted. An abrupt reduction of the resistance indicates membrane rupture, while a smooth progressive decrease over 5 to 20 minutes to a value <50 MΩ indicates successful perforation. In the former case, cells should be discarded. In the latter, the pipette is withdrawn slowly, with the amplifier switched to current-clamp mode with zero resting current, in order to prevent large current flows that might disrupt the resealing of the membrane and the vesicle formation. Subsequently, the amplifier is switched back to voltage-clamp mode to record single-channel activity. Successful vesicle formation can be confirmed visually at 400 × magnification, and the presence of cytoplasm by preloading the cells with a fluorescent dye, such as 40 µM carboxyfluorescein diacetate, a membrane-permeant non-fluorescent compound. The presence of organelles can be detected with dyes, like rhodamine 123, which partitions into functioning mitochondria and produces a characteristic punctate staining that can be detected in GH3 cells and perforated vesicles (Axon Guide, 1993).

## 4.3. Application of Perforated Patch Technique in High-Throughput Electrophysiology Screening

The perforated patch technique by polyene antibiotics has been used in high-throughput electrophysiology platforms, which allow simultaneous recordings from several cells using automated systems. In the IonWorks™ HT system (Molecular Devices, Sunnyvale, CA), recordings are made using a planar, multiwell substrate, containing 384 small wells arranged in an 8 × 48 array (PatchPlate™) (Schroeder et al., 2003). Each well allows positioning of one cell into a tiny aperture (1–2 μm in diameter), separating two isolated fluid chambers: the upper, extracellular compartment, and the intracellular fluidics system at the bottom. Then a high-resistance seal (but typically less than 1 GΩ) is formed in between the cell membrane and the peripheral orifice of the aperture onto which the cell sits, by differential pressure applied across the two chambers.

The system gains access to the cell interior by perforated patch, after application of nystatin or amphotericin B into the intracellular fluidics system. The intracellular fluidics system contains also the Ag/AgCl electrode, from which recordings are carried out after perforation of the membrane. $R_a$ values between 5 and 10 MΩ are achieved in few minutes, allowing stable voltage clamping, and despite sub-GΩ seal resistance, current recordings can be carried out. A parallel format in many wells allows voltage-clamp recordings of up to 384 individual cells to be made in minutes. Schroeder et al. (2003) have reported success rates of 60% to 80% for recordings from hKv1.5 and hNav1.3, human Kv3.2, Kv2.1, and HERG (human ether-a-go-go potassium) channels expressed in CHO and HEK cells, translating into 3500 successful patches per 8-hour day. In a recent modification, such that each recording well contained 64 apertures for cells instead of one, Finkel et al. (2006) used a population patch-clamp technique, wherein a single voltage-clamp amplifier sums the whole-cell currents from multiple cells at once, after each one is sealed to a separate aperture in the planar substrate well. Electrical access to the cell is established by amphotericin B perforated patch, in a similar manner. The perforating solution is prepared from preweighted, 5-mg amphotericin B aliquots, stored at 4°C, after solubilization in 180 μL DMSO under sonication and subsequent dilution into 50 mL of internal buffer by vortexing for 1 minute. The final concentration of amphotericin is 100 μg/mL.

Seal resistance values are also sub-G$\Omega$ (average 120M$\Omega$). Recordings using these techniques under sub-G$\Omega$ seals should not be considered as a substitute for conventional patch recording but as a much higher throughput form of screening using a direct electrophysiological assay.

## 5. Problems and Troubleshooting of the Perforated Patch Configuration

Shortcomings of the perforated patch technique include the following:

1. It is technically more complex, more demanding, more expensive, and more time-consuming compared to the conventional whole–cell technique. A much longer time period is needed to achieve adequately low $R_a$ in perforated patches compared to the standard whole-cell patch. However, $\beta$-escin seems to function better in this context.

2. Sometimes perforated patches fail. This may be secondary to failure to obtain a G$\Omega$ seal (as a result of interference with the seal), premature rupture of the membrane (due to a high concentration of ionophore or mechanical factors), spontaneous conversion to the standard whole-cell mode, inability to establish perforation, or prolonged delay to establish low $R_a$. In case of membrane rupture, recordings are similar to those in a standard whole-cell mode, but dialysis and penetration of the ionophore into the cell may have unpredictable effects. Loss of potency of solutions with time or after prolonged storage of stocks, exposure to light or heat, inadequate concentration, or excessive prefilling distance in the tip with ionophore-free solution may result in failure, and can be easily identified and corrected by specific sequential adjustments.

3. Perforated patch does not allow the study of effects of relatively large modulating compounds acting at cytoplasmic, submembrane sites, since they may be impermeable to the pores.

4. Although one advantage of the technique is the lack of any specific effects of the perforating agents on currents under investigation and the signaling mechanisms that modulate them, it has been reported that nystatin inhibits the K$^+$ channel Kv1.3 in a time- and concentration-dependent fashion, even at concentrations much lower than those required for perforation (Hahn

et al., 1996). Amphotericin B also inhibited Kv1.3 currents, but was less potent compared to nystatin.

5. Chloride fluxes, which may develop if the pipette-filling solution does not contain impermeant anions and Cl⁻ at concentrations matching those in the cell, may result in changes in voltage and cell volume. Chloride shifts may occur through polyene antibiotic-induced perforated patches, but not with gramicidin. Another way to bypass the problem is by adding an impermeant anion that matches the effective valence of the intracellular impermeant anions. Thus, 20 to 25 mM Cl⁻ and 125 to 130 mM methanesulfonate in the pipette solution appears to keep cell volume reasonably stable (Axon Guide, 1993).

6. Finally, one problem is $R_a$ considerably higher than that obtained with the conventional whole-cell technique, introducing significant voltage errors. In current clamp mode, also, action potential recordings via perforated patches may be erroneous with standard patch-clamp amplifiers. In these amplifiers, alterations in the recorded voltage, secondary to current dissipation by their input section, yield distorted current clamp results in the presence of high $R_a$. Conventional patch-clamp amplifiers are not able to follow rapid events, such as action potentials (Magistretti et al., 1996), especially under high $R_a$. The problem can be bypassed by employing special microelectrode bridge amplifiers, although these are limited by suboptimal performance at voltage clamp. Alternatively, the more advanced patch-clamp amplifiers, such as the EPC 8 or 9, C version (HEKA Elektronik, Lambrecht, Germany), or the Axopatch, 200A or 200B (Molecular Devices), offer a "fast current clamp" mode, which acts as a capacitance neutralization adjustment, neutralizing the capacitance of the electrode and to some extent of the amplifier, allowing better recording of rapid changes in membrane action potential, relatively free of voltage distortions. Other amplifiers suitable for fast or high-speed current clamping include the EPC 10 and the MultiClamp 700B (Molecular Devices).

7. High $R_a$ may also introduce errors in voltage-clamp due to voltage drop, proportional to the passing current. The $R_a$ depends on the resistance of the pipette itself, and the resistance of the perforated patch of membrane, which is determined by the total channel number and the single-channel conductance of the channels formed by the perforating agent. The number of

the channels depends on the density of channels formed in a unit surface area times the patch area. So, in order to minimize $R_a$, pipettes with tips as wide as possible and suitable geometrical shape should be pulled, as described previously.

With regard to the perforating agents, these need to be selected taking into consideration several factors, including the type of cells and channels under examination. Regarding polyene antibiotics, amphotericin B has produced the lowest $R_a$—something that may be related to a wider pore, with single-channel conductance twice that of nystatin, at least in measurements made in salts after double-sided application. We have found that β-escin produces perforated patches with even lower $R_a$, comparable to those after the standard patch, in addition to having several other advantages that would make it an ideal agent for perforation. Finally, attention to details and meticulous careful preparation and storage of solutions is essential. Solutions of polyene antibiotics and β-escin are highly sensitive to light, so during experiments we wrap the syringes containing the diluted pipette-filling solutions with aluminum foil and keep them in a container with ice. Ideally, even for the short time required to back-fill pipettes, solutions should be protected from light exposure. Careful dispersion of solutions with an ultrasonicator or vortex mixer may significantly improve solubilization, as does the addition of Pluronic F127 and fluorescein. These steps may improve efficacy and reduce the delay for permeabilization. Finally, considering the presence of significant variability in recording conditions from cell to cell and current to current, investigators should determine which agent is more suitable to their own conditions and at what concentrations, based on initial pilot experiments.

## Acknowledgment

This work has been supported by a grant from the National Institute of Neurological Disorders and Stroke (KNS049420A).

## References

Akagi, K., and Nagao, T., et al. (1999) Responsiveness of beta-escin-permeabilized rabbit gastric gland model: effects of functional peptide fragments. *Am. J. Physiol.* **277**, G736–744.

Akaike, N. and Harata, N. (1994) Nystatin perforated patch recording and its applications to analyses of intracellular mechanisms. *Jpn. J. Physiol.* **44**, 433–473.

Andreoli, T. E., Bangham, J. A., et al. (1967) The formation and properties of thin lipid membranes from HK and LK sheep red cell lipids. *J. Gen. Physiol.* **50**, 1729–1749.

Armstrong, D. and Eckert, R. (1987) Voltage-activated calcium channels that must be phosphorylated to respond to membrane depolarization. *Proc. Natl. Acad. Sci. USA* **84**, 2518–2522.

Arnould, T., Janssens, D., et al. (1996) Effect of aescine on hypoxia-induced activation of human endothelial cells. *Eur. J. Pharmacol.* **315**, 227–233.

Axon Guide. (1993) *Advanced Methods in Electrophysiology.* Axon Instruments, Molecular Devices Corp., Sunnyvale, CA.

Bangham, A. D., Horne, R. W., et al. (1962) Action of saponin on biological cell membranes. *Nature*, **196**, 952–955.

Becq, F. (1996) Ionic channel rundown in excised membrane patches. *Biochim. Biophys. Acta* **1286**, 53–63.

Belles, B., Hescheler, J., et al. (1988a) A possible physiological role of the Ca-dependent protease calpain and its inhibitor calpastatin on the Ca current in guinea pig myocytes. *Pflugers Arch.* **412**, 554–556.

Belles, B., Malecot, C. O., et al. (1988b) "Run-down" of the Ca current during long whole-cell recordings in guinea pig heart cells: role of phosphorylation and intracellular calcium. *Pflugers Arch.* **411**, 353–360.

Bezanilla, F., and Caputo, C., et al. (1986) Potassium conductance of the squid giant axon is modulated by ATP. *Proc. Natl. Acad. Sci. USA* **83**, 2743–2745.

Bremerich, D. H., Hirasaki, A., et al. (1997a) Halothane attenuation of calcium sensitivity in airway smooth muscle. Mechanisms of action during muscarinic receptor stimulation. *Anesthesiology* **87**, 94–101.

Bremerich, D. H., Warner, D. O., et al. (1997b) Role of protein kinase C in calcium sensitization during muscarinic stimulation in airway smooth muscle. *Am. J. Physiol.* **273**, L775–781.

Byerly, L. and Hagiwara, S. (1982) Calcium currents in internally perfused nerve cell bodies of Limnea stagnalis. *J. Physiol.* **322**, 503–528.

Chad, J. E. and Eckert, R. (1986) An enzymatic mechanism for calcium current inactivation in dialysed Helix neurones. *J. Physiol.* **378**, 31–51.

Chad, J., Kalman, D., et al. (1987) The role of cyclic AMP-dependent phosphorylation in the maintenance and modulation of voltage-activated calcium channels. *Soc. Gen. Physiol. Ser.* **42**, 167–186.

Ciorba, M. A., Heinemann, S. H., et al. (1997) Modulation of potassium channel function by methionine oxidation and reduction. *Proc. Natl. Acad. Sci. USA* **94**, 9932–9937.

Cotero, B. V., Rebolledo-Antunez, S., et al. (1998) On the role of sterol in the formation of the amphotericin B channel. *Biochim. Biophys. Acta* **1375**, 43–51.

D'Ambrosio, R. *Perforated Patch-Clamp Technique*, In: Patch-Clamp Analysis: Advanced Techniques 1st ed. (Walz, W., ed.). Humana Press, Totowa, NJ, Chapter 6.

de Kruijff, B. and Demel, R. A. (1974) Polyene antibiotic-sterol interactions in membranes of Acholeplasma laidlawii cells and lecithin liposomes. 3. Molecular structure of the polyene antibiotic-cholesterol complexes. *Biochim. Biophys. Acta* **339**, 57–70.

de Kruijff, B., Gerritsen, W. J., et al. (1974a) Polyene antibiotic-sterol interactions in membranes of Acholeplasma laidlawii cells and lecithin liposomes. I. Specificity of the membrane permeability changes induced by the polyene antibiotics. *Biochim. Biophys. Acta* **339**, 30–43.

de Kruijff, B., Gerritsen, W. J., et al. (1974b) Polyene antibiotic-sterol interactions in membranes of Acholesplasma laidlawii cells and lecithin liposomes. II. Temperature dependence of the polyene antibiotic-sterol complex formation. *Biochim. Biophys. Acta* **339**, 44–56.

Doroshenko, P. A., Kostyuk, P. G., et al. (1982) Intracellular metabolism of adenosine 3′,5′-cyclic monophosphate and calcium inward current in perfused neurones of Helix pomatia. *Neuroscience* **7**, 2125–2134.

Ebihara, S., Shirato, K., et al. (1995) Gramicidin-perforated patch recording: GABA response in mammalian neurones with intact intracellular chloride. *J. Physiol.* **484** (pt 1), 77–86.

Eckert, R., Chad, J. E., et al. (1986) Enzymatic regulation of calcium current in dialyzed and intact molluscan neurons. *J. Physiol. (Paris)* **81**, 318–324.

Ermishkin, L. N., Kasumov, K. M., et al. (1976) Single ionic channels induced in lipid bilayers by polyene antibiotics amphotericin B and nystatine. *Nature* **262**, 698–699.

Fan, J. S. and Palade, P. (1998) Perforated patch recording with beta-escin. *Pflugers Arch.* **436**, 1021–1023.

Fernandez, J. M., Fox, A. P., et al. (1984) Membrane patches and whole-cell membranes: a comparison of electrical properties in rat clonal pituitary (GH3) cells. *J. Physiol.* **356**, 565–585.

Finkel, A., Wittel, A., et al. (2006) Population patch clamp improves data consistency and success rates in the measurement of ionic currents. *J. Biomol. Screen.* **11**, 488–496.

Finkelstein, A. and Holz, R. (1973) Aqueous pores created in thin lipid membranes by the polyene antibiotics nystatin and amphotericin B. *Membranes* **2**, 377–408.

Forscher, P. and Oxford, G. S. (1985) Modulation of calcium channels by norepinephrine in internally dialyzed avian sensory neurons. *J. Gen. Physiol.* **85**, 743–763.

Furukawa, T., Yamane, T., et al. (1996) Functional linkage of the cardiac ATP-sensitive K+ channel to the actin cytoskeleton. *Pflugers Arch.* **431**, 504–512.

Gruszecki, W. I., Gagos, M., et al. (2003) Organization of antibiotic amphotericin B in model lipid membranes. A mini review. *Cell. Mol. Biol. Lett.* **8**, 161–170.

Hahn, S. J., Wang, L. Y., et al. (1996) Inhibition by nystatin of Kv1.3 channels expressed in Chinese hamster ovary cells. *Neuropharmacology* **35**, 895–901.

Hamill, O. P., Marty, A., et al. (1981) Improved patch-clamp techniques for high-resolution current recording from cells and cell-free membrane patches. *Pflugers Arch.* **391**, 85–100.

Hilgemann, D. W. and Ball, R. (1996) Regulation of cardiac Na+, Ca2+ exchange and KATP potassium channels by PIP2. *Science* **273**, 956–959.

Hladky, S. B. and Haydon, D. A. (1970) Discreteness of conductance change in bimolecular lipid membranes in the presence of certain antibiotics. *Nature* **225**, 451–453.

Hladky, S. B. and Haydon, D. A. (1984) Ion movements in gramicidin channels. *Curr. Top. Membr. Transp.* **21**, 327–371.

Holz, R. and Finkelstein, A. (1970) The water and nonelectrolyte permeability induced in thin lipid membranes by the polyene antibiotics nystatin and amphotericin B. *J. Gen. Physiol.* **56**, 125–145.

Horn, R. and Korn, S. J. (1992) Prevention of rundown in electrophysiological recording. *Methods Enzymol.* **207**, 149–155.

Horn, R. and Marty, A. (1988) Muscarinic activation of ionic currents measured by a new whole-cell recording method. *J. Gen. Physiol.* **92**, 145–159.

Hoshi, T. (1995) Regulation of voltage dependence of the KAT1 channel by intracellular factors. *J. Gen. Physiol.* **105**, 309–328.

Iizuka, K., Dobashi, K., et al. (1997) Receptor-dependent G protein-mediated $Ca^{2+}$ sensitization in canine airway smooth muscle. *Cell. Calcium* **22**, 21–30.

Kasumov, K. M., Borisova, M. P., et al. (1979) How do ionic channel properties depend on the structure of polyene antibiotic molecules? *Biochim. Biophys. Acta* **551**, 229–237.

Kerrick, W. G. and Hoar, P. E. (1994) Relationship between ATPase activity, $Ca^{2+}$, and force in alpha-toxin- and beta-escin-treated smooth muscle. *Can. J. Physiol. Pharmacol* **72**, 1361–1367.

Kleinberg, M. E. and Finkelstein, A. (1984) Single-length and double-length channels formed by nystatin in lipid bilayer membranes. *J. Membr. Biol.* **80**, 257–269.

Konishi, M. and Watanabe, M. (1995) Molecular size-dependent leakage of intracellular molecules from frog skeletal muscle fibers permeabilized with beta-escin. *Pflugers Arch.* **429**, 598–600.

Korn, S. J. and Horn, R. (1989) Influence of sodium-calcium exchange on calcium current rundown and the duration of calcium-dependent chloride currents in pituitary cells, studied with whole cell and perforated patch recording. *J. Gen. Physiol.* **94**, 789–812.

Korn, S. J. and Weight, F. F. (1987) Patch-clamp study of the calcium-dependent chloride current in AtT-20 pituitary cells. *J. Neurophysiol.* **58**, 1431–1451.

Kupper, J., Bowlby, M. R., et al. (1995) Intracellular and extracellular amino acids that influence C-type inactivation and its modulation in a voltage-dependent potassium channel. *Pflugers Arch.* **430**, 1–11.

Kurachi, Y., Asano, Y., et al. (1990) Voltage-dependent inhibition of the delayed outward potassium current by OPC-8490, a novel positive inotropic agent, in isolated atrial myocytes of guinea-pig heart. *Naunyn Schmiedebergs Arch. Pharmacol.* **341**, 324–330.

Kyrozis, A. and Reichling, D. B. (1995) Perforated-patch recording with gramicidin avoids artifactual changes in intracellular chloride concentration. *J. Neurosci. Methods* **57**, 27–35.

Launikonis, B. S. and Stephenson, D. G. (1999) Effects of beta-escin and saponin on the transverse-tubular system and sarcoplasmic reticulum membranes of rat and toad skeletal muscle. *Pflugers Arch.* **437**, 955–965.

Le Foll, F., Castel, H., et al. (1998) Gramicidin-perforated patch revealed depolarizing effect of GABA in cultured frog melanotrophs. *J. Physiol.* **507** (pt 1), 55–69.

Levitan, E. S. and Kramer, R. H. (1990) Neuropeptide modulation of single calcium and potassium channels detected with a new patch clamp configuration. *Nature* **348**, 545–547.

Li, M., West, J. W., et al. (1992) Functional modulation of brain sodium channels by cAMP-dependent phosphorylation. *Neuron* **8**, 1151–1159.

Liem, L. K., Simard, J. M., et al. (1995) The patch clamp technique. *Neurosurgery* **36**, 382–392.

Lindau, M. and Fernandez, J. M. (1986) IgE-mediated degranulation of mast cells does not require opening of ion channels. *Nature* **319**, 150–153.

Lucero, M. T. and Pappone, P. A. (1990) Membrane responses to norepinephrine in cultured brown fat cells. *J. Gen. Physiol.* **95**, 523–544.

Magistretti, J., Mantegazza, M., et al. (1996) Action potentials recorded with patch-clamp amplifiers: are they genuine? *Trends Neurosci.* 19, 530–534.

Marom, S., Goldstein, S. A., et al. (1993) Mechanism and modulation of inactivation of the Kv3 potassium channel. *Receptors Channels* **1**, 81–88.

Marty, A. and Finkelstein, A. (1975) Pores formed in lipid bilayer membranes by nystatin, Differences in its one-sided and two-sided action. *J. Gen. Physiol.* **65**, 515–526.

Marty, A. and Neher, E. (1995) *Tight-Seal Whole-Cell Recordings,* Plenum Press, New York.

Miloshevsky, G. V. and Jordan, P. C. (2004) Gating gramicidin channels in lipid bilayers: reaction coordinates and the mechanism of dissociation. *Biophys. J.* **86**, 92–104.

Mueller, P. and Rudin, D. O. (1967) Development of K+-Na+ discrimination in experimental bimolecular lipid membranes by macrocyclic antibiotics. *Biochem. Biophys. Res. Commun.* **26**, 398–404.

Neher, E. and Sakmann, B. (1976) Single-channel currents recorded from membrane of denervated frog muscle fibres. *Nature* **260**, 799–802.

Rae, J., Cooper, K., et al. (1991) Low access resistance perforated patch recordings using amphotericin B. *J. Neurosci. Methods* **37**, 15–26.

Rosenmund, C. and Westbrook, G. L. (1993) Rundown of N-methyl-D-aspartate channels during whole-cell recording in rat hippocampal neurons: role of Ca2+ and ATP. *J. Physiol.* **470**, 705–729.

Sarantopoulos, C., McCallum, B., et al. (2002) Gabapentin decreases membrane calcium currents in injured as well as in control mammalian primary afferent neurons. *Reg. Anesth. Pain Med.* **27**, 47–57.

Sarantopoulos, C., McCallum, J. B., et al. (2004) Beta-escin diminishes voltage-gated calcium current rundown in perforated patch-clamp recordings from rat primary afferent neurons. *J. Neurosci. Methods* **139**, 61–68.

Schroeder, K., Neagle, B., et al. (2003) Ionworks HT: a new high-throughput electrophysiology measurement platform. *J. Biomol. Screen* **8**, 50–64.

Sutton, L. E. (1965) *Tables of Interatomic Distances and Configuration in Molecules and Ions,* The Chemical Society, London.

Tang, X. D. and Hoshi, T. (1999) Rundown of the hyperpolarization-activated KAT1 channel involves slowing of the opening transitions regulated by phosphorylation. *Biophys. J.* **76**, 3089–3098.

Tosteson, D. C., Andreoli, T. E., et al. (1968) The effects of macrocyclic compounds on cation transport in sheep red cells and thin and thick lipid membranes. *J. Gen. Physiol.* **51**(suppl), 373S.

Venegas, B., Gonzalez-Damian, J., et al. (2003) Amphotericin B channels in the bacterial membrane: role of sterol and temperature. *Biophys. J.* **85**, 2323–2332.

Williams, B. A., Dickenson, D. R., et al. (1999) Kinetics of rate-dependent shortening of action potential duration in guinea-pig ventricle; effects of IK1 and IKr blockade. *Br. J. Pharmacol.* **126**, 1426–1436.

Yawo, H. and Chuhma, N. (1993) An improved method for perforated patch recordings using nystatin-fluorescein mixture. *Jpn J. Physiol.* **43**, 267–273.

Zeidler, U., Barth, C., et al. (1995) Radiation-induced and free radical-mediated inactivation of ion channels formed by the polyene antibiotic amphotericin B in lipid membranes: effect of radical scavengers and single-channel analysis. *Int. J. Radiat. Biol.* **67**, 127–134.

# 9

# Fast Drug Application

## Manfred Heckmann and Stefan Hallermann

## 1. What Do We Gain from Fast Drug Application?

"The determination of the three-dimensional structures of protein molecules showed for the first time in detail the construction of the molecular 'machines' of the life cycle. If we want to learn how these 'machines' work, it is not sufficient only to know their construction. We actually have to see them at work, and this requires dynamical studies, i.e. penetration into the dimension of time" (Eigen, 1968).

The dimension of time is of particular relevance at synapses. Synaptic events are usually short, on the order of milliseconds, and their exact duration and timing are of prime relevance. "For many synaptic receptors/channels, the 'natural' mode of transmitter application is a short, steep pulse, and many of these systems show rapid desensitisation" (Dudel et al., 1992). This quotation is from the first review of a technique that allows very fast application of drugs to receptor channels in outside-out patches. This chapter provides practical tips and additional information for setting up a fast drug application system. For further information about this technique, see Jonas (1995) and Sachs (1999).

## 2. Material and Equipment

Chapters 1 to 5 provided general information about the patch-clamp equipment. This chapter focuses on the aspects that are relevant for fast drug application. We describe our system, consisting of a piezo, a monoluminal application pipette, and an application chamber.

From: *Neuromethods, Vol. 38: Patch-Clamp Analysis: Advanced Techniques, Second Edition*
Edited by: W. Walz @ Humana Press Inc., Totowa, NJ

## 2.1. The Microscope

In principle, the technique of fast drug application does not require a microscope. However, a microscope is helpful for patch-clamping in most preparations. We tried different types of microscopes and found upright microscopes more convenient than inverted microscopes. Differential interference contrast (DIC) optics facilitate visualizing the liquid filament, and a video camera with a monitor is also helpful. We use upright microscopes with a fixed stage (Fig. 1A). The micromanipulator is mounted on the table. If, instead, a microscope with a mobile stage is used, the micromanipulator should be mounted on the stage. With respect to the micromanipulator, it is important to recognize that one needs to cover relatively long distances (up to 10 mm) with the electrode during an experiment with fast drug application. Because electrodes need to be replaced frequently, mounting the head-stage or the whole micromanipulator on a hinge might help. In the course of an experiment, a patch electrode is lowered onto the preparation, an outside-out patch is excised (Hamill et al., 1981), and, with a fixed stage and the micromanipulator mounted on the table (Fig. 1A), the stage is moved relative to the electrode to reach the application chamber. Thereby the electrode tip remains in the optical field.

---

Fig. 1. A piezo-driven application system on the stage of an upright microscope. (**A**) The photograph and the schematic drawing show components of the system: the piezo device (1), which can be fixed in its frame (2) with a screw (3), and its power supply (4); a platelet (5) to hold the application pipette (6) connected to a tube (7); in- (8) and out-flow (9) of the application chamber (10); in- (11) and out-flow (12) of the bath (13); objective (14); head-stage of an Axopatch 200A amplifier (15); patch-clamp electrode (16); fixed stage (17) of an upright microscope. (**B**) The application pipette is made from borosilicate glass with an outer diameter of 0.5 mm and an inner diameter of about 0.3 mm. To obtain the desired angle, a longer piece of glass (5 to 10 cm long) is heated in the flame of a common lighter. A scratch with a diamond knife on the convex side of the bend facilitates breaking the glass as close to the bend as desired. The open end of the application pipette can then be fire-polished in the lighter flame. (**C**) The application pipette is glued into a fiberglass platelet. A tube whose inner diameter fits tightly the outer diameter of the pipette is then put over the other end. Finally, the tool is mounted to the piezo with a screw.

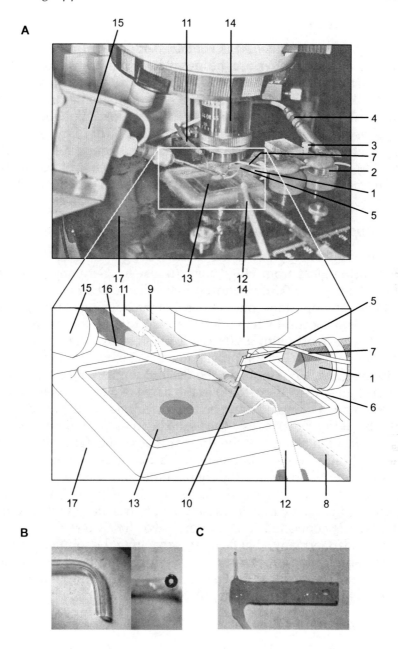

## 2.2. The Piezo

The piezo (1 in Fig. 1A) is contained in a metal tube and pre-stressed by a spring. We use the low-voltage minitranslator P810.30 element from Physik Instrumente (Waldbronn, Germany). It lengthens by 45 μm on application of 100 V. It needs to be driven with sufficient power to overcome its capacitive load. Suitable power supplies for this piezo element are available from Physik Instrumente. To fix the piezo device on the stage, we use a frame (2 in Fig. 1A), held by strong magnets. The piezo can be rotated within the frame to allow easy access to the application pipette. A plastic screw (3 in Fig. 1A) facilitates fixing the piezo.

## 2.3. Application Pipets and Tubing

Our application pipettes are made from glass tubes. We found borosilicate glass from Hilgenberg (Malsfeld, Germany), with an outer diameter of 0.5 mm and an inner diameter of about 0.3 mm suitable. An important constraint is that the opening of the application pipette needs to fit under the objective of the microscope. Usually we work with a 20× LD Zeiss objective (14 in Fig. 1A). Therefore, the application pipette needs to be bent to fit below the objective (Fig. 1B). To obtain the desired angle, we heat a longer piece of glass (5 to 10 cm long) in the flame of a common lighter. A scratch with a diamond knife on the convex side of the bend facilitates breaking the glass as close to the bend as desired. One end of the application pipette is then fire-polished in the lighter flame and the application pipette is glued into a fiberglass platelet. A tube whose inner diameter fits tightly the outer diameter of the pipette is placed over its other end. Silicon might help to seal this connection. Finally, the platelet is mounted to the piezo with a screw. The tube is connected to a six-port valve (HVX 86915 Hamilton, Darmstadt, Germany) that facilitates changing the solution flowing to the application pipette rapidly. Syringes are used as solution reservoirs. The syringes are sealed airtight and connected to compressed air with a pressure regulator.

## 2.4. Application Chamber

The tip of the application pipette is placed in an application chamber (10 in Fig. 1A) that is separated from the rest of the bath. We make our application chamber from a Kimax glass tube (outer diameter 3 mm). An opening of about 120 degrees and about 1 cm in length is made in the tube. This opening is the access to the chamber for both the application pipette and the patch electrode. The application chamber is glued into the bath chamber. The ends of the glass tube, that is, the application chamber, are then connected to a solution reservoir and a pump. It is important to obtain a very steady flow of the solution in the application chamber. We use the hydrostatic pressure depending on the height of the solution reservoir and an adjustable pump to fine-tune the flow as precisely as possible. If the pump is not ideal, an air chamber (i.e., a large syringe, half full with water and connected parallel to the tube) may help to steady the flow. We adjust the pump so that a little solution is always sucked from the bath to avoid contamination of the bath with the solution in the application chamber.

## 2.5. Patch Electrodes

Many electrodes are used in experiments with outside-out patches and a rapid application system. We prefer using electrodes with a resistance of about $10\,M\Omega$ when filled with a standard intracellular saline. We produce these electrodes on-line during the experiments, which has the advantage that the tip size and the geometry of the electrode can always be fine-tuned as desired. For example, by adjusting the tip diameter, one can try to increase or decrease the numbers of channels in a patch. We found the DMZ Universal Puller by Zeitz Instruments (München, Germany), suitable, but other commercial pullers may also be appropriate. Zeitz Instruments also offers a quartz-glass puller for low-noise recordings (Dudel et al., 2000; Hallermann et al., 2005).

## 2.6. Computers and Software

We use commercially available computers and software for the recording and evaluation of our data. A pulse generator like the Master8 from AMPI (Jerusalem, Israel), or the Max21 from Zeitz Instruments are useful to control the piezo device. Usually, the recording software triggers the pulse generator. Short current traces (100 to 3000 ms) are recorded at intervals of 0.01 to 60 seconds directly onto the hard drive of the computer. To make sure that periodic noise sources are likely to be out of phase, we use an interval of, say, 1.001 second instead of exactly 1 second. The most important functions of the data analysis program are averaging, subtracting, and fitting current traces. Jitter or imprecise timing of the traces (inevitable in earlier times with videotapes or date-recorders) precludes precise subtraction. On good days up to 2,000 sweeps have to be processed per patch, which can be a cumbersome procedure with some programs. We found the ISO2 software (MFK, Taunusstein, Germany) more useful for this purpose than PClamp from Axon or PulseFit from HEKA. However, using ISO2 is and more difficult on new computers due to outdated hardware requirements. To facilitate data evaluation, the processor hardware must perform optimally.

## 3. Procedures, Applications, and Results

### 3.1. Testing the System

After the equipment has been set up, it needs to be tuned. The first step is to optimize the solution flow from the application pipette. One needs a steady stream of solution with a sharp interface (Fig. 2A). This can easily be judged with the microscope and the video monitor if there is a difference in osmolarity between the solutions. Turn on the solution flow, focus on the liquid filament, and check that it remains steadily in focus for, say, half an hour or the period you want to record from a patch. To measure the time course of the solution exchange, monitor the liquid junction potential. A reasonable exchange rate is shown in Fig. 2C and D. Keep in mind that inevitably the exchange at an intact patch is somewhat slower than at an open pipette.

The position of the patch electrode relative to the liquid filament is relevant for the solution exchange (Fig. 2E,F). It is important to map out the best position and to practice how to find it. With a real patch, there is no time for searching for the position, and the process is more difficult than with an open pipette. It is advisable to control the performance of the system routinely after experiments (use pressure to blow away the patch).

### 3.2. Working with Real Patches

Artifacts that superimpose with the currents of interest arise when the piezo is turned on and off. To illustrate the problem, we picked an extreme example (Fig. 3). We cannot give definitive advice about how to reduce these artifacts; you have to try several approaches to find the one that works. Small changes in the setup often have large effects. Particularly important are the mounting of the piezo on the stage, the size of the application pipette, its fixation

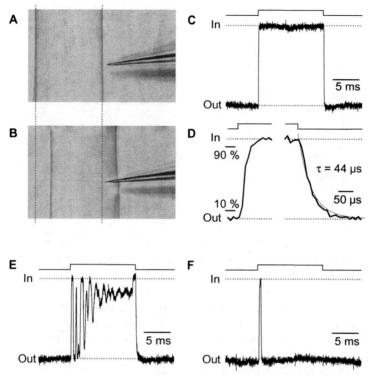

Fig. 2. Testing the performance of the system. (**A**) The liquid filament with reduced osmolarity close to a patch-clamp electrode is visible in differential interference contrast. (**B**) In response to a voltage pulse (thin lines above the current traces), the liquid filament moves 45 μm toward the electrode. After the pulse, the liquid filament returns to its initial position (dotted lines). Recording the liquid junction current during such a movement with an open patch electrode facilitates measuring the speed of the solution exchange (**C**). In this case a 12.5-ms pulse was applied to the piezo. (**D**) The rise and decay phase of the trace shown in **C** are shown on an expended time scale. The 10% to 90% rise time is about 30 μs, and the decay of the current is fitted with a monoexponential function with a time constant of 44 μs. (**E,F**) Effects with less favorable positions of the same patch-clamp electrode. Due to unavoidable vibrations of the system, the tip is not necessarily always immersed for the whole pulse duration.

to the piezo, and the type of electrode glass. We have the impression that quartz glass electrodes, which have otherwise superior noise performance, tend to ring more than usual borosilicate electrodes in our rapid application system. Some artifact will always

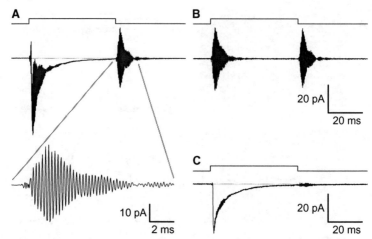

Fig. 3. Vibration artifacts. (**A**) Average current response from an outside-out patch from a CA3 pyramidal neuron of a rat hippocampal slice. The patch was held at –80 mV, and 10 mM glutamate was applied for 50 ms. Large artifacts occur at the beginning and the end of the pulse (thin lines above the current traces). This is a common problem, but an extreme example is shown here. (**B**) Average current response from the same outside-out patch recorded a few seconds earlier without glutamate flowing through the application pipette. (**C**) Subtraction of the trace shown in **B** from the one in **A** purges the current.

remain, which can be eliminated by subtraction of blank traces (Fig. 3C). Usually it is possible to start off with a much smaller vibration artifact than the one shown in Fig. 3 in a real recording, and it is then often possible to obtain clean average currents even at a very high gain (see, for example, the trace with 0.2 mM glutamate in Fig. 4 of Heckmann and Dudel, 1997).

### 3.3. Data Analysis

We review here the data for one particular receptor channel (Heckmann et al., 1996), to briefly illustrate how we obtain kinetic information from average current responses with a fast drug application system. The interested reader should also refer to Colquhoun et al. (1992), Dilger and Brett (1990), Dudel et al. (1990), and Franke et al. (1991).

Figure 4A shows data from a recording with an outside-out patch. In this case recombinant kainate receptors were exposed to various glutamate concentrations. A first step might be to measure

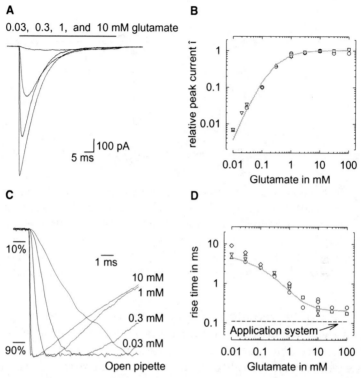

Fig. 4. From current traces to the rate constants of channel activation. The data shown here and in Fig. 6 are from outside-out patches with recombinant, homomeric kainate receptors (GluR6), expressed in HEK 293 cells. (**A**) Averaged current traces from a patch, which was held at –60 mV, are shown superimposed. The peak current amplitude depends on the glutamate concentration. Each trace is the average of three to seven single responses to 0.03, 0.3, 1, or 10 mM glutamate. The horizontal bar above the traces indicates the application (pulse duration 50 ms; interval 5 s). (**B**) Double logarithmic plot of the peak current î versus the glutamate concentration. î was normalized to the î value with 10 mM glutamate. The data from five patches are marked by different symbols. The gray line is a fit of the data with the kinetic mechanism and the rate constants given in Fig. 5. The peak current amplitude is half-maximal with about 0.5 mM (ratio of $k_-/k_+$). For glutamate concentrations $K_m$, the slope is about 1.3, giving a lower limit of 2 for the number of binding steps. (**C**) Dependence of current rise time on glutamate concentration. The rising phase of the traces in **A** are shown normalized. In addition, the response of the open electrode recorded at the end of the experiment is shown. (**D**) Double logarithmic plot of the rise time (10–90%) versus the glutamate concentration. The data from five patches are shown. The gray line is a fit of the data with the kinetic mechanism.

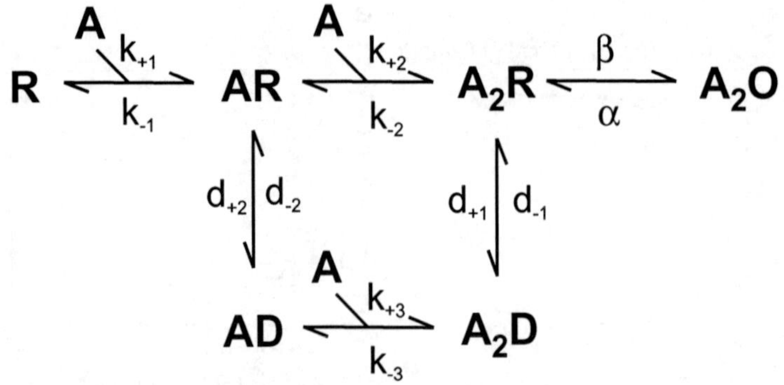

$$R \underset{k_{-1}}{\overset{A \atop k_{+1}}{\rightleftarrows}} AR \underset{k_{-2}}{\overset{A \atop k_{+2}}{\rightleftarrows}} A_2R \underset{\alpha}{\overset{\beta}{\rightleftarrows}} A_2O$$

$$d_{+2} \Big\updownarrow d_{-2} \qquad\qquad d_{+1} \Big\updownarrow d_{-1}$$

$$AD \underset{k_{-3}}{\overset{A \atop k_{+3}}{\rightleftarrows}} A_2D$$

Fig. 5. Kinetic reaction scheme. $R$ represents a closed channel; $A$, agonist molecules; and $AR$, $A_2R$, and $A_2O$, a closed and an open channel with one or two agonist molecules bound. $D$ represents desensitized states. For the computer simulations, the following rate constants were chosen: $k_{+1}$ $7.2 \cdot 10^6$, $k_{-1}$ 1000, $k_{+2}$ $3.6 \cdot 10^6$, $k_{-2}$ 2000, $k_{+3}$ $3.6 \cdot 10^6$, $k_{-3}$ 80, $\alpha$ 1500, $\beta$ 7000, $d_{+1}$ 800, $d_{-1}$ 0.08, $d_{+2}$ 400, $d_{-2}$ 1 (units: $s^{-1}$ and $M^{-1}s^{-1}$ for $k_{+n}$).

the affinity of the receptors. High densities of receptors are preferable in these experiments because one needs to measure current responses also at low agonist concentrations. Dose–response curves for the peak current amplitude and the current rise times are shown in Fig. 4B and D. The half-maximal activating concentration ($K_m$) is about 0.5 mM (Fig. 4B). For glutamate concentrations well below the $K_m$, the logarithmic slope is above 1.3 for these receptors. This provides a lower limit of 2 for the number of agonist binding steps (Fig. 5). To estimate the $k_+$ and $k_-$ rate constants and the channel opening ($\beta$) and closing rate constants ($\alpha$), we fitted the scheme to the two dose-response curves (Fig. 5). Commercial programs like ChanneLab from Synaptosoft Inc. (http://www.synaptosoft.com) or Berkeley Madonna (http://www.berkeleymadonna.com) are user-friendly environments in which to do these simulations. More information about the shape of the rise time curves is found in Franke et al. (1991), Heckmann et al. (1996), and von Beckerath et al. (1995).

As is apparent in Fig. 4A, the current decays in the presence of glutamate. This is due to desensitization of the receptors, and it demands at least one desensitized stated ($A_2D$ in Fig. 5) that is reached either from $A_2R$ or $A_2O$. Experiments like the ones shown in Fig. 6C and D provide more information about the kinetics of desensitization. The experiments shown in Fig. 6A and B measure

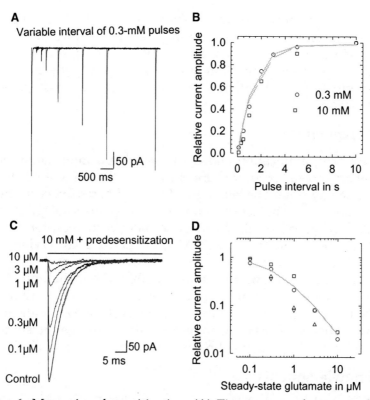

Fig. 6. Measuring desensitization. (**A**) Time course of recovery from desensitization. As shown in Fig. 4A, the currents desensitize. To measure the recovery from desensitization, here a control pulse was followed by a test pulse after 0.1, 0.3, 0.5, 1, 2, 3, and 5 s. Averaged currents of three to five single responses are shown. (**B**) Plot of the relative current amplitudes with 0.3 and 10 mM glutamate versus pulse interval. The solid and dashed gray lines are the results of computer simulations of these experiments for the two concentrations. (**C**) To quantify steady-state desensitization by low glutamate concentrations, responses to test pulses with a saturating glutamate concentration were recorded as controls. The other responses were recorded in the presence of 0.1 to 10 µM glutamate, as indicated on the left side of the traces. The traces are averages of four to seven single responses from one patch. (**D**) Double logarithmic plot of the relative peak current versus steady-state glutamate concentration. The data from four patches are shown. The gray line is the result of a computer simulation of this experiment with the model given in Fig. 5.

recovery from desensitization. Simulations facilitate testing reaction schemes (Fig. 5) and estimating the relevant rate constants.

In addition to the purely agonist elicited responses, the technique of fast drug application can be used to study the mechanism of competitive inhibition, ion-channel block, and allosteric modification.

## 4. Perspectives

Fast drug application allows one to obtain information about many aspects of receptor channel function that cannot be studied in the steady state. The technique described here is very useful for basic research, characterization of new pharmacological compounds (Krampfl et al., 2006) or mutations in receptor channel subunits (Krampfl et al., 2005; Maljevic et al., 2006; Wyllie et al., 2006). With the modification described by Tour et al. (1995), even more complex pulse protocols are possible. The main limitation in experiments with fast drug application is the stability and quality of the recording, which requires training. A challenge for the future is to advance with fast drug application from average current traces to low-noise single-channel recordings (Hallermann et al., 2005).

## Acknowledgments

This work is supported by the Deutsche Forschungsgemeinschaft.

## References

Colquhoun, D., Jonas, P., and Sakmann, B. (1992) Action of brief pulses of glutamate on AMPA/kainate receptors in patches from different neurones of rat hippocampal slices. *J. Physiol.* **458**, 261–287.

Dilger, J. P. and Brett, R. S. (1990) Direct measurement of the concentration- and time-dependent open probability of the nicotinic acetylcholine receptor channel. *Biophys. J.* **57**, 723–731.

Dudel, J., Franke, C., and Hatt, H. (1990) Rapid activation, desensitization, and resensitization of synaptic channels of crayfish muscle after glutamate pulses. *Biophys. J.* **57**, 533–545.

Dudel, J., Franke, C., and Hatt, H. (1992) High-speed application of drugs with a piezo drive, in *Practical Electrophysiological Methods* (Kettenmann H., Grantyn R., eds.), Wiley-Liss, New York, pp. 141–145.

Dudel, J., Hallermann, S., and Heckmann, M. (2000) Quartz glass pipette puller operating with a regulated oxy-hydrogen burner. *Pflügers Arch.* **441**, 175–180.

Eigen, M. (1968) New looks and outlooks on physical enzymology. *Q. Rev. Biophys.* **1**, 3–33.

Franke, C., Hatt, H., Parnas, H., and Dudel, J. (1991) Kinetic constants of the ace-tylcholine (ACh) receptor reaction deduced from the rise in open probability after steps in ACh concentration. *Biophys. J.* **60**, 1008–1016.

Hallermann, S., Heckmann, S., Dudel, J., and Heckmann, M. (2005) Short openings in high resolution single channel recordings of mouse nicotinic receptors. *J. Physiol.* **563**, 645–662.

Hamill, O. P., Marty, A., Neher, E., Sakmann, B., and Sigworth, F. J. (1981) Improved patch-clamp techniques for high-resolution current recording from cells and cell-free membrane patches. *Pflügers Arch.* **391**, 85–100.

Heckmann, M., Bufler, J., Franke, C., and Dudel, J. (1996) Kinetics of homomeric GluR6 glutamate receptor channels. *Biophys. J.* **71**, 1743–1750.

Heckmann, M. and Dudel, J. (1997) Desensitization and resensitization kinetics of glutamate receptor channels from Drosophila larval muscle. *Biophys. J.* **72**, 2160–2169.

Jonas, P. (1995) Fast application of agonists to isolated membrane patches, in *Single-Channel Recording*, 2nd ed. (Sakmann B. and Neher E., eds.), Plenum, New York, pp. 231–243.

Krampfl, K., Maljevic, S., Cossette, P., et al. (2005) Molecular analysis of the A322D mutation in the GABA receptor alpha-subunit causing juvenile myoclonic epilepsy. *Eur. J. Neurosci.* **22**, 10–20.

Krampfl, K., Schlesinger, F., Cordes, A. L., and Bufler, J. (2006) Molecular analysis of the interaction of the pyrazine derivatives RPR119990 and RPR117824 with human AMPA-type glutamate receptor channels. *Neuropharmacology* **50**, 479–490.

Maljevic, S., Krampfl, K., Cobilanschi, J., et al. (2006) A mutation in the GABA(A) receptor alpha$_{(1)}$-subunit is associated with absence epilepsy. *Ann. Neurol.* **59**, 983–987.

Sachs, F. (1999) Practical limits on the maximal speed of solution exchange for patch-clamp experiments. *Biophys. J.* **77**, 682–690.

Tour, O., Parnas, H., and Parnas, I. (1995) The double-ticker: an improved fast drug-application system reveals desensitization of the glutamate channel from a closed state. *Eur. J. Neurosci.* **7**, 2093–2100.

von Beckerath, N., Adelsberger, H., Parzefall, F., Franke, C., and Dudel, J. (1995) GABAergic inhibition of crayfish deep extensor abdominal muscle exhibits a steep dose-response relationship and a high degree of cooperativity. *Pflügers Arch.* **429**, 781–788.

Wyllie D. J., Johnston A. R., Lipscombe D., and Chen P. E. (2006) Single-channel analysis of a point mutation of a conserved serine residue in the S2 ligand binding domain of the NR2A NMDA receptor subunit. *J. Physiol.* **574**(Pt 2), 477–489.

# 10

# Pipette Internal Perfusion: Methods and Applications

*Srinivas M. Tipparaju and Aruni Bhatnagar*

## 1. Introduction

The patch-clamp technique has revolutionized the understanding of ion channel physiology (Hamill et al., 1981; Neher and Sakmann, 1976). This technique facilitates the measurement of currents and voltage in cells under voltage-clamp and current-clamp conditions. Cells can be patched in a variety of configurations. Whole-cell currents and voltage can be measured by forming a high resistance seal between the cell membrane and the rim of the glass or quartz patch pipette. In the whole-cell configuration the seal is ruptured to gain access to the cell interior. Resistive and capacitive currents are dissociated by instantaneously changing the voltage between the cell interior and the bath electrode, and changes in current could be recorded using the classical voltage-clamp approach (Hodgkin and Huxley, 1952). Current injections can be used to generate action potentials from excitable cells (Bhatnagar, 1997). In addition, single-channel activity can be recorded from both cell-attached and excised patches (Hamill et al., 1981).

For studies on ion channel regulation, it is often necessary to change the composition of solutions across cell membrane or to add drugs and metabolites to the external medium or to the cell interior. A variety of approaches have been adopted to accomplish rapid changes in the composition of fluids on the outside or the inside of cells. These approaches enhance the utility of the patch-clamp technique, which, in combination with perfusion techniques, can be used to address a wider range of questions related to the regulation of ion channels by drugs, metabolites, and second messengers. In

From: *Neuromethods, Vol. 38: Patch-Clamp Analysis: Advanced Techniques, Second Edition*
Edited by: W. Walz @ Humana Press Inc., Totowa, NJ

the whole-cell configuration, for instance, direct access to the cell interior makes it possible to add substances directly to the cytosol. Although this could be accomplished by simply recording currents with patch pipettes containing internal solutions of different composition, it is usually necessary to change the pipette solution composition by internal perfusion after recordings with control internal solution have been made. Several such internal perfusion methods have been developed that can be used with whole-cell, cell-attached, or excised-patch recordings.

The internal solution used to fill patch pipettes usually contains ions at their cytosolic concentration, although low concentrations of calcium-chelating agents are sometimes added to prevent the influx of calcium during patch rupture. Cells that have high metabolic rates are usually patched with adenosine triphosphate (ATP) to prevent its depletion during prolonged whole-cell recordings. In addition, impermeable or expensive drugs, chemicals, and metabolites can be directly included in the patch-pipette solution. The main advantage of this approach is that it is simple and straightforward. Usually if the substance does not basally affect ion conductances, it is possible to include it in the internal solution. Changes in current elicited by a stimulus are then recorded in the presence or absence of substance in the patch pipette. For instance, to examine the role of antioxidants, Vogalis and Harvey (2003) first recorded currents in the presence of hydrogen peroxide from intestinal neurons patched with a pipette containing ionic solution only and then with pipettes containing catalase and glutathione. Similarly Vargas and Lucero (2002) added the catalytic subunit of protein kinase A (PKA) to the internal solution to study its effect of hyperpolarization-activated current in rat olfactory receptor neurons.

Although direct addition of substances to the internal solution of the recording pipette is simple, it has two major disadvantages. First, no control recordings are possible, and currents in the absence of the drug or the metabolite cannot be recorded from the same cell. Current recordings from a different cell patched without the drug or metabolite in the patch-pipette solution could be used as controls, but due to cell-to-cell variations, many more recordings are required to demonstrate statistically significant differences. Second, inclusion of surface active chemicals can prevent the formation of a high-resistance seal. Proteins or peptides dissolved in the patch-pipette solution usually diffuse from the pipette solution and coat the rim of the electrode, thereby making the formation of

GΩ seals difficult. To circumvent these changes, cells are patched with normal internal solution (containing only ions, chelators, and ATP), and then the internal solution of the pipette is changed using internal perfusion techniques.

The choice of a specific patch-clamp configuration and perfusion method depends on the requirements of the experiments. It is important to consider also the mode of action of the compound of interest. If the drug or metabolite binds directly to the channel without the requirement of secondary mediation or other cellular events, then it is possible to use either whole-cell or cell-attached patches. If, on the other hand, the compound binds only to the cytosolic part of the channel, then the whole-cell mode or the inside-out configuration can be used, depending on whether ensemble or single-channel recordings are desired. If the compound acts only on the exterior of the membrane, then cell-attached patch recordings are necessary. If the compound works only after metabolic transformation, then only the whole-cell configuration is useful. The whole-cell configuration is also suitable for recording the effects of proteins or metabolites that do not alter channel function directly but alter membrane currents by affecting other components of cell metabolism or signaling. In each case, the concentration of the compound of interest can be rapidly changed by internal perfusion techniques.

## 2. Methods of Internal Perfusion

Several methods have been used to perfuse the recording pipette internally to change the composition of its solution. The most intuitive approach is to insert a single or a multibarrel perfusion pipette inside the patch pipette and deliver the solution of interest to the tip of the recording electrode. Initially this was accomplished by mounting the perfusion pipette on a separate micromanipulator, which was used to position the perfusion assembly near the tip of the patch electrode. Solution in the perfusion pipette was injected into the patch pipette using a hypodermic needle. An alternative approach was to build a suction port and pipette solution waste chamber within the pipette holder (Soejima and Noma, 1984), which can also be made to hold the pipette solution reservoir (Lapointe and Szabo, 1987). Another approach was to use a double-barrel glass to perfuse from one barrel and withdraw the solution from the other (Makielski et al., 1987), but this has found limited application because of the technical difficulties of forming high resistance seals with multibarrel electrodes.

Most current methods of internal perfusion involve a remote reservoir containing the perfusion solution that is introduced into the patch pipette via a perfusion tube or pipette. The capillary for internal perfusion is inserted into the patch pipette, 100 to 300 μm behind the tip. The capillary is usually made of polymer-coated quartz. For solution exchange, negative pressure is applied to the back end of the pipette (Irisawa and Kokubun, 1983) or both positive and negative pressures are used to allow smooth exchange of fluids near the tip of the patch pipette. Perfusion pipettes can be used with single or dual (Verrecchia et al., 1999) patch pipette recordings.

## 3. Experimental Setup and Practical Procedures

For patch-clamp recording, a conventional electrophysiological setup is required. Optical recording capabilities are helpful in monitoring the efficacy and speed of internal perfusion, but are not always necessary. We use an inverted microscope connected to an electrical recording system. The microscope is placed on a vibration-free table and covered with a dark cage to minimize electrical and optical interference. Electrical recordings are made using Axopatch. For optical recordings, we have built a microfluorometer (single excitation, dual emission) around the inverted microscope, equipped with an epifluorescence attachment and two photomultiplier tubes (PMTs). The cells are placed on the bottom of a circular tissue chamber, which is temperature-controlled by a Peltier element. The cells are illuminated by a 150-W xenon lamp and light from the lamp is collimated by a beam probe and delivered to the filter assembly with a dichroic mirror installed beneath the microscope stage. The fluorescence is collected through the side port of the microscope. To minimize collection of stray light, a rectangular shutter is used to mask the area not covered by the cell. The light collected is split by a beam splitter, which is connected to two identical photomultiplier tubes. Specific filters are installed on the face of the PMTs. Data are collected by a set of concatenated counters using LabView software. This setup is used to establish the efficiency of internal perfusion. Fluorescent dye is included in the injection reservoir, and uniform distribution of the dye is imaged on the cell.

Fluorescence imaging is not necessary to establish efficacy. Changes in cell currents could be monitored to ensure adequate and uniform perfusion. The combination of internal perfusion and fluorescence monitoring may be useful; however, for applications that require monitoring of intracellular ions, pH, and reactive oxygen species, use fluorescent dyes. The internal perfusion technique is particularly useful in uniformly and reproducibly loading cells with dyes. Often, variable dye loading in cells is a significant problem when quantifying changes in the fluorescence of cell-permeable dye esters using single emission and excitation wavelengths. Internal perfusion is also the technique of choice when using cell-impermeant dyes or fluorescent proteins.

## 3.1. Electrical Recordings with Internal Perfusion

In an ideal internal perfusion setup, it should be possible to form GΩ seals with the internal solution alone, so that control recordings could be made before changing the composition of the pipette solution by internal perfusion. The assembly should be pressurized, and upon activation, positive pressure should be applied to rapidly deliver the substance into the pipette tip. Moreover, the negative pressure should balance the positive pressure, so that the cell does not sense changes in pressure applied for exchanging solutions at the tip of the recording pipette. Overall, the setup should allow smooth exchange of solutions inside the pipette, without large mechanical movements (which could destroy the patch) and without increasing the electrical noise (which could compromise electrical recordings).

Several commercial systems meet these conditions with variable success. Bioscience Tools (San Diego, CA) sells computerized solution delivery systems that can be used as an attachment with a conventional pipette holder. The delivery system is housed in a metal box. Opening of a valve drains a small reservoir system near the tip of the recording pipette. The system can be adapted for using multiple solution changes. The PLI-100 system supplied by Harvard Apparatus (Holliston, MA) works on the principle of pressure injection and is compatible with several single or multiple channel cell injectors such as the PM-4/PM-8 multiinjectors (Warner Instruments, Hamden, CT) that come with their own control circuitry and accessories. We have used the 2PK+ system (Adams and List) from ALA Scientific Instruments (Westbury, NY). This system is based on the most generic approach. It represents a refinement

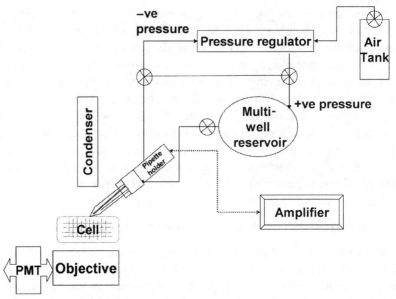

Fig. 1. Schematic diagram of a patch-clamp setup with internal pipette perfusion. The schematic shows a setup for internal pipette perfusion during patch-clamp recording and cell imaging. A multiwell reservoir is used to deliver solutions of different composition to the quartz capillary so that recordings with several different solutions are possible with the same cell. Solid arrows indicate the flow of perfusion pressure, which facilitates solution exchange at the tip of the recording patch pipette. The air tank is the pressure source. Dotted line indicates electrical connections between the glass patch-pipette holder and the patch-clamp amplifier. Valves for pressure flow are marked with a circled ×. Note that the tip of the glass patch pipette is tapered. PMT; photomultiplier tube; negative (−ve) and positive (+ve) pressures are indicated.

of earlier designs in which a quartz microcapillary was inserted into a glass recording pipette and the solution was changed near the tip of the patch pipette (Lapointe and Szabo, 1987). This system and its use are described above (Fig. 1).

### 3.2. Fabrication of Quartz Perfusion Pipettes

A quartz microcapillary or perfusion pipette is used to deliver the perfusate. The quartz capillary of inner diameter 170 μm and outer diameter 250 µm (Polymicro Technology, Phoenix, AZ, or Scientific Instruments, New York, NY) is used. The capillary is coated with polyamide, which provides good mechanical resistance and flexibility, and the quartz interior has excellent low noise characteristics and chemical resistance. The glass perfusion pipette is pulled to a diameter of 20 to 60 µm over a microflame. Usually a diameter near 40 µm is ideal. The most useful procedure is to suspend the quartz capillary vertically using alligator clips and melt the capillary near the middle using a butane torch or a microflame. Breaking the tubing leaves a portion of the quartz surface exposed. The carbon deposited on the surface is cleaned with sandpaper or a microknife. The tubing is cleaned thoroughly by sonication and is then washed six times with water followed by methanol. The tubing should be completely dried in air before use.

The choice of the geometry of the perfusion pipette is determined by the lag time that is acceptable within the experimental design. The relationship between the time course of solution exchange and tip diameter (determined by the appearance of dye from the tip pipette and the time needed for the new solution to diffuse to the tip of the patch pipette) is nonlinear (Lapointe and Szabo, 1987). For a given negative pressure, decreasing the tip diameter leads to a decrease in the time required for the new solution to produce 90% of the total conductance change. Overall, there is a compromise. Fine-tip dimensions allow the perfusion pipette to be placed close to the mouth of the patch pipette. Decreasing tip dimensions, however, decrease flow rate, which can prevent fast fluid exchange. Moreover, if the perfusion pipette is very sharp, it can diminish the flow of its own solution if pushed too far into the patch pipette. We have found that it is easier to form seals with patch pipettes that are not completely jammed by the perfusion pipette.

The microcapillary is connected with short polyethylene tubing to a positive pressure vessel. First, negative pressure is applied to the pipette to draw the solution from the microcapillary. Positive pressure is used to inject fluid into the pipette and from there to the cell. The pipette is then opened to air to allow dialysis of the cell interior. Although cleanliness is paramount, it is not necessary to make a fresh capillary for each new experiment. The capillary, however, should be thoroughly cleaned and dried before reusing. Quartz is durable, and it provides excellent chemical resistance, and hence the capillary can be used as long as the polyamide coating on the quartz capillary remains intact.

### 3.3. Fabrication of Patch Pipettes

Shape, resistance, and noise characteristics of the patch pipette are key determinants of the quality of electrophysiological data that can be recorded. The shape of the patch pipette is particularly important for internal perfusion experiments. In general, soft glass pipettes have low series resistance and could be fashioned into more favorable tip geometry. They have high capacitance, which could be reduced by fire-polishing and coating with elastomer (Sylgard, Dow Corning, Midland, MI). Hard glass has excellent electrical properties, but hard glass pipettes have sharp ends, high series resistance, and low cone angles (Rae and Levis, 1992). For internal perfusion, the use of borosilicate glass is recommended. In our laboratory, glass pipettes are pulled using a multistage horizontal puller (P97 Flaming/Brown micropipette puller, Sutter Instruments, Novato, CA). When filled with normal pipette solution, these pipettes have a resistance of 1 to $3\,M\Omega$. For low-noise recordings, it is advisable to use thick-walled short pipettes. The pipettes should be fire-polished and coated with low-loss elastomer and, as far as possible, shallow depths of immersion should be used to further increase the quality of the recordings (Levis and Rae, 1998).

The efficiency of internal perfusion can be further enhanced by pressure polishing. For this, the pipette is connected to pressurized gas during polishing. Under pressure, the glass walls of the pipette expand during heating, resulting in a tip with higher cone angles. Both soft and hard glass pipettes can be pressure polished. Several different types of pipette tips can be fashioned (Fig. 2). Because the resistance of the pipette is primarily located near the tip (Levis and Rae, 1998), for the same tip diameter, pressure-polished pipettes have much lower resistance than conventionally fabricated pipettes (Goodman and Lockery, 2000). For internal perfusion, bullet-shaped pipettes are preferred. Several microfuges with pressure-polishing capabilities are commercially available (from GlasswoRx, St. Louis, MO; ALA Scientific, Westbury, NY; Flyion, GmbH; and WPI, Sarasota, FL). The major advantage of pressure polishing is that it increases the cone angle, which facilitates solution exchange while at the same time reducing the distributed resistive and capacitative (RC) noise (Goodman and Lockery, 2000).

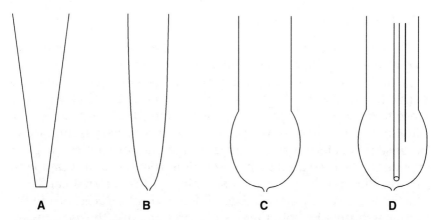

Fig. 2. Examples of pipette tip geometry. Pipettes with tip shapes shown in **A** to **C** can be generated by conventional electrode pullers and fire polishers. Pipettes with broader shanks (**B**) are useful for recordings with viscous internal solution or cells with an extensive extracellular matrix. Pipette shapes shown in **C** and **D** can be obtained by pressure polishing and may be more suitable for internal perfusion applications. The bullet-like or the bulb-like shape of such pipettes can readily accommodate an internal quartz perfusion capillary as well as an Ag-AgCl electrode (solid line). The broad shank allows the perfusion pipette capillary to be placed close to the mouth of the recording patch pipette without occluding its lumen. This facilitates pressurized delivery of the solution right at the orifice connecting the pipette to the cell.

### 3.4. Setup

Components of the perfusion pipette have to be carefully machined. If a sophisticated machine shop is not available, the use of a commercial system is highly recommended. The pressure and vacuum need to be precisely controlled, and therefore high fidelity pumps are needed. These pumps should be powered with a DC power supply. To prevent electrical interference, the power supply should be placed outside the Faraday cage and the output should be filtered. Specialized pipette holders are needed. The pipette holder should have a side port for the suction tube and an additional port where the quartz capillary can enter and easily access the pipette lumen. It should be possible to lock the system to allow for negative and positive pressure applications.

### 3.5. Experimental Procedure

The microreservoir is filled with filtered solutions and connected to the quartz capillary. The capillary is filled by using positive pressure so that no air bubbles are trapped. Once the quartz capillary is filled, it is inserted in the glass patch pipette by placing it very close to the tip of the pipette and tightening the other end of the pipette holder. The positioning of the quartz capillary is important, because this determines the rate at which the solution in the glass pipette can be displaced by new solution from the quartz capillary. Figures 2D and 3 show the positioning of the internal capillary. Once the quartz capillary is back-filled and introduced inside the glass patch-pipette, both the positive pressure and the vacuum are kept in the off position, so that the tip of the pipette does not sense changes in pressure. This also prevents any solutions from leaking from the quartz capillary into the patch pipette.

Depending on the experiment, cells can be placed in the recording chamber and equilibrated and washed with the external solu-

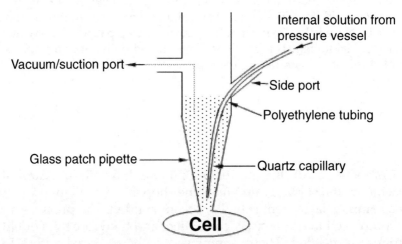

Fig. 3. Configuration for pipette internal perfusion. The schematic shows the position of the quartz capillary in the glass pipette. The quartz capillary is connected to the perfusion solution in the pressure vessel (multiwell reservoir) via polyethylene tubing that enters through the side port. The solution from the quartz capillary is injected at positive pressure into the lumen of the patch pipette near its tip. Negative pressure is applied through the suction/vacuum port to withdraw the patch pipette solution in the shank.

tion. After selecting a suitable cell, the patch pipette can be advanced into the recording bath and the pipette offset can be compensated. After establishing the gigaohm seal, the capacitive transient is compensated and the membrane is ruptured to get whole-cell access. Depending on the type of cell used, it is recommended to slightly lift (20–30 μm) the cell away from the bottom of the dish. This minimizes the chance of transmitting mechanical disturbances to the cell or impaling the cell with the electrode should the micromanipulator drift under gravity. Variations in this procedure could be used for cell-attached and excised-patch recording (Hamill et al., 1981).

After recording currents with the "control" solution, the solution from the quartz capillary can be delivered to the patch pipette in a gradual manner by increasing the pressure up to 40 mm Hg. Precision adjustment controls are needed for a stepwise increase in the applied positive pressure or the vacuum. The positive pressure and the vacuum cannot be measured simultaneously; therefore, positive pressure must be released by releasing the vacuum on the negative side. This process allows smooth exchange of solutions at the tip of the patch pipette (Fig. 3). The time for the exchange and arrival of solution into the cell depends on the geometry of the setup. An efficient procedure should exchange fluids in less than 1 minute. Pressure polished pipettes can give exchange rates of less than 50 ms. Successful perfusion is documented by monitoring changes in dye distribution or in the cell current parameter of interest.

## 4. Applications

Internal perfusion has been used to address many scientific questions relating to the role of ions, second messengers, metabolites, proteins, and signaling molecules in regulating ion channel function. For instance, internal perfusion has been used to suppress specific ionic conductances so as to isolate their contribution to overall membrane conductance, or to unmask other weaker currents. A combination of internal perfusion and voltage clamp was used by Lee and coworkers (1978) to completely block $K^+$ currents by internally perfusing nerve cell bodies with $Cs^+$. Similarly, an early version of the internal pipette perfusion technique was used by Ishizuka et al. (1984) to eliminate outward current in dorsal root ganglion cells, in which inward currents are difficult to analyze because of contamination by outward $K^+$ currents that cannot be completely suppressed by the $I_K$ blockers tetraethylammonium (TEA) and 4-aminopyridine (4-AP). Hence, internal perfusion with $Cs^+$ (in combination with externally acting TEA and 4-AP) was used to abolish $I_K$ and unmask inward currents. For the same reason, internal perfusion with fluoride was used to irreversibly suppress L-type $Ca^{2+}$ current to allow measurements of the T-type $Ca^{2+}$ current in hippocampal CA1 pyramidal neurons (Takahashi et al., 1991). Hattori et al. (1984) also used internal perfusion and current clamp to study γ-aminobutyric acid (GABA)-induced depolarization in frog dorsal root ganglion. Internal and external ion substitutions were used to establish the ionic basis of the response. Interestingly, internal perfusion was adequately demonstrated by showing that the membrane behaved as a simple chloride electrode when the internal and external chloride concentrations were changed. A relatively simple method for internal perfusion has been described by Lee and coworkers (1980), who used a suction pipette designed specifically to aspirate neurons (50–150 μM in diameter) in the cells using a negative pressure of about −30 mm Hg. With proper precautions, spatial control and temporal resolution on the order of $10^{-4}$ second was possible. Moreover, as the authors show, this method could be used to manually sever axons from the cell body, thereby avoiding enzymatic disruption techniques that could alter the membrane properties.

The combination of optical and electrophysiological recordings coupled with pipette internal perfusion represents a very powerful approach for studying ion physiology. For instance, in a study by Mollard and associates (1989), loading cells with Indo-1 by internal perfusion allowed uniform dye-loading and calibration and simultaneous measurements of calcium current and $[Ca^{2+}]_i$. A combination of calcium imaging and voltage-clamp recordings was also used by Osipchuk and coworkers (1990), who studied the effects of acetylcholine, cholecystokinin, internally applied guanosine triphosphate (GTP)-γ-s, inositol 1,4,5-triphosphate ($IP_3$), and $Ca^{2+}$ on cytoplasmic free $Ca^{2+}$ concentration. Changes in $[Ca^{2+}]_i$ were measured by microfluorimetry (Fura-2) and changes in $Ca^{2+}$-dependent $Cl^-$ currents. Two plastic tubes (20-μm open diameter) were used to exchange solutions at the tip of the patch pipette. Control currents were recorded in the whole-cell mode, and the effect of GTP-γ-S was studied by internal perfusion; pulses of $Ca^{2+}$ release were observed in both the current and microfluorimetric recordings. Similarly, by simultaneously recording both voltage and

$[Ca^{2+}]_i$, Noel and associates (1992) were able to establish the role of inositol signaling in $\alpha_1$-adrenoceptor evoked changes in $[Ca^{2+}]_i$. Internal perfusion with monoclonal anti-phosphatidylinositol 4,5-bisphosphate antibody was used to prevent agonist-induced changes in $[Ca^{2+}]_i$ (Noel et al., 1992).

The internal perfusion technique has also been used to determine the role of second messengers and metabolites in regulating ion conductances. In isolated neonatal rabbit ventricular myocytes, after recording $I_{Ca}$ with "control" internal solution, Kumar et al. (1997) used internal perfusion to demonstrate that cyclic guanosine monophosphate (cGMP), specifically its nonhydrolyzable analogue 8-Br-cGMP, increases L-type calcium currents. By using the pipette internal perfusion technique, these authors were able to demonstrate developmental differences in basal cGMP-dependent L-type calcium currents. Similar studies have been reported by Vogalis and coworkers (2003), who showed that internal perfusion of enteric neurons with ATP-γ-S suppresses slow afterhyperpolarization. Internal perfusion with the catalytic subunit ($PKA_{cat}$) caused a significantly decrease in the slow afterhyperpolarization current. Similarly, internal perfusion of CA1 hippocampal neurons with a constitutive active fragment of protein kinase C (PKC) (PKM) was used by Xiong et al. (1998) to demonstrate that PKC phosphorylation regulates *N*-methyl-D-aspartate (NMDA) receptor function.

We have used internal perfusion to study the role of nucleotides in regulating voltage-gated $K^+$ currents (Tipparaju et al., 2005). We found that inactivation of Kv currents in COS-7 cells cotransfected with Kvα1.5 and Kvβ1.3 could be prevented by internally perfusing the pipette with 1 mM of the oxidized form of nicotinamide adenine dinucleotide ($NAD^+$) but not reduced nicotinamide adenine dinucleotide phosphate (NADPH), suggesting that that the channel complex responds differently to oxidized and reduced nucleotides. Many other applications are possible. The technique requires only that stable electrical recordings should be possible with the cell of interest and that an adequate perfusion system could be installed and used.

## 5. Conclusion

Pipette internal perfusion is a sophisticated technique that provides greater cellular access during electrophysiological and optical recordings. The technique is based on an intuitive design that is simple to implement and use. As with any other patch-clamp technique, it requires dedicated time and personal attention. Although used mostly to study ionic changes, internal perfusion has been adapted to deliver enzymes, antibodies, and metabolites into cells. This allows the cell to be used as a veritable test tube in which reactants could be added at will to change the behavior of a single cell or a single molecule (channel). Few other techniques offer such possibilities. Nevertheless, the approach has been relatively underutilized, but as the technology becomes easier to use and commercial perfusion pipettes become widely available, the promise of the approach may be more fully realized.

## Acknowledgments

Work in the authors' laboratory is supported by the American Heart Association (AHA) and the National Heart, Lung, and Blood Institute (NHLBI).

## References

Bhatnagar, A. (1997). Contribution of ATP to oxidative stress-induced changes in action potential of isolated cardiac myocytes. *Am. J. Physiol* **272**, H1598–H1608.

Goodman, M. B. and Lockery, S. R. (2000). Pressure polishing: a method for re-shaping patch pipettes during fire polishing. *J. Neurosci. Methods* **100**, 13–15.

Hamill, O. P., Marty, A., Neher, E., Sakmann, B., and Sigworth, F. J. (1981). Improved patch-clamp techniques for high-resolution current recording from cells and cell-free membrane patches. *Pflugers Arch.* **391**, 85–100.

Hattori, K., Akaike, N., Oomura, Y., and Kuraoka, S. (1984). Internal perfusion studies demonstrating GABA-induced chloride responses in frog primary afferent neurons. *Am. J. Physiol* **246**, C259–C265.

Hodgkin, A. L. and Huxley, A. F. (1952). A quantitative description of membrane current and its application to conduction and excitation in nerve. *J. Physiol.* **117**, 500–544.

Irisawa, H. and Kokubun, S. (1983). Modulation by intracellular ATP and cyclic AMP of the slow inward current in isolated single ventricular cells of the guinea pig. *J. Physiol.* **338**, 321–337.

Ishizuka, S., Hattori, K., and Akaike, N. (1984). Separation of ionic currents in the somatic membrane of frog sensory neurons. *J. Membr. Biol.* **78**, 19–28.

Kumar, R., Namiki, T., and Joyner, R. W. (1997). Effects of cGMP on L-type calcium current of adult and newborn rabbit ventricular cells. *Cardiovasc. Res.* **33**, 573–582.

Lapointe, J. Y. and Szabo, G. (1987). A novel holder allowing internal perfusion of patch-clamp pipettes. *Pflugers Arch.* **410**, 212–216.

Lee, K. S., Akaike, N., and Brown, A. M. (1978). Properties of internally perfused, voltage-clamped, isolated nerve cell bodies. *J. Gen. Physiol.* **71**, 489–507.

Lee, K. S., Akaike, N., and Brown, A. M. (1980). The suction pipette method for internal perfusion and voltage clamp of small excitable cells. *J. Neurosci. Methods* **2**, 51–78.

Levis, R. A. and Rae, J. L. (1998). Low-noise patch-clamp techniques. *Methods Enzymol.* **293**, 218–266.

Makielski, J. C., Sheets, M. F., Hanck, D. A., January, C. T., and Fozzard, H. A. (1987). Sodium current in voltage clamped internally perfused canine cardiac Purkinje cells. *Biophys. J.* **52**, 1–11.

Mollard, P., Guerineau, N., Audin, J., and Dufy, B. (1989). Measurement of $CA^{2+}$ transients using simultaneous dual-emission microspectrofluorimetry and electrophysiology in individual pituitary cells. *Biochem. Biophys. Res. Commun.* **164**, 1045–1052.

Neher, E. and Sakmann, B. (1976). Single-channel currents recorded from membrane of denervated frog muscle fibres. *Nature* **260**, 799–802.

Noel, J., Fukami, K., Hill, A. M., and Capiod, T. (1992). Oscillations of cytosolic free calcium concentration in the presence of intracellular antibodies to phosphatidylinositol 4,5–bisphosphate in voltage-clamped guinea pig hepatocytes. *Biochem. J.* **288**, 357–360.

Osipchuk, Y. V., Wakui, M., Yule, D. I., Gallacher, D. V., and Petersen, O. H. (1990). Cytoplasmic $Ca^{2+}$ oscillations evoked by receptor stimulation, G-protein activation, internal application of inositol trisphosphate or $Ca^{2+}$: simultaneous microfluorimetry and $Ca^{2+}$ dependent $Cl^-$ current recording in single pancreatic acinar cells. *EMBO J.* **9**, 697–704.

Rae, J. L. and Levis, R. A. (1992). Glass technology for patch clamp electrodes. *Methods Enzymol.* **207**, 66–92.

Soejima, M. and Noma, A. (1984). Mode of regulation of the ACh-sensitive K-channel by the muscarinic receptor in rabbit atrial cells. *Pflugers Arch.* **400**, 424–431.

Takahashi, K., Ueno, S., and Akaike, N. (1991). Kinetic properties of T-type $Ca^{2+}$ currents in isolated rat hippocampal CA1 pyramidal neurons. *J. Neurophysiol.* **65**, 148–155.

Tipparaju, S. M., Saxena, N., Liu, S. Q., Kumar, R., and Bhatnagar, A. (2005). Differential regulation of voltage-gated $K^+$ channels by oxidized and reduced pyridine nucleotide coenzymes. *Am. J. Physiol. Cell Physiol.* **288**, C366–C376.

Vargas, G. and Lucero, M. T. (2002). Modulation by PKA of the hyperpolarization-activated current (Ih) in cultured rat olfactory receptor neurons. *J. Membr. Biol.* **188**, 115–125.

Verrecchia, F., Duthe, F., Duval, S., Duchatelle, I., Sarrouilhe, D., and Herve, J. C. (1999). ATP counteracts the rundown of gap junctional channels of rat ventricular myocytes by promoting protein phosphorylation. *J. Physiol.* **516** (pt 2), 447–459.

Vogalis, F. and Harvey, J. R. (2003). Altered excitability of intestinal neurons in primary culture caused by acute oxidative stress. *J. Neurophysiol.* **89**, 3039–3050.

Vogalis, F., Harvey, J. R., and Furness, J. B. (2003). PKA-mediated inhibition of a novel K⁺ channel underlies the slow after-hyperpolarization in enteric AH neurons. *J. Physiol.* **548**, 801–814.

Xiong, Z. G., Raouf, R., Lu, W. Y., et al. (1998). Regulation of N-methyl-D-aspartate receptor function by constitutively active protein kinase C. *Mol. Pharmacol.* **54**, 1055–1063.

# 11

# Loose-Patch-Clamp Method

## Héctor G. Marrero and José R. Lemos

## 1. Introduction

One of the basic tenets of experimental science has been the search for techniques that give relevant information from intact systems while causing minimal effects. This concept is particularly important in the physiological branch of the biological sciences. In general, reductionist methods assume that the net behavior of a system is the result of the summation of individual parameters, an assumption that is not always true in the biological sciences. On the other hand, in electrophysiology, reductionist methods have been successful, especially using voltage and current patch-clamp techniques. These, however, have practical limitations in that in general they require special processing of cells and tissues for their application. The effects of special processing become more notable with the need for having "clean" surfaces for the formation of tight seal patches. Although voltage- and current-clamp techniques with tight seals have given good information on the biophysical characteristics of ionic currents (and channels), such information from intact or in situ systems is more difficult, because these are minimally altered preparations. Information obtained with tight-patch techniques is also limited by the changes caused in the internal environment of cells (whole-cell configuration) or isolation from external effects (i.e., isolation of the patch under the cell-attached configuration).

The loose-patch-clamp technique, based on the use of pipette probes, which do not form tight (high resistance) seals on the tested surfaces, is an alternative to circumvent some of the practical and conceptual limitations of tight-patch techniques. In general, it does not require special processing of tissues or cells and offers the

From: *Neuromethods, Vol. 38: Patch-Clamp Analysis: Advanced Techniques, Second Edition*
Edited by: W. Walz @ Humana Press Inc., Totowa, NJ

ability to test multiple areas of a surface using the same probe (pipette). Furthermore, alteration of the surface and internal contents of tissues and cells (Milton and Caldwell, 1990; Roberts et al., 1990) is minimal or nonexistent (permitting recording for long periods of time), and the effects of the in situ environment can be taken into account in many cases.

Due to the nature of the loose seal, the area under study (patch) is also amenable to artificial external pharmacological changes. Pharmacology of ion channels can thus be reliably tested by either changing the external environment of the pipette (and waiting for internal pipette equilibration, as in the case of very low resistance seals) or changing both internal (pipette) and external (bath) solutions. Loose-patch techniques have been used to study ionic currents from the surface of muscles (Almers et al., 1983, 1984; Antoni et al., 1988; Caldwell et al., 1986; Caldwell and Milton, 1988; Eickhorn et al., 1990; Körper et al., 1998; Lupa et al., 1995; Milton and Behforouz, 1995; Milton et al., 1992; Roberts, 1987; Roberts and Almers, 1984; Ruff, 1999; Wolters et al., 1994; Yee Chin et al., 2004), neurons (Carta et al., 2004; Delay and Restrepo, 2004; Garcia et al., 1990; Nunemaker et al., 2003; Nygård et al., 2005; Smith and Otis, 2003), pituitary glands (Marrero and Lemos, 2003, 2005), whole nerves (Blanco et al., 1993; Marrero et al., 1989; Marrero and Orkand, 1993; Ortiz et al., 1988), for detecting activity at neuromuscular junctions (Beam et al., 1985; Re et al., 2003, 2006; Ruff, 1996), and for characterizing specific synaptic components (Dunant and Muller, 1986; Forti et al., 1997; Re et al., 2006). Techniques for the study of the biophysical properties of ion channels have already been developed for loose-patch-clamp approaches (see reviews by Anson and Roberts, 2002; Roberts and Almers, 1992), mostly on individual cells.

This chapter differentiates as much as possible the theoretical considerations from the methods, in order to facilitate its practical use for the reader. A general, brief description of loose-patch voltage-clamp methods is given in a later section. It is highly recommended that the reader consult previous reviews for a more detailed rationale of the loose-patch technique. More detailed examples of the technique used in the study of surface responses from whole tissues are given in a later section. These results were obtained from loose-patch-clamp recordings from large surfaces with many components (cells, axons and glia, for example) and, therefore, responses from large populations in situ. Furthermore, the importance of these examples is in their potential for future development of techniques for ion channel studies in vivo, with minimal invasiveness.

## 2. General Considerations

The loose-patch-clamp configuration, a form of voltage clamp in its original concept, is an analogue of the cell-attached voltage-clamp method. In this manner, voltages applied to the pipette affect the surface under its opening, and summations with the membrane potential (underneath the pipette opening, at the patch) must be taken into account with the proper convention: positive potentials in the pipette hyperpolarize and negative ones depolarize the membrane. The main difference between the two techniques is due to seal resistance considerations. In tight-seal cell-attached patches, the seal resistance is very large ($\geq 1\,G\Omega$, with small currents through the seal), thus making the potential applied at the patch (pipette tip) close in value to the applied pipette potential, and detection of elicited currents is mostly limited to the area covered by the pipette opening.

In contrast, with loose patches the seal resistance is considerably smaller ($K\Omega$ to $M\Omega$ range), with a significant amount of current flowing through the seal, thereby affecting the potential at the tip of the pipette (Fig. 1) (see Practical Methods, below, for typical pipette sizes and seals). If not accounted for, or compensated for in some manner (see Procedures for Patching, in Practical Methods, below), the deviations caused by the low seal resistance (usually referred to as errors) also cause uncertainty in the determination of the magnitude of membrane currents. The situation is complicated even more when considering preparations (e.g., whole tissues) that contain interstitial spaces where additional currents flow into and out of the tissue volume, equivalent to additional leak currents (Fig. 1).

The goal in loose-patch voltage clamp, as in other voltage-clamp techniques, is to be able to control the voltage across the membrane patch ($V_m$) under study. This voltage is given by $V_m = V_{int} - V_p$, where $V_{int}$ is the voltage inside the membrane and $V_p$ is the voltage

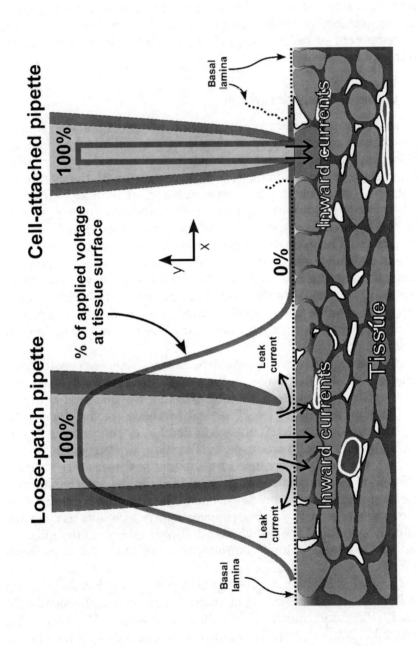

Fig. 1. Idealized illustration of typical loose-patch (left half) and cell-attached tight-patch (right half) pipettes. A large (15–30 µm diameter) loose-patch pipette at the surface of intact tissue is shown covering many components (basal lamina, cells, etc.) underlying the patch. The currents generated in the loose patch are separated into those flowing through the seal (leak currents) and those flowing into the tissue (inward currents). The basal lamina of the intact tissue is shown for the purpose of illustrating its integrity in cases where the loose patch is used. Note that, in comparison, the basal lamina has to be disrupted in the standard cell-attached case in order to achieve high-resistance seals (with a smaller ~1-µm-diameter pipette). In this case, the currents are considered to flow mostly through the patch membrane (arrows, inward currents). Also shown is the expected effect of the applied voltage at the surface of the tissue (transparent gray trace), for cases where there are *no* active (ion) currents elicited. A voltage applied inside the pipettes is experienced at the surface at a certain strength (y axis, percent of applied voltage at tissue surface) and depends on its relative position to the pipette along the surface of the tissue (x axis, in this two-dimensional illustration). In the loose-patch case, some of the applied voltage affects the surface of the tissue outside of the pipette due to the low-resistance seal (voltage spread, i.e., incomplete space clamp). The voltage drops gradually away from the pipette on either side. In contrast, the applied voltage in the tight seal is constrained to the inside of the pipette. The value of the voltages in each pipette is not necessarily the same. The relative size of the loose-patch pipette, as well as the voltage spread, has been exaggerated for the purpose of illustration.

---

at the pipette tip. There is at present no direct means to determine the internal potential with the loose-patch method. There are, however, some approaches that can be used to circumvent this difficulty, and they are enumerated in the Practical Methods section below. Still, certain measures have to be made for the correct determination of the pipette tip potential, which, as mentioned above, depends on the magnitude of the seal resistance. Simple model circuit analysis (Fig. 2A) shows that when having a small seal resistance (without yet considering the effect of surface ionic currents) the voltage at the pipette tip ($V_p$) is given by

$$V_p = V_{cmd}\left(\frac{R_s}{R_s + R_p}\right) = V_{cmd} \cdot A \tag{1}$$

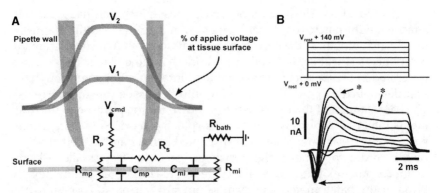

Fig. 2. Effects observed with loose-patch-clamp method on intact tissues. (**A**) Schematic of the model circuit of the loose patch. Resistances are shown for the pipette ($R_p$), patch membrane ($R_{mp}$), seal ($R_s$), input ($R_{mi}$), and bath ($R_{bath}$); see their use in the text. The capacitances are for the patch membrane ($C_{mp}$) and the membrane outside of the patch ($C_{mi}$). Also shown is the percent of applied voltage (see also Fig. 1) at the surface of the tissue (transparent gray traces) for two different pipette voltages ($V_1$ and $V_2$), showing the increase of spread with increased applied voltage. In this scheme, the minimum voltage needed for detectable sodium currents (for example) moves away from the pipette as the voltage is increased. This helps explain the behavior of the voltage dependence for the sodium currents from the surface of intact tissues when using large-diameter pipettes; there will be regions where the effective voltage is smaller than the applied (command) voltage as well as new regions included when the voltage is increased (recruitment). (**B**) An example of this behavior for responses obtained from the surface of an intact posterior pituitary using a pipette of 20- to 25-μm-diameter opening and pulses of about 8-ms duration (square traces on top, depolarizing voltages are expressed as $V_{rest} + V_{applied}$ since the resting membrane potential is not known). Here, sodium responses increase to a maximum but show little or no indication of reversal with further increases in applied voltage (saturation of responses; arrow), in contrast to the outward potassium currents where the reversal potential is not in this range (asterisks with arrows).

where $R_p$, $R_s$, and $V_{cmd}$ are the pipette resistance, seal resistance, and command voltages, respectively. Under the conditions of the loose patch, where $R_s$ is of the order of magnitude of $R_p$, the voltage at the pipette tip is thus reduced from the command voltage by a factor of $A$ (note that when $R_s \gg R_p$, then $V_p \approx V_{cmd}$, as in the tight-seal case). If currents arising from the surface ($I_{patch}$, flowing

**C**

**D**

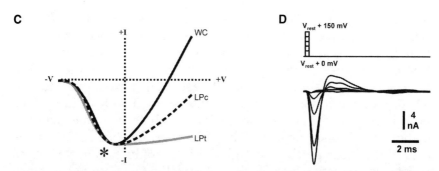

Fig. 2 (*Continued*) (**C**) Simulation of the behavior of voltage dependence of inward currents (e.g., sodium) obtained from whole cell (WC), loose patch on individual cells (LPc), and loose patch with large pipette openings (see **B**) on intact tissue (LPt). The values of the axis are omitted in order to emphasize a general behavior. In general, responses to voltages below initiation of reversal show similar behavior (asterisk, loose-patch voltages are assumed to be corrected), but reversal potentials are much harder to determine using large pipettes in intact tissue than in individual cells or with whole-cell configuration. While a hindrance in the determination of channel properties, such a loss of voltage control (bad space clamp) is an advantage when testing for local action potentials elicited by short-duration (≤0.4 ms) pulses applied to the patch (**D**). These responses have the properties of action potential currents (Marrero and Lemos, 2003). The current saturation properties are different from those with long pulses, consistent with increases in bad space-clamping when considering long-lasting depolarizations; short pulse depolarizations elicit responses after the cessation of the stimulus, when space-clamp errors would be diminished. The magnitude of the peak (inward) currents represents a population of channels being recruited by the short pulse rather than channels that are directly influenced by any part of the applied voltage.

through the series and seal resistances) are included, their effect is to add their "effective" voltage to the pipette tip as

$$V_p = V_{cmd} \cdot A \pm |I_{patch}| R_p \cdot A \qquad (2)$$

(see also Anson and Roberts, 2002; Stühmer et al., 1983), where the "±" sign emphasizes the effect of currents flowing into or out of the patch and $A$ is as defined in equation 1. Thus, with the loose-patch conditions, applied voltages are not only reduced but also the voltage errors associated with series resistance are affected (again, notice the approximation to the tight-patch case when $R_s \gg R_p$). As with standard patch-clamp methods, series resistance effects (the term $I_{patch}R_p$ in equation 2) may also produce errors in the determination of the onset of voltage-gated currents. For example, with a large series resistance that has not been well compensated for, an I-V plot of the sodium currents would show a sudden jump (i.e., negative dip) in the curve near the initial values of voltage depolarization (see Armstrong and Gilly, 1992; Sherman et al., 1999). In the case of the loose patch, although there is a reduction in this series resistance correction by a factor of $A$ with the low pipette resistance also taken into account, the *precision* is still affected. Ways to deal with this are given in the Practical Methods section, below.

Another source of error is the capacitance from the pipette walls and from the patch under study. Voltage (command) changes induce current transients that, if not compensated for, are the cause of the same type of errors present in standard patch-clamp methods. The transients usually reach values well above the applied voltages in addition to decaying exponentially with time, as indicated by $I_{capacitive} = C\partial V/\partial t$, where $C$ is any of the capacitances under the influence of the voltage change. Although capacitance effects due to the pipette can be dealt with in many efficient ways (see Practical Methods, below), in many cases the uncertainties arising from membrane and cell capacitances are not easily controllable or accounted for. Of major concern is the effect that membrane, or cell, capacitance has on the voltage *induced* inside the cell ($V_{int}$ in general; see Anson and Roberts, 2002; Roberts and Almers, 1992; Strickholm, 1995) if not properly accounted for or compensated. More specifically, for an applied voltage pulse that induces a rectangular current $I_m$ through the membrane, the change in membrane voltage ($\Delta V_I$) would be given by

$$\Delta V_I = I_m R_{mi} (1 - e^{-t/\tau}), \quad \tau = R_{mi} C_m \tag{3}$$

where $R_{mi}$ is the passive membrane resistance (sometimes referred to as input resistance, or lumped membrane resistance), $C_{mi}$ is the membrane capacitance, $\tau$ is the membrane time constant, and $t$ is the duration of the current. For long durations ($t \rightarrow \infty$), then $\Delta V_I \approx I_m R_{mi}$, while for short durations ($t \ll \tau$) then $(1 - e^{-t/\tau}) \approx t/\tau$ and $\Delta V_I \approx I_m R_{mi} \cdot t/\tau$. The significance of this is that, due to the time it takes to charge the membrane, the change in voltage inside the cell will lag behind the applied voltage step pulse and will also depend on the pulse duration ($t$). In the standard whole-cell configuration the effects of charging the membrane are detected (as current transients) and can be minimized to a reasonable degree by manual compensation (analog feedback from the amplifier or with a pulse that is scaled using the values set for the membrane capacitance and series resistance; Strickholm, 1995). With loose-patch, similar procedures can be used to automatically cancel current transients (even using the set values of series resistance obtained manually; see below). However, the compensations are to be made with the assumption that the detected transients contain only some information on the charging state of the membrane, since the internal voltage is not known. This voltage change would represent an addition to the uncertainty of $V_{int}$, and thus in $V_m$ (since $V_m = V_{int} - V_p$). It can clearly be seen that, in order to minimize the errors in the uncertainty $\Delta V_l$, the loose-patch method should be used in cells with low $R_{mi}$ or stable intracellular potentials.

In addition to the voltage and current deviations introduced by the relatively small seal resistance, one must consider that the current flowing through the seal is going to form part of the detected parameters in experiments. Namely, since the loose-patch-clamp is mostly used in voltage-clamp mode, the resulting currents detected are a mixture of membrane currents (the parameter under study) and seal currents (usually referred to as "leakage" currents and a hindrance in the detection of membrane responses). It is estimated that the currents flowing through the relatively low seal resistance (as compared to the pipette) are on the order of 10 times the magnitude of the membrane currents (Anson and Roberts, 2002; Roberts and Almers, 1992), thus interfering with the detection of the smaller ionic responses.

Finally, space clamp errors can arise from the "spreading" of uncontrolled voltage beyond the limit of the pipette opening (Fig. 1). These errors, again, are associated with the low-resistance seal and are traditionally thought of as arising mostly from beneath the rim of the pipette (Almers et al., 1984; Roberts and Almers, 1984), where there is a transition from the inside of the pipette to the outside bath. Although, with good compensation, the biophysical characteristics of detectable currents can be reliably obtained at low applied voltages (depending on the preparation and pipette, this range would usually be below 100 to 120 mV depolarizations from resting potential), it is, nevertheless, often difficult to obtain either proper reversal potentials or a classically behaved I-V plot at higher applied potentials. For example, when studying systems with classical fast sodium (inward) currents, the responses show the initiation of the reversal potential somewhere between −30 and −10 mV (calculated $V_m$). At voltages close to the reversal potential, and higher, the currents show reversed sodium currents but with an additional "inward" component. It is, therefore, generally thought that the difficulty in obtaining correct reversal potentials is due to the sum of responses from well-clamped areas and those from lower voltage areas (not well clamped) lying under the pipette rim (see examples in Almers et al., 1983; Antoni et al., 1988). This concept of "pipette rim responses" is generally based on the assumption that at the contact area the total seal resistance is a summation of series resistances, spread (usually linearly and radially) from the inner to the outer diameter of the rim. With this in mind, the "rim effects," therefore, would be improved by reducing the rim thickness. However, not all situations show an improvement or dependency on the pipette rim thickness (Almers et al., 1983) and the understanding of "space clamping" might be limited by the assumption of a linear behavior of seal resistance. At this point, there is no clear understanding on the dependence of the seal resistance with position within the pipette annulus.

The understanding of space-clamp problems is further limited when assuming space clamp effects to be constrained to the rim's outer limit. In fact, unlike model circuit approaches, there would be reason to believe that space-clamp spread would extend beyond the pipette rim limits when considering the electric fields emanating from the open spaces of the loose patch. A situation where pipette field effects could be involved is when loose-patch voltage clamp is used with large-opening (15–30 μm) pipettes on intact tissues (see previous section), where many components can be considered to form part of the patch (depending on the size of the components; glia + axons + neurons, for example). Effects from the interstitial spaces would also be observed in such cases and would be interpreted as an extra seal leakage spread throughout the immediate volume underlying the patch (Fig. 1). As such, the effects of the pipette potential could extend beyond the pipette's outer rim, thus enhancing the space clamp errors (Fig. 2A; a thorough approach has been made for similar situations when considering external stimulation of cardiac tissues; see Le Guyader et al., 2001; Newton et al., 1999; Plonsey and Barr, 1987; Pollard and Barr, 2006; Sepúlveda et al., 1989). In such cases, it is often found that progressive depolarization of the patch induces currents (sodium, for example) that start to reverse only slightly (or not at all) as the applied pulse is increased (Fig. 2B, arrow). When plotted against the corrected voltages, the plots show little change as voltage is increased beyond the optimum (asterisk in Fig. 2C). Thus, with large pipette openings, at low applied voltages, the results with whole tissues show consistency with standard patch-clamp methods and with loose-patch applied to continuous surfaces (e.g., individual cells), but show more dramatic deviations at higher depolarizations. It is thought that this is due to either the effects of space clamping extending beyond the pipette rim (see below) or currents flowing along the interstitial spaces into the volume of the tissue. Although this effect, which might be present to some degree even at low applied voltages, would translate as a cause of "incomplete clamping," it would still represent an advantage when studying tissue responses elicited by very short pulses (see Fig. 2D and examples, below): electrically active cells underlying the pipette would be stimulated to produce local action potentials that are not reduced or abolished by voltage clamping since the voltage is not well controlled.

## 3. Practical Methods

### 3.1. Making Pipettes

With some modifications in procedure, loose-patch pipettes can be made using the same equipment as with standard patch-clamp pipettes. Vertical or horizontal pipette pullers can thus be used (usually with a two-step pull protocol) with standard patch-clamp glass capillaries. For large-opening (15–30 µm) pipettes, we found useful, capillaries with 0.2- to 0.25-mm wall thickness since reproducibility of openings was high, perhaps due to the relative low heat required (this also helps extend the lifetime of the puller's filament). The tip of the pipettes is then fire-polished using a standard micro-forge. This helps also in having better control of the pipette opening as well as the tip rim thickness. To minimize pipette capacitance effects and to avoid the bath solution creeping up the pipette (which would change the pipette capacitance with time), the outside of the pipettes can be coated with different agents offered on the market for such purposes. Coating with Sigmacote® (Sigma-Aldrich, St. Louis, MO) is preferred over Sylgard® (Dow Corning, Midland, MI) since it is less work-intensive or skill-requiring; a polished unfilled pipette is mounted into a holder, attached to a syringe, and positive pressure is applied while dipping the pipette briefly into the Sigmacote. The coating should be dried also while the pipette is still mounted and while applying positive pressure, in order to avoid the Sigmacote's getting inside the pipette.

### 3.2. Electrode Wires

As with standard patch-clamp, silver (Ag) electrode wires, coated with AgCl, can be used efficiently for the inside of the pipette, and Ag/AgCl pellets can be used for the external references (bath electrodes). We found that although there were increases in noise when using agar bridges to connect the reference electrodes to the bath, the practice was still useful for avoiding Ag contamination of samples. The pipette electrodes can be chlorinated using many different means, such as bathing in sodium hypochlorite solutions or electrolysis (preferred method) in a 2M KCl solution: "+" side of a DC source at 20 to 50 V for about 2 to 5 seconds. While electrolysis usually renders the surface of the Ag wires well chlorinated, the mechanical stability of the AgCl coat is rather fragile. The need for repeatedly recoating becomes an annoyance but in addition, pieces of the coating sometimes fall inside to the tip of the pipette. To avoid this, it is recommended that the coated wire first be washed with distilled water (to remove any remaining KCl solution) and then be "fixed" by gentle heating. This procedure melts the AgCl coat onto the wire and renders it more mechanically sturdy. Excessive heat must be avoided in order not to "evaporate" the AgCl layer.

### 3.3. Solutions

Unless a tight seal is possible, the solutions used inside the pipette and in the bath should be equal in ionic composition. Although in certain cases different compositions have been used successfully (with an estimated seal resistance set to 10 times the pipette resistance; see Garcia et al., 1990), it is not recommended as a standard practice since the degree of mixing at the tip of the pipette is difficult to assess, especially when considering tests made on intact tissues.

### 3.4. Equipment

With the exception of amplifiers, most of the standard equipment used in tight-seal experiments is also used with loose-patch experiments. Patching individual cells and thin tissues can be done with the aid of an inverted microscope while with intact thick tissues it is more adequate to use a high-power dissection microscope. The amplifiers should have the capability of compensating small pipette and seal resistances, in addition to "fast" and "slow" capacitances. There are many choices already on the market (EPC9 and EPC10 from HEKA Electronik, Lambreaht/Pfalz, Germany, and PC-ONE with loose-patch head-stage, from Dagan Corp., Minneapolis, MN), some with specialized features suitable for loose-patch recording, which can compensate 100% of seal and series resistances. In cases where electrode polarization is a concern (although chlorination should take care of that; see above), it is better to use an amplifier that provides for separate pipette electrodes for stimulation and for recording, respectively (each one with its own bath electrode). Data are usually handled on-line by computer programs, or amplifiers, that can provide leak subtraction protocols (see below) in addition to the standard stimulation protocols. Standard manipulator systems can also be used with the popular configuration of a micromanipulator mounted on a macromanipulator (hand driven). We found that hydrostatic (oil or water) as well as motor-driven (inch-worm DC) micromanipulators are more advantageous than piezoelectric micromanipulators, which, although very precise, have a small range of movement. This is important if multiple tests will be done from different sites of a large sample (cell, tissue, or explant).

### 3.5. Procedures for Patching

This section emphasizes the importance of practical minimization of the errors associated with the determination of the pipette and seal resistances as well as procedures for leak subtraction. As mentioned above, and in contrast with tight-patch techniques, errors in the values of $R_p$ and $R_s$ in a loose-patch-clamp lead to significant deviations in the voltages and measured currents. While most amplifiers or programs used for loose-patch have the capability of automatically scaling the applied voltages (voltage compensation), not all of them scale the resultant currents automatically. In these cases the resulting currents can be corrected either off-line or before the start of an experiment by setting the amplifier's gain factor to the appropriate value. Nevertheless, since in most cases what are sought are the relative changes in current responses, what needs to be known with better accuracy is the effectiveness of the applied voltages (i.e., the corrected voltages). The first step after placing a pipette in the bath, and canceling "junction" (or baseline) currents, is to determine the resistance of the pipette ($R_p$). This can be done by measuring the current of a small voltage step (usually about 5 mV, mono- or bipolar with transient capacitances canceled), for $R_p = V_{step}/I_{step}$, and setting it in the amplifier (or computer program, as the setup requires) for series resistance compensation. However, since there is a finite resistance of the bath (its value depending on solution composition and pipette geometry; Roberts, 1987; Sakmann and Neher, 1983) in series with the true pipette resistance, this ratio overestimates the pipette resistance. For example, a pipette with cylindrical geometry (i.e., resembling a cylinder in form) would have a "bath" resistance (usually called field convergence resistance) given by $R_{bath} = 0.37\rho/d$, where $\rho$ is the resistivity of the solution at 55 to 65 $\Omega$·cm and $d$ is the diameter of the pipette opening. For a 15-μm-diameter pipette, this gives a range for $R_{bath}$ of 14 to 16 kΩ, which represents 9% to 14% of the measured $R_p$ (usually 110–150 kΩ). As a standard, it is safe to assume an overestimate of ≈10%, so that the true pipette resistance is set as = $0.9 \cdot R_p$.

The next step is to establish the value of the seal resistance ($R_s$). Two methods can be used for this purpose. One is to first bring the pipette to the cell (or tissue) until a decrease in the step pulse current is observed, indicating contact with the surface. The pipette is then approached further, stopped, and the seal resistance compensation is increased until the step pulse current equals that of the baseline. At this point, both the series resistance and the *A* factor (equations 1 and 2) are automatically compensated for. Further capacitance compensation and baseline zeroing can be done at this point.

Another variation of this method, and the one we prefer for the study of intact tissues, is to set the seal resistance to a desired value (always after determining the pipette resistance) and approach the tested surface until the step pulse current equals the baseline current (i.e., nulled). This has the advantage of having the seal resistance as an unchanged parameter when multiple sites are tested, and saves some time in its adjustment. Again, further capacitance compensation and baseline zeroing can be done as needed. With either method, there are some other practical considerations still to be taken into account. To minimize the errors, it is taken as a rule that the seal resistance be higher than twice the pipette resistance ($R_s > 2R_p$; Anson and Roberts, 2002). With cells, this can be achieved by either increasing the pipette pressure manually (approaching further) or by suctioning through the pipette. With intact tissues, where up to $R_s \approx 5R_p$ is used with large-opening pipettes, suctioning is difficult to achieve successfully, usually due to the mechanical properties of the samples, and it may lead to the irreversible distortion of the surface. In either case, the stability of the seal must be constantly monitored and adjusted (compensated) since it may change during the course of an experiment run. When made by mechanical manipulation, the re-tuning (as well as the initial seal) must be done with care so as not to affect the tested surface irreversibly, especially in the case of fragile whole tissues.

Since resistance and capacitance compensations are not absolutely accurate, leakage currents could still form part of the detected signals, and further leak subtraction is usually needed. The correction can be made by the amplifier itself (a feature in some models) or by setting a $P/N$ protocol in the patch-clamp computer program (more popularly used). The purpose of using $P/N$ protocols is to obtain from small pulses the current component that is purely ohmic (i.e., containing solely components from static capacitances as well as series and seal resistances, and usually called control pulses or subpulses) and subtract them from those that contain elicited ionic currents (test pulses). The subpulses have a magnitude equal to the value of the test pulse ($\pm P$) divided by an integer ($N$), with $N$ of them being applied before or after the test pulse, averaged (a feature in some software), then multiplied by $N$, and finally subtracted from the test pulse current (or added if negative subpulses were chosen). In some instances, a single subpulse of magnitude $P/N$ is used, scaled by $N$, and subtracted. Thus, caution should be taken when planning these protocols; it must be determined that the magnitude of the resulting subpulses does not elicit membrane ionic currents. This is particularly important since in most automatic methods the $P/N$ value is that of the corresponding test pulse and not a constant for all applied pulses. For example, in a situation where a series of increasing test pulses will be applied as 10 positive step pulses in increments of 10 mV, starting from 0 mV with $N$ chosen as 2, the smallest nonzero pulse will have $P/N$ = 5 mV (a reasonable value), whereas the highest subpulse will be 45 mV, a value that most probably elicits ionic currents. In this particular example it is clear that the $N$ must be increased in order to avoid eliciting ionic currents with the subpulses. On the other hand, increasing the number of subpulses to be equal to $N$ and then averaging and scaling increases the noise (by $\sqrt{N}$; the average, multiplying the noise by $1/\sqrt{N}$, has to be rescaled, multiplying it by $N$), and is not improved when subpulses are subtracted from test pulses (due to error propagation). Using a single $P/N$ subpulse (with $N > 1$) increases the noise more dramatically, since it has to be multiplied by $N$ before subtraction.

Note that, in comparison, the noise is less affected with programs that simply do an addition of all subpulses before subtraction from the test pulse. It is likely that, due to the low seal resistance, noise problems will be encountered especially in the cases where there are small membrane responses (recall that the currents of interest are around 10% of the total). For these reasons, it is recommended that the subpulse currents be stored independently from the test pulse responses, whether or not processed data are displayed on-line during an experiment (i.e., to display $P/N$-subtracted currents). This gives the experimenter the opportunity to later choose scaled subpulses that may be more suitable for off-line processing.

In the cases where there is the need to have a large number of subpulses (usually above the $N = 4$ taken as standard), the noise can be reduced by averaging multiple runs (noise is reduced by a factor of $1/\sqrt{n}$, where $n$ is the number of runs). Finally, the timing between each subpulse and between the group of subpulses and the test pulse must be chosen carefully. While it is desired that the subpulses be as close as possible to the test pulse (in order to account for possible changes between test pulses), the decay of each response is not instantaneous; pulses still exhibit a decay "tail" since the compensations are not perfect. The timing between each pulse must be such that their decay tails do not interfere with each other. The presence of decay tails is more noticeable when using large pipettes in intact tissues. Since the time to decay to baseline is dependent on the pulse magnitude (although with the same time-constant), some programs offer the feature of choosing the timing between subpulses independently of the timing between the subpulse group and the test pulse. The situation is worse when, as explained above, very long pulses are used and internal potentials are changed considerably (again, more noticeable in intact tissues when using large pipettes). Thus, when there is an even longer decay of baseline current due to the internal voltage changes, the timing between all pulses must be monitored and corrected if necessary. In some cases, the problems arising from this, as well as those from eliciting active currents with prepulses, can be resolved by setting the holding potential of the subpulses independently from the holding potential of the test pulses (a feature present in most $P/N$-able programs). Care must be taken with the timing of the start and end of the prepulse holding potential period, since the change in holding potential may represent a step pulse by itself.

As was mentioned in the introduction, the patch under the pipette is sensitive to pharmacological changes in the external (bath) environment. This is due to the low seal resistance of the loose-patch where there would be diffusion from the outside to the inside of the pipette. Although changes can be readily seen with many applied substances that are effective in the micromolar range (toxins and ion channel blockers), in practice more care should be taken when testing the pharmacology of ionic currents. Namely, this approach usually requires application of pharmacology at higher concentrations than needed, for example, in whole cells, with relatively long times before observing any effect, and the concentrations at the membrane surface area of the patch are not precisely known (i.e., unknown degree of mixing inside the pipette). It was found most useful and reliable if the pharmacology is changed inside the pipette to the same degree as in the bath. Two methods for achieving this are suggested. One is using a pipette holder that permits the solution exchange of the pipette. The exchange should be made with little or no pressure changes between the pipette and outside. Otherwise, caution should be taken in not disrupting the tested surface during the exchange, and it is recommended that the pipette be lifted slightly during the exchange and resealed at the original location. As a control, the procedure should be done with no changes in the media, in order to test possible "mechanical" effects (i.e., tests in procedure reproducibility). This method is not recommended for unprotected tissues, as it may perturb some of the surface components. The other method is to lift the pipette from the patch, and exchange its contents with the bath media (assuming that the pharmacology has already been changed in the bath) by repeatedly alternating suction and pressure. The amount of lift is usually determined as that which does not disturb the test surface during the exchange. The pipette resistance should be monitored during the procedure and checked for deviations from its original value at the end of the exchange (for example, checking for air bubbles left inside the pipette). Tests for reproducibility should also be done with unaltered media when using this method.

### 3.6. Methods for Circumventing the Unknown $V_{int}$

To determine the membrane potential ($V_m$), the intracellular potential ($V_{int}$) should be known. There is at present no direct or reliable means for determining, or controlling, the potential inside cells using solely loose-patch methods. The best way, although not applicable to all systems tested, is to measure the intracellular potential using intracellular electrodes or whole-cell configurations (Antoni et al., 1988; Carta et al., 2004; Delay and Restrepo, 2004; Forti et al., 1997; Garcia et al., 1990; Körper et al., 1998; Marrero et al., 1989; Nygård et al., 2005; Re et al., 2003; Ruff, 1996, 1999; Wolters et al., 1994; Yee Chin, et al., 2004). Different approaches can be used, however, to *estimate* the internal potential of cells studied with loose-patch-clamp. Most approaches use known characteristics of sodium currents as a guide. For example, a crude, yet quick, way is to generate an I-V curve (V as the applied command values) and finding the voltage at which the maximum occurs (see, for example, Fig. 2C). Adding this value to the one obtained from an I-V using a more precise method (from, say, intracellular recording) would yield, approximately, the value of the internal potential. This approach assumes that the basic current-voltage dependence is not different between the well-clamped and loose-patch methods, however, and, therefore, is not a reliable means of accurately determining the internal potential. It is, nevertheless, still useful when searching for relative changes in internal potentials and ionic currents. In another more precise variation, protocols are applied for testing the activation or inactivation properties of sodium currents in the loose-patch configuration. The protocols are the same as those used in the determination of these biophysical properties when tested using tight-patch methods (Almers et al., 1983, 1984; Antoni et al., 1988; Beam et al., 1985; Caldwell et al., 1986; Milton and Behforouz, 1995; Ruff, 1996, 1999). For example, the half-height voltage for inactivation, obtained with the values of the *applied* loose-patch voltages, is corrected (added) with the known values by other means (intracellular recordings, for example). This yields an estimated intracellular voltage that relies on the assumption that biophysical properties of the sodium channels are unchanged between recording methods. Nevertheless, it must be remembered that in the not well-clamped approach of the loose-patch, currents flowing across the membrane will affect the internal voltage. Thus, at best, especially in cells with high impedance, these methods for estimating internal voltage yield information on the internal potential at the maximum of (for example) the sodium currents, and not at the resting (unstimulated) state. With either of the above approaches there is the danger of underestimating the effect of membrane currents on the internal voltage, thus yielding possible overestimates (more negative) in the assessment of the resting potential at peak current values. Such considerations should still be kept in mind even with systems known to be stable in their resting potential (i.e., muscle fiber cells).

Methods for circumventing the difficulty in determining and controlling the membrane potential have already been developed to be used with the loose patch. These are mostly based on the simultaneous use of whole-cell (or intracellular) and loose-patch techniques (Almers et al., 1983; Anson and Roberts, 1998; Roberts et al., 1990). In practice, the methods require the use of two types of pipettes, one for direct intracellular application of voltages or currents, and one in the loose-patch configuration. In this way, intracellular voltages can be measured and controlled while using the loose patch either for stimulating specific areas or for recording currents localized at the position of the loose-patch pipette (Anson and Roberts, 1998), thus extending the use of loose-patch techniques to cells with high-input impedance. The roles of each method can be switched, whenever it would be more reliable to stimulate through the loose patch and record from the whole-cell configuration (such as when intracellular voltage control is not effective). The methods are used in cells large enough to accommodate both approaches and, in certain configurations, have the advantage of being practically insensitive (as compared to loose-patch alone) to errors in series and seal resistances. The combination of loose-patch and intracellular voltage control has also been successfully used in mapping ion channel distribution on the surface of cells with more precision than with loose-patch alone (Anson and Roberts, 1998).

### 3.7. Space-Clamp Problems

Using the combination of intracellular voltage control together with the loose patch, it was possible to prove that space-clamp problems are due to uncontrolled voltages beyond the tip of the pipette opening (Almers et al., 1983). Thus, space-clamp problems are reduced when the loose patch is used simultaneously with methods that control the internal potential (intracellular, whole-cell, etc., see above).

In another approach that uses the loose patch alone, space-clamp problems can also be minimized through the use of concentric pipettes (Almers et al., 1984; Roberts and Almers, 1984). With this arrangement, the voltage inside both pipettes is controlled to the same values, while recording only from the central pipette. In this way, currents induced at the central pipette opening and beyond would have the same magnitude. So, even when recordings include components outside the main central pipette, currents correspond to known applied voltages. Although the pipette preparation is rather difficult, the method solves fairly well the problem of the space clamp over continuous surfaces (*i.e.*, individual cells).

## 4. Applications of the Loose-Patch-Clamp: Recordings from the Surface of Intact Tissues

The inherent problems present in the loose-patch technique are actually advantages when considering experiments made on the surface of intact tissues. Many tissue preparations are covered with a thin basal lamina (see Fig. 1) that may permit efficient stimulation of the underlying cells (or components). In others, a thick sheath layer is encountered as a barrier and presents more complicated problems when using the loose patch. In some cases, the porosity of these electrically inert layers, such as those composed of colla-gen, acts as an additional seal leakage that may enhance the errors in voltage and current determination. In addition, the large thick-ness of the layers encountered in many preparations makes it very difficult to determine the "effective" applied voltage at the surface of the underlying cells.

A more difficult situation is encountered when the layer is formed by "protective" nonporous components, such as with glial sheaths. In this case, the layers may either act as an effective, dielectric, thick barrier, or, worse, may have the property of channeling cellular currents through less resistive paths under the sheath and, there-fore, of not being detected at the loose patch. Although in many instances further processing of the tissues is made in order to eliminate surface barriers, the processing may defeat the purpose of examining intact tissues. With the exception of situations with such "channeled" currents, the problems related to thick layers can be minimized by using pipettes with an opening diameter much bigger than the thickness of the layer. This solution, however, may result in patches being made over surfaces covering many components (cells, axons, etc.). If this is the case, the interstitial spaces between multiple components might become an additional problem in the determination of voltage dependence of responses, since they would represent an additional seal leakage (see Fig. 1) and would exacerbate space clamping (it would be expected that the space clamping would be extended into the *volume* of the tissue due to the nonzero resistance of the interstitial spaces). This approach, therefore, is not recommended for studying the biophysical characteristics of ion channels when tested in the classical way, that is, using standard voltage-activation and -inactivation stimulus protocols. When used, however, there must be some assessment of the conditions of the surface to be tested as well as the composition of the volumes immediately below the desired tested areas. As with any other approach using the loose patch, it is also recommended that I-V plots be made as part of any experiment in order to assess the degree of deviation from standard behavior.

Large pipettes in intact tissues are very useful for the pharmacological characterization of ion channels and for comparing responses from different areas in whole, intact systems (Marrero and Lemos, 2003), as well as for the study of nerve stimulation effects on elicited currents (Marrero et al., 1989; Marrero and Lemos, 2005). For example, using pulses of 4-ms duration or more, loose-patch responses from the intact posterior pituitary of rats exhibit fast inward currents followed by fast and slow outward components (Fig. 2B). Pharmacological characterization reveals that they are, respectively, sodium [tetrodotoxin (TTX) sensitive], fast potassium "A" [4-aminopyridine (4-AP) sensitive], and slow potassium B-K [tetraethylammonium (TEA) sensitive] currents. The posterior pituitary is composed, mostly, of secretory terminals belonging to magnocellular neurons in the supraoptic (SON) and paraventricular (PVN) nuclei of the brain. When tested with the loose patch, the SON and PVN areas show that these currents are also present,

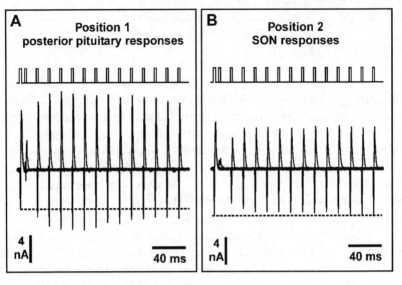

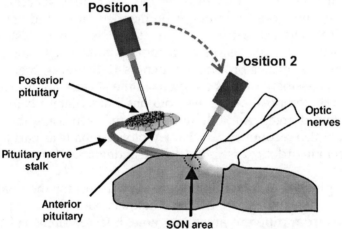

Fig. 3. Comparison of loose-patch responses from different regions of the hypothalamic-neurohypophysial system (HNS) of the rat. Using a whole HNS explant (drawing at bottom half), loose-patch responses are obtained from the neurohypophysis (NH, inverted in order to gain access to its surface, responses in **A**) with the pipette at position 1, or from the surface of the hypothalamic supraoptic nuclei area (SON, responses in **B**) at position 2. The same protocol was used in both instances, where a depolarizing pulse (+50 mV, 4-ms duration, square line on top) is followed by an identical one sometime later. These experiments consisted of a run of 14 episodes, each with the time between the two pulses increased by a constant value. In the neurohypophysis or pituitary, the response from the second pulse shows an increase over the first response (control, dashed line) as the time between pulses is increased (**A**). This behavior is not seen in the responses from the hypothalamic areas of SON (**B**) or paraventricular nuclei (not shown). The records also show differences between the two areas, by the proportion of types of currents in individual responses: faster outward ("A" type) potassium current in the pituitary terminals (NH) than in the cell body (SON) regions (compare control of each set).

although not in the same proportion (i.e., the relative magnitude of the potassium current to the sodium currents) as in the posterior pituitary (Fig. 3). Thus, the loose-patch method uniquely allows direct comparison between neuronal cell body and terminal electrophysiology. All of these responses have very similar pharmacological characteristics to those in isolated SON cells and neurohypophysial terminals (Marrero and Lemos, 2003). The proportion of currents from the SON, although characteristic for each area, is not the same as with isolated neurons, perhaps due to differences between the in situ and isolated conditions.

Two effects are observed that are definitively associated with conditions in situ, since they are not observed in isolated terminals or in perturbed preparations. One is the detection of local action potential currents within the loose-patch responses (see Marrero and Lemos, 2003, 2005), more distinguishable when pulses of short duration are used (Fig. 2D). It is believed that this is the result of incomplete clamping on excitable tissue; pulses excite the underlying tissue where, not being well-clamped, the internal voltage fluctuates freely as in an elicited action potential. The loose patch, therefore, can detect some of the associated currents of these local excitations elicited by stimulation through the loose-patch pipette. This represents an advantage over other loose-patch procedures where action potential currents are also detected, but they would be either spontaneous or induced by using a second electrode (Carta et al., 2004; Delay and Restrepo, 2004; Nunemaker et al., 2003; Re et al., 2003; Smith and Otis, 2003). This advantage is even more notable when considering that, within the limits of the action potential duration, the action potential firing rate thus can be controlled. The other unique effect observed in the in situ cases is the potentiation of currents by previous stimulation. This is observed in the posterior pituitary of the rat (Marrero and Lemos, 2005; Fig. 3A) and in the optic nerve of *Rana pipiens* (Marrero et al., 1989; Fig. 4B). In the posterior pituitary, the effect is dependent on the activity of calcium and fast potassium channels ("A"), and is the result of activity-induced calcium depletion in the narrow interstitial spaces between terminals in situ; the lower calcium concentrations favor the enhancement of sodium currents. The potentiation is not observed, however, in the SON areas (Fig. 3B). In *Rana pipens*, the effect, also dependent on external calcium concentrations, is the result of nerve activity influencing glial responses (Fig. 4B), as has been corroborated using whole-cell patch-clamp configurations

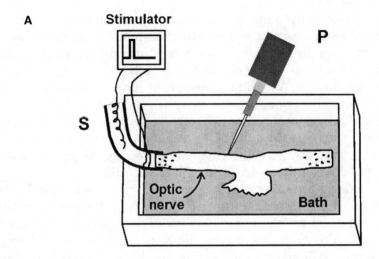

Fig. 4. Effect of whole nerve stimulation on the loose-patch responses from the surface of the optic nerve of *Rana pipiens*. (**A**) Experimental setup showing a frog optic nerve mounted on a suction electrode (S) and the placement of the loose-patch pipette (P). (**B**) Changes in response amplitude for a loose-patch depolarizing pulse of 50 mV (relative to resting potential: $V_{rest}$) and of 5-ms duration. Shown are responses for control ("before") and 20 ms after nerve stimulation ("after," nerve stimulation at "stimulus"). Stimulation of the nerve is evident as a large inward deflection indicating currents from a propagating action potential (usually referred to as action potential field currents) originating at the distal end (suction end) of the nerve. Responses elicited after nerve stimulation show an increase in currents 15 ms after the action potential peak current (arrow with *). The thin line is zero current baseline and the dashed line is control amplitude, extended to "after" response for comparison. This potentiation effect in the optic nerve (called "facilitation" in the original experiments) is affected by external calcium concentrations (Marrero and Orkand, 1993) in a similar manner as in the pituitary (Marrero and Lemos, 2005).

on surface glia from mechanically de-sheathed optic nerves (Marrero and Orkand, 1993). Decreases in intensity of the potentiation in de-sheathed nerves are explained by the exposure of the surface glia to the controlled ion concentrations of the exterior (bath). The unperturbed conditions in the intact tissue thus would explain the inability to observe this phenomenon in isolated terminals or cells. Without the surface protection and limited interstitial

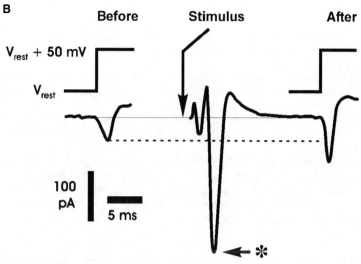

Fig. 4 (*Continued*)

space present in intact tissue, the external conditions in isolated terminals (i.e., concentration of ions) would not be subject to changes due to channel activity since they are kept constant by the "infinite" reservoir of the bath.

## 5. Conclusion

The examples cited in this chapter, using different types of intact tissues, demonstrate that the loose-patch-clamp method is an applicable approach that, in some instances, is superior to the tight-patch-clamp method. With intact tissues, even in cases where there is much uncertainty as to controllable loose-patch parameters, it can still be used to determine "novel" properties (e.g., action potential currents) not demonstrated, or difficult to detect, with standard tight-seal methods. Application of the loose-patch-clamp method to other intact tissues opens new avenues for the investigation of their properties in situ. Of particular importance is the ability to examine electrophysiological effects where homeostasis, cell interactions, and tissue architecture are normal or, at least, intact. It is even possible to directly compare the functional properties of different compartments of intact cells/systems. Therefore, the loose-patch-clamp method should be a powerful approach to apply to many intact nervous systems in vivo.

# References

Almers, W., Roberts, W. M., and Ruff, R. L. (1984) Voltage clamp of rat and human skeletal muscle: measurements with an improved loose-patch technique. *J. Physiol.* **347**, 751–768.

Almers, W., Stanfield, P. R., and Stühmer, W. (1983) Lateral distribution of sodium and potassium channels in frog skeletal muscle: measurements with a patch-clamp technique. *J. Physiol.* **336**, 261–284.

Anson, B. D. and Roberts, W. M. (1998) A novel voltage clamp technique for mapping ionic currents from cultured skeletal myotubes. *Biophys. J.* **74**, 2963–2972.

Anson, B. D. and Roberts, W. M. (2002) Loose-patch voltage clamp technique, in *Neuromethods: Patch Clamp Analysis: Advanced Techniques* (Walz, W., Boulton, A. A., and Baker, G. B., eds.) Humana Press, Totowa, NJ, pp. 265–286.

Antoni, H., Böcker, D., and Eickhorn, R. (1988) Sodium current kinetics in intact rat papillary muscle: measurements with the loose-patch-clamp technique. *J. Physiol.* **406**, 199–213.

Armstrong, C. M. and Gilly, W. F. (1992) Access resistance and space clamp problems associated with whole-cell clamping, in *Methods in Enzymology* (Rudy, B. and Iverson, L., eds.) Academic Press, New York, pp. 100–122.

Beam, K. G., Caldwell, J. H., and Campbell, D. T. (1985) Na channels in skeletal muscle concentrated near the neuromuscular junction. *Nature* **313**, 588–590.

Blanco, R. E., Marrero, H., Orkand, P. M., and Orkand, R. K. (1993) Changes in ultrastructure and voltage-dependent currents at the glia limitans of the frog optic nerve following retinal ablation. *Glia* **8**, 97–105.

Caldwell, J. H., Campbell, D. T., and Beam, K. G. (1986) Na channel distribution in vertebrate skeletal muscle. *J. Gen. Physiol.* **87**, 907–932.

Caldwell, J. H. and Milton, R. L. (1988) Sodium channel distribution in normal and denervated rodent and snake skeletal muscle. *J. Physiol.* **401**, 145–161.

Carta, M., Mameli, M., and Valenzuela, C. F. (2004) Alcohol enhances GABAergic transmission to cerebellar granule cells via an increase in Golgi cell excitability. *J. Neurosci.* **24**, 3746–3751.

Delay, R. and Restrepo, D. (2004) Odorant responses of dual polarity are mediated by cAMP in mouse olfactory sensory neurons. *J. Neurophysiol.* **92**, 1312–1319.

Dunant, Y. and Muller, D. (1986) Quantal release of acetylcholine evoked by focal depolarization at the Torpedo nerve-electroplaque junction. *J. Physiol.* **379**, 461–478.

Eickhorn, R., Weirich, J., Hornung, D., and Antoni, H. (1990) Use dependence of sodium current inhibition by tetrodotoxin in rat cardiac muscle: influence of channel state. *Pflügers Arch.* **416**, 398–405.

Forti, L., Bossi, M., Bergamaschi, A., Villa, A., and Malgaroli, A. (1997) Loose-patch recordings of single quanta at individual hippocampal synapses. *Nature* **388**, 874–878.

Garcia, U., Grumbacher-Reinert, S., Bookman, R., and Reuter, H. (1990) Distribution of $Na^+$ and $K^+$ currents in soma, axons and growth cones of leech Retzius neurones in culture. *J. Exp. Biol.* **150**, 1–17.

Körper, S., Wink, M., and Fink, R. H. (1998) Differential effects of alkaloids on sodium currents of isolated single skeletal muscle fibers. *FEBS Lett.* **436**, 251–255.

Le Guyader, P., Trelles, F., and Savard, P. (2001) Extracellular measurement of anisotropic bidomain myocardial conductivities. I. Theoretical analysis. *Ann. Biomed. Eng.* **29**, 862–877.

Lupa, M. T., Krzemien, D. M., Schaller, K. L., and Caldwell, J. H. (1995) Expression and distribution of sodium channels in short- and long-term denervated rodent skeletal muscles. *J. Physiol.* **483**, 109–118.

Marrero, H., Astion, M. L., Coles, J. A., and Orkand, R. K. (1989) Facilitation of voltage-gated ion channels in frog neuroglia by nerve impulses. *Nature* **339**, 378–380.

Marrero, H. G. and Lemos, J. R. (2003) Loose-patch clamp currents from the hypo-thalamo-neurohypophysial system of the rat. *Pflügers Arch.* **446**, 702–713.

Marrero, H. G. and Lemos, J. R. (2005) Frequency-dependent potentiation of voltage-activated responses only in the intact neurohypophysis of the rat. *Pflügers Arch.* **450**, 96–110.

Marrero, H. and Orkand, R. K. (1993) Facilitation of sodium currents in frog neu-roglia by nerve impulses: dependence on external calcium. *Proc. Biol. Sci.* **253**, 219–224.

Milton, R. L. and Behforouz, M. A. (1995) Na channel density in extrajunctional sarcolemma of fast and slow twitch mouse skeletal muscle fibres: functional implications and plasticity after fast motoneuron transplantation on to a slow muscle. *J. Muscle Res. Cell Motil.* **16**, 430–439.

Milton, R. L. and Caldwell, J. H. (1990) How do patch clamp seals form? A lipid bleb model. *Pflügers Arch.* **416**, 758–762.

Milton, R. L., Lupa, M. T., and Caldwell, J. H. (1992) Fast and slow twitch skeletal muscle fibres differ in their distribution of Na channels near the endplate. *Neurosci Lett.* **135**, 41–44.

Newton, J. C., Knisley, S. B., Zhou, X., Pollard, A. E., and Ideker, R. E. (1999) Review of mechanisms by which electrical stimulation alters the transmem-brane potential. *J. Cardiovasc. Electrophysiol.* **10**, 234–243.

Nunemaker, C. S., DeFazio, R. A., and Moenter, S. M. (2003) A targeted extracel-lular approach for recording long-term firing patterns of excitable cells: a practical guide. *Biol. Proced. Online* **5**, 53–62.

Nygård, M., Hill, R. H., Wikstrom, M. A., and Kristensson, K. (2005) Age-related changes in electrophysiological properties of the mouse suprachiasmatic nucleus in vitro. *Brain Res. Bull.* **65**, 149–154.

Ortiz, S., Rodriguez, O., Orkand, P. M., Orkand, R. K., and Marrero, H. (1988) Voltage-gated currents recorded from the surface of the frog optic nerve. *P. R. Health Sci. J.* **7**, 141–143.

Plonsey, R. and Barr, R. C. (1987) Interstitial potentials and their change with depth into cardiac tissue. *Biophys. J.* **51**, 547–555.

Pollard, A. E. and Barr, R. C. (2006) Cardiac microimpedance measurement in two-dimensional models using multisite interstitial stimulation *Am. J. Physiol. Heart Circ. Physiol.* **290**, H1976–H1987.

Re, L., Corneli, C., Sturani, E., et al. (2003) Effects of Hypericum extract on the acetylcholine release: a loose patch clamp approach. *Pharmacol. Res.* **48**, 55–60.

Re, L., Rossini, F., Re, F., et al. (2006) Prion protein potentiates acetylcholine release at the neuromuscular junction. *Pharmacol. Res.* **53**, 62–68.

Roberts, W. M. (1987) Sodium channels near end-plates and nuclei of snake Skeletal Muscle *J. Physiol.* **388**, 213–232.

Roberts, W. M. and Almers, W. (1984) An improved loose patch voltage clamp method using concentric pipettes. *Pflügers Arch.* **402**, 190–196

Roberts, W. M. and Almers, W. (1992) Patch voltage clamping with low-resistance seals: loose patch clamp. *Methods Enzymol.* **207**, 155–176.

Roberts, W. M., Jacobs, R. A., and Hudspeth, A. J. (1990) Colocalization of ion channels involved in frequency selectivity and synaptic transmission at presynaptic active zones of hair cells. *J. Neurosci.* **10**, 3664–3684.

Ruff, R. L. (1996) Sodium channel slow inactivation and the distribution of sodium channels on skeletal muscle fibres enable the performance properties of different skeletal muscle fibre types. *Acta Physiol. Scand.* **156**, 159–168.

Ruff, R. L. (1999) Effects of temperature on slow and fast inactivation of rat skeletal muscle Na(+) channels. *Am. J. Physiol.* **277**, C937–C947.

Sakmann, B. and Neher, E. (1983) Geometric parameters of pipettes and membrane patches, in *Single Channel Recording*, 1st ed. (Sakmann, B. and Neher, E., eds.), Plenum Press, New York, pp. 37–51.

Sepúlveda, N. G., Roth, B. J., and Wikswo, J. P., Jr. (1989) Current injection into a two-dimensional anisotropic bidomain. *Biophys. J.* **55**, 987–999.

Sherman, A. J., Shrier, A., and Cooper, E. (1999) Series resistance compensation for whole-cell patch-clamp studies using a membrane state estimator. *Biophys. J.* **77**, 2590–2601.

Smith, S. L. and Otis, T. S. (2003) Persistent changes in spontaneous firing of Purkinje neurons triggered by the nitric oxide signaling cascade. *J. Neurosci.* **23**, 367–372.

Strickholm, A. (1995) A single electrode voltage, current- and patch-clamp amplifier with complete stable series resistance compensation. *J. Neurosci. Methods* **61**, 53–66.

Stühmer, W., Roberts, W. M., and Almers, W. (1983) The loose patch clamp, in *Single-Channel Recording* (Sakmann, B. and Neher, E., eds.), Plenum, New York, pp. 123–132.

Wolters, H., Wallinga, W., Ypey, D. L., and Boom, H. B. (1994) Ionic currents during action potentials in mammalian skeletal muscle fibers analyzed with loose patch clamp. *Am. J. Physiol.* **267**, C1699–C1706.

Yee Chin, J., Matthews, H. R., Fraser, J. A., Skepper, J. N., Chawla, S., and Huang, C. L. (2004) Detubulation experiments localise delayed rectifier currents to the surface membrane of amphibian skeletal muscle fibres. *J. Muscle Res. Cell Motil.* **25**, 389–395.

# 12

# Recording Currents from Channels and Transporters in Macropatches

*Guiying Cui, Matthew D. Fuller,
Christopher H. Thompson, Zhi-Ren Zhang,
and Nael A. McCarty*

## 1. Introduction

This chapter describes methods for the study of ion channels and transporters by recording from membrane macropatches. While investigators have made use of many different cell types for such experiments, we focus here on studies of these proteins expressed exogenously in *Xenopus* oocytes. We rely on this model system in our laboratory for a number of reasons, including the fact that we are able to obtain seals of very high resistance, typically >150 GΩ. Where possible, we draw comparisons with the study of the same channels by other macroscopic recording techniques; where possible, we also compare results from macropatch experiments with results of similar experiments using single-channel recording. We provide examples of experiments with the following proteins: the human cystic fibrosis transmembrane conductance regulator (CFTR), the rabbit ClC-2 voltage-gated chloride channel, and a $Na^+/Ca^{2+}$ exchanger from *Drosophila melanogaster* (Calx1.2).

## 2. Preparation of Oocytes

Adult female *Xenopus laevis* (Xenopus One, Ann Arbor, MI, or Xenopus Express, Woods Hole, MA) are placed under anesthesia by whole-body immersion in ice-cold tricaine methanesulfonate (MS-222, 0.2% in deionized water). The depth of anesthesia is monitored by pinching the skin in the web of a hindfoot. Once ready, the frog is laid on a bed of ice, belly up. A small, 1.5–2 cm

From: *Neuromethods, Vol. 38: Patch-Clamp Analysis: Advanced Techniques, Second Edition*
Edited by: W. Walz @ Humana Press Inc., Totowa, NJ

laparoscopic incision is made into the skin and underlying musculature of the abdomen, and a portion of the ovary is explored. Several hundred oocytes of varying stages are removed and placed in $Ca^{2+}$-free Barth's solution (containing, in mM: 88 NaCl, 7.5 Tris base, 1 KCl, 2.4 $NaHCO_3$, 0.82 $MgSO_4$; pH 7.5) with 2 mg/mL type 1A collagenase, and incubated with gentle agitation for 2.5 hours at room temperature. Once the cells have become dispersed, stage V to VI cells are isolated and washed three times in Barth's solution with $Ca^{2+}$ (Barth's solution plus: 0.33 mM $Ca(NO_3)_2$ and 0.41 mM $CaCl_2$; pH 7.5). Oocytes lacking follicle cells are selected and are incubated until needed at 18°C in half-strength Liebovitz's L-15 medium with L-glutamine (Invitrogen, Carlsbad, CA) supplemented with HEPES (22 mM, pH 7.5), gentamicin (50 μg/mL), penicillin (10 U/mL), streptomycin (100 μg/mL), and 2.5% to 5% horse serum (Quick et al., 1992). Postsurgery, the incision is closed using a 4-0 Ethilon monofilament suture, and the frog is allowed to recover for 2 hours in isolation in a shallow amount of tank water. The frog is then returned to the colony and observed over the remainder of the day. Upon recovery from surgery, frogs behave in a manner indistinguishable from the others.

Oocytes are injected with in vitro transcribed complementary RNA (cRNA) (mMessage mMachine kit, Ambion, Inc., Austin, TX), in a volume of 50 to 100 nL in diethylpyrocarbonate (DEPC)-treated water. For most experiments using either single-channel recording or two-electrode voltage clamp (TEVC), cRNA is prepared from traditional expression vectors such as pSP64poly(A), pAlter, pBlue-Script, and pSPORT. For some experiments using macropatch recording, where a large number of channels are needed or the single-channel conductance is low, cRNA is prepared from constructs in the pGEMHE vector, which includes the 5' and 3' untranslated regions from the *Xenopus* β-globin gene for greatly improved expression (Liman et al., 1992). For macropatch experiments, oocytes are injected with 25 to 100 ng of cRNA for CFTR, 50–100 ng of cRNA for ClC-2, or 30–50 ng of cRNA for Calx1.2. For single-channel and TEVC experiments, oocytes are injected in a range of 5 to 100 ng of CFTR cRNAs, or 5 to 25 ng of ClC-2 cRNAs. For TEVC experiments with CFTR, we also inject 0.4 ng of cRNA for the $\beta_2$-adrenergic receptor ($\beta_2$-AR), allowing activation of CFTR by exposure to isoproterenol in the bathing solution. Recordings are made 24 to 72 hours after the injection of cRNAs. For macropatch, giant patch, and single-channel patch experiments, oocytes are shrunk in a hypertonic solution [in mM: 200 monopotassium aspartate, 20 KCl, 1 MgCl$_2$, 10 ethylene glycol-bis(2-aminoethylether)-N,N,N',N'-tetraacetic acid (EGTA), and 10 HEPES-KOH, pH 7.2] and the oocyte vitelline membrane is removed manually using fine forceps.

## 3. Preparation of Electrodes

We use borosilicate glass from Sutter Instruments (Novato, CA; catalog number BF-150-10) for all of our electrodes. The filament fused to the inner face of the glass tubing facilitates the removal of bubbles when filling with electrode solution. All of our electrodes are pulled using a Sutter Instruments P-2000 laser-based puller. The following are pull programs for each type of electrode; these should be used only as starting conditions, as the numbers will need to be tweaked for each puller/glass combination.

For macropatch recordings:

Line 1: HEAT = 430, FIL = 5, VEL = 40, DEL = 230, PUL = 0
Line 2: HEAT = 410, FIL = 4, VEL = 35, DEL = 230, PUL = 0.

The pull is completed in three loops, resulting in tip diameters of ~12 μm and 8 to 9 μm before and after fire-polishing, respectively. Pipette resistance is 3 to 4 MΩ after fire-polishing (with 150 mM salt solutions in pipette and bath). Seal resistances on oocyte membranes for inside-out macropatches are typically >200 GΩ and for outside-out macropatches are typically >150 GΩ. For outside-out macropatches, our best recordings come from pipettes that do not narrow down to the tip at an angle, but that instead have a region of cylindrical shape at the very tip. This seems to promote stronger interaction between the membranes and the inner surface of the glass. For inside-out macropatches, better seals are formed with pipettes that do not have such a region at the tip.

For single-channel patch recordings:

Line 1: HEAT = 500, FIL = 4, VEL = 45, DEL = 200, PUL = 0.

The pull is completed in four loops, resulting in tip diameters of ~1.5 to 2.5 μm and 1 μm before and after fire-polishing, respectively. Pipette resistance is ~10 MΩ after fire-polishing (with 150 mM salt solutions in pipette and bath). Seal resistances for inside-out or outside-out single-channel patches on oocyte membranes are 200 to 500 GΩ. Our best recordings with long-lasting, high resistance seals come from pipettes that have a region of cylindrical shape at the very tip.

For giant patch recordings:

Line 1: HEAT = 420, FIL = 5, VEL = 35, DEL = 220, PUL = 0.

The pull is completed in six loops, resulting in tip diameters of ~20 µm and 18 µm before and after fire-polishing, respectively. Pipette resistance is 0.1 to 0.5 MΩ after fire-polishing (with 150 mM salt solutions in pipette and bath). Seal resistances for inside-out giant patches on oocyte membranes are typically 25 to 50 GΩ. Pipettes with a region of cylindrical shape at the tip seem to work best for giant patch recordings.

Pulled electrodes for single-channel recording are coated with Sylgard (Dow Corning, Midland, MI) before use to reduce fast noise. However, this may not be necessary for macropatch recording if current densities are high. All electrodes are fire-polished immediately before use, which helps avoid problems due to the accumulation of dust at the tip. For patches on membranes of some cell types, including muscle cells and many small cells, coating the electrode tip with a hydrocarbon mixture containing light and heavy mineral oils plus Parafilm (American National Can, Neenah, WI) is usually required for macropatch or giant patch recording (Hilgemann, 1995). We have not found it necessary to coat electrodes for use on *Xenopus* oocyte membrane patches in any configuration.

## 4. Seal Formation

### 4.1. Seals for Single-Channel Patches

Fill the electrode just enough to contact the silver wire when mounted in the electrode holder. Connect the syringe (1 cm³ size) to the suction port and apply 0.3 to 0.5 cm³ positive pressure to keep the tip clean as it goes through the air–water interface. Position the electrode in the bath solution adjacent to the oocyte. Measure the pipette resistance in response to the step command provided by the amplifier. Open the syringe to atmospheric pressure. Using the micromanipulator, slowly approach the oocyte (dark side) at a location where the electrode is normal to the membrane surface; stop moving when the pipette current decreases by 25%. Wait ~1 minute, then very slowly and smoothly apply ~0.05 to 0.2 cm³ negative pressure until the deflection of the pipette current nearly disappears. Release the negative pressure and remove the syringe from the suction line, and then wait a few minutes for the seal to strengthen and stabilize. If the seal does not form completely within 2 to 3 minutes, it will be tempting to repeat application of negative pressure, although this does not usually promote successful seal formation. To form an excised, inside-out patch, rapidly move the manipulator axially away from the membrane surface. To form an excised, outside-out patch, first apply hard, fast negative pressure to rupture the cell membrane, going briefly to the whole-cell configuration. Then slowly move the manipulator axially away from the membrane surface. The membrane will rupture and then re-seal almost immediately.

## 4.2. Seals for Macropatches

Fill the electrode just enough to contact the silver wire. Connect the syringe to the suction port only just prior to lowering the electrode into the bath solution, applying no positive pressure. Position the electrode in the bath adjacent to the oocyte and measure the pipette resistance in response to the step command on the amplifier. Using the micromanipulator, slowly approach the oocyte (dark side) at a location where the electrode is normal to the membrane surface; stop moving the electrode when the oocyte membrane just begins to indent where the electrode is touching it. Then apply ~0.05 to 0.2 cm³ negative pressure very slowly and smoothly to form a seal. For inside-out macropatches, release the negative pressure, remove the syringe and wait a few minutes for the seal to strengthen and stabilize. To form an excised, inside-out macropatch, rapidly move the micromanipulator axially away from the membrane surface. For outside-out macropatches, upon seal formation, release the negative pressure but do not remove the syringe from the suction port; allow the seal to strengthen for a few minutes. Next, apply a strong, fast negative pressure (>0.5 cm³) to rupture the cell membrane within the pipette. Release the negative pressure and remove the syringe. The "Zap" function on the Axoclamp 200B amplifier (Molecular Devices, Sunnyvale, CA), can also be used to rupture the cell membrane, but this is not recommended, as it significantly reduces the seal strength. Finally, slowly move the micromanipulator axially away from the membrane surface. The membrane will rupture and then re-seal almost immediately.

### *4.3. Seals for Giant Patches*

Fill the electrode quite full, since the solution will be leaking out due to the large tip diameter. Mount the electrode in the holder. Connect the syringe to the suction port without applying positive pressure (again, due to the large tip diameter). Position the electrode in the bath solution adjacent to the oocyte. Manually move the tip to the surface of the oocyte until a dimple is visible on the surface; this may appear as a change in color from dark to light at the point of contact. The initial approach must be done visually, instead of by monitoring electrode current, since the resistance is so low. The seal will be formed with very gentle suction ($<0.05\,cm^3$). This is the tricky step for giant patches, as the suction must be applied with low negative pressure but must be applied rapidly.

## 5. Why Use Macropatches?

The macropatch configuration offers many advantages over other macroscopic recording methods such as conventional whole-cell recording in small cells or TEVC in *Xenopus* oocytes; several examples are listed here. An obvious benefit, common to all forms of excised patch recording, is the ability to control the composition of solutions bathing both sides of the membrane. For instance, cytoplasmic [Cl$^-$] in an oocyte under normal conditions is approximately 30 mM, which is too low to support large inward currents over the typical duration of a voltage pulse. Using excised macropatches, one can set $[Cl^-]_{in} = [Cl^-]_{out}$, facilitating the study of chloride channels under symmetrical conditions. Since osmotic effects are not usually a problem with macropatches, one can use fairly high symmetrical [Cl$^-$]; we often use 300 mM [Cl$^-$]. For studying channels with very low single-channel conductance, we often use asymmetrical [Cl$^-$] conditions. For example, the R334C mutant of CFTR exhibits single-channel conductance of ~1.5 pS for the full-conductance state, compared to 7.6 pS in the wild-type channel (Zhang et al., 2005a). Currents carried by this channel can be enhanced by increasing the driving force for chloride; for inside-out single-channel recordings of this mutant, we use a pipette solution containing 30 mM *N*-methyl-D-glucamine (NMDG)-Cl and 270 mM NMDG-aspartate, and a bath (cytoplasmic) solution containing 300 mM NMDG-Cl (Zhang et al., 2005a), and measure currents at $V_M = -100\,mV$.

Macropatch recordings are also useful for studying ion selectivity patterns from both sides of the membrane. It is clear, at least in the case of CFTR, that the relative permeabilities for substitute anions depend on which side of the membrane the substitute anions are presented. For example, TEVC experiments tell us that large anions such as glutamate and gluconate do not permeate the CFTR channel when applied to the outside (McCarty and Zhang, 2001). However, gluconate is capable of permeating the channel when applied to the cytoplasmic face (Linsdell and Hanrahan, 1998). These data are consistent with the presence, slightly less than 50% of the distance into the channel from the extracellular end, of a narrow "selectivity filter," which controls both binding and permeability of chloride and other anions (McCarty, 2000; McCarty and Zhang, 2001). The macropatch technique is uniquely adapted to studying the properties of this region, in wild-type channels and channels bearing point mutations, by virtue of the ability to generate large currents in both inside-out and outside-out configurations, allowing the rapid substitution of anions at the cytoplasmic or extracellular face of the membrane, respectively.

Whole oocytes studied under TEVC do not well tolerate voltage pulses of long duration, partly due to activation of background conductances. In contrast, macropatches can be held at extreme voltages for many hundreds of milliseconds. This allows time-dependent processes to run to completion, enabling better quantitation of kinetic rates.

Some ion channels interact better with pore blockers that are applied to the cytoplasmic end of the channel rather than the extracellular end (Machaka et al., 2002). This likely reflects the presence of a physical barrier that prohibits the approach of the blocker to its binding site from the extracellular end of the channel. This is certainly the case for CFTR, which is blocked by a variety of small organic molecules exclusively from the cytoplasmic side, such as diphenylamine-2-carboxylic acid (DPC), 5-Nitro-2-(3-phenylpropylamino)benzoic acid (NPPB), glibenclamide, and 4,4'-diisothiocyanatostilbene-2,2'-disulfonic acid (DIDS) (Linsdell and Hanrahan, 1996; McCarty et al., 1993; McDonough et al., 1994; Zhang et al., 2000). In whole-cell experiments, the first three of these compound blocks block CFTR slowly when applied to the bath, as the compounds become protonated and are able to permeate the plasma membrane. Once inside the cell, the charged forms are able to block the channel pore. However, one never knows the true blocker concentration under these conditions. In contrast, the excised, inside-out macropatch configuration allows one to accurately control the cytoplasmic drug concentration.

We recently compared the block of wild-type CFTR by DPC, NPPB, and glibenclamide in TEVC experiments with block by the same drugs in the excised, inside-out macropatch configuration (Zhang et al., 2004a). Fundamental aspects of block by each drug were similar between the two configurations, such as voltage dependence on inhibition. Both DPC and NPPB are rapid blockers, leading to time-independent inhibition at hyperpolarizing potentials in both TEVC and macropatch configurations. However, glibenclamide blocks CFTR channels with complex kinetics including both time-dependent and time-independent components (Zhang et al., 2004a). In TEVC recordings, CFTR currents at strongly hyperpolarizing potentials exhibit slight time dependence in the absence of glibenclamide, most likely due to the aforementioned low cytoplasmic chloride concentration. The inherent time dependence of CFTR currents in this configuration precludes analysis of the time-dependent decay of current induced by glibenclamide-mediated block. In contrast, wild-type CFTR currents in excised, inside-out macropatches show no time-dependence in the presence of symmetrical [Cl$^-$] (Fig. 1). In the presence of cytoplasmic glibenclamide, macropatch currents exhibited time-dependent decay at hyperpolarizing potentials following an initial phase of time-independent block (Zhang et al., 2004a). The time constant describing the relaxation was concentration-dependent but not voltage-dependent. The time-independent component of block observed in macropatches reflects the interaction of drug with binding sites that confer brief drug-induced closed states in single-channel patches (Zhang et al., 2004b); the forward rates to those states are very high. The time-dependent component observed in macropatches reflects the interaction of drug with a binding site that is characterized, at the single-channel level, by both a low forward rate and a low reverse rate. The site underlying the time-dependent component of block exhibited greater affinity for glibenclamide than did the site(s) underlying the time-independent block. This is consistent with the differences in affinities for interaction with the states generating fast vs. slow block of single-channel currents, as well.

Upon stepping to depolarizing potentials, where glibenclamide blocks CFTR currents very poorly, a time-dependent increase in current is observed, reflecting the relief from pore block (Zhang et al., 2004a). In the absence of added blocker, inside-out macropatch currents at depolarizing potentials exhibited no time-dependence (Fig. 1). However, in the presence of cytoplasmic glibenclamide, outward currents increased rapidly upon stepping to $V_M = +100\,mV$, representing rapid relief from block at the sites that underlie the brief glibenclamide-induced blocked states in single channels. This was followed by a slower time-dependent phase, which represents relief from block at the site that underlies the longer glibenclamide-induced blocked states in single channels (Zhang et al., 2004a). The rate constant for this relaxation was not sensitive to glibenclamide concentration, but did exhibit significant voltage dependence. Hence, the kinetics of block in the macropatch configuration reflect the kinetics of block studied at the single-channel level. This example shows how macropatch experiments can be used to study

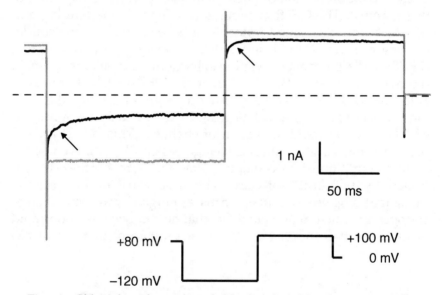

Fig. 1. Glibenclamide-mediated block of wild-type cystic fibrosis transmembrane conductance regulator (CFTR) currents in an inside-out macropatch. The gray line indicates currents measured in the absence of blocker. The solid black line indicates currents measured in the presence of 50 µM cytoplasmic glibenclamide. The dashed line indicates zero-current level. Pipette solution contained (in mM): 150 N-methyl-D-glucamine (NMDG)-Cl, 5 MgCl$_2$, 10 N-tris(Hydroxymethyl)methyl-2-aminoethanesulfonic acid (TES) (pH 7.5). Bath/intracellular solution contained: 150 NMDG-Cl, 1.1 MgCl$_2$, 2 EGTA, 10 TES, 1 Mg adenosine triphosphate (ATP), 50 U/mL protein kinase A (PKA), and 0 to 50 µM glibenclamide. The voltage protocol used is shown below. The arrow at the left indicates time-dependent block; the arrow at the right indicates time-dependent relief from block.

macroscopic consequences of microscopic kinetics (although certain caveats exist; see Colquhoun and Hawkes, 1995); macropatch experiments are relatively easy to perform compared to single-channel experiments. We are currently using macropatch recordings from oocytes expressing wild-type CFTR and many pore-domain mutants to identify the binding sites within the channel pore that underlie the complex glibenclamide-induced kinetics observed in single channels (Cui et al., 2004).

Another major benefit of the macropatch approach is the ability to study chemical reactions in a mass action sense, by virtue of the large numbers of channels present in the macropatch. An example is the activation of ligand-gated channels upon rapid exposure to their agonist. The CFTR channel is activated by the binding and hydrolysis of adenosine triphosphate (ATP) at its two nucleotide-binding domains (NBD1 and NBD2) following protein kinase A (PKA)-mediated phosphorylation of its globular regulatory domain (Ikuma and Welsh, 2000; Riordan et al., 1989; Welsh et al., 1992). After removal of PKA and ATP from the solution bathing the cytoplasmic face of a macropatch bearing CFTR channels, reapplication of ATP leads to rapid activation of channels. This is best accomplished using a fast solution exchange system (Fig. 2). We use the system from Warner Instruments (Hamden, CT; model SF-77B) controlled by pCLAMP software. The time constant for solution exchange using this system is <25 ms, as judged by the activation of endogenous calcium-activated Cl⁻ channels in oocytes upon rapid exposure of the cytoplasmic face to a solution containing 10 mM

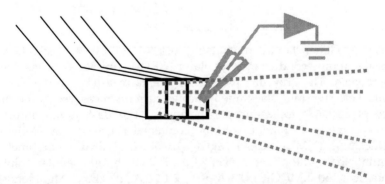

Fig. 2. Fast perfusion system. We use the system from Warner Instruments (model SF-77B), which is operated by a stepper motor and allows one to position one of three outflow tubes in front of the pipette. Solution exchange time at the surface of the patch is <25 ms.

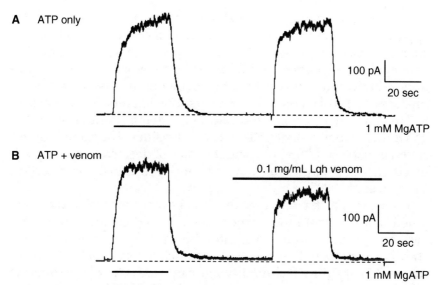

Fig. 3. Activation of CFTR channels in an excised, inside-out macro-patch by rapid exposure to ATP. (**A**) Two sequential responses to rapid exposure of the cytoplasmic surface of the patch to solution containing 1 mM ATP. Note that maximal current amplitudes are nearly identical. (**B**) Exposure of a patch to a solution containing ATP alone or ATP plus 0.1 mg/mL equivalent of venom from the scorpion *Leiurus quinquestriatus hebraeus* (Lqh venom). Maximal current is reduced in the presence of venom, although this only occurs when venom is applied to the patch following washout of ATP (Fuller et al., 2005). Pipette solution was (in mM): 150 NMDG-Cl, 5 MgCl$_2$, and 10 TES (pH 7.4, adjusted with Tris). Bath/intracellular solution was (in mM): 150 NMDG-Cl, 1.1 MgCl$_2$, 2 Tris-EGTA, 1 ATP, and 10 TES (pH 7.4, adjusted with Tris), and 0 or 0.1 mg/mL equivalent of Lqh venom. Patches were held at $V_M = -80$ mV. The dashed line indicates the zero-current level. Channels were phosphorylated by exposure to an intracellular solution containing 50 U/mL PKA prior to the start of the record; washout of PKA and ATP led to a substantial reduction of maximal ATP-stimulated current due to the activity of membrane-associated phosphatases.

[Ca$^{2+}$] (Fuller et al., 2004). We and others have been able to measure the apparent rates of ATP-dependent channel opening by rapid application of ATP to the cytoplasmic face of macropatches pulled from oocytes expressing CFTR (Fig. 3) (see also Vergani et al., 2003, 2005). This has been useful in our efforts to determine the mechanism of action of a novel peptide toxin inhibitor of CFTR (Fuller et al., 2005); we were able to show that the toxin does not alter apparent channel opening rate. Because the toxin inhibits CFTR channels in a strongly state-dependent manner, it is important to be able to change solution rapidly in order to avoid dephosphoryla-tion-induced rundown by the action of membrane-associated phos-phatases. Control records (Fig. 3, top) show that repeated exposures to 1 mM ATP, using this rapid solution exchange system, led to the generation of equivalent CFTR current densities. The same configu-ration is also useful for studying the rates of deactivation of ligand-gated channels upon washout of ligand (see Gadsby et al., 2006; Vergani et al., 2003).

The macropatch configuration also is superb for the study of the kinetics of covalent modification. Using CFTR channels bearing a cysteine residue engineered into the outer pore vestibule, we showed previously that covalent modification by methanethiosul-fonate (MTS) sulfhydryl-modifying reagents such at [2-(Trimethyl ammonium)ethyl] methanethiosulfonate (MTSET$^+$) and (2-sulfo-natoethyl) methanethiosulfonate (MTSES$^-$) alters single-channel conductance without changing open probability (Smith et al., 2001; Zhang et al., 2005a). As described above, R334C-CFTR exhibits a greatly reduced single-channel conductance compared to wild-type CFTR. Covalent modification of the engineered cysteine by exposure to MTSET$^+$ led to the deposition of positive charge at this position, which increased conductance most likely by increasing the local concentration of chloride at the mouth of the pore. In those single-channel experiments, the very tip of the electrode was filled with our usual pipette solution but the pipette was back-filled with the same solution containing 200 µM MTSET$^+$. Through time, the MTSET$^+$ diffused to the tip where it interacted with R334C-CFTR channels there, depositing positive charge. Upon covalent modifi-cation, the amplitude of the full conductance state increased in a single step 2.1- to 2.3-fold (Smith et al., 2001; Zhang et al., 2005a). The same effect was seen when R334C-CFTR channels were modi-fied by bath-applied MTSET$^+$ in outside-out macropatches (Zhang et al., 2005a,b). In these experiments, pipette tips were filled with intracellular solution with 1 mM ATP but lacking PKA; the elec-trode shaft was back-filled with the same solution containing PKA (inclusion of PKA at the electrode tip inhibits seal formation, likely due to the glycerol component of the enzyme storage buffer). Over the course of 30 to 90 minutes, PKA diffused to the tip and phos-phorylated the CFTR channels, leading to steady-state chloride current. The electrode was then positioned in front of a flowing

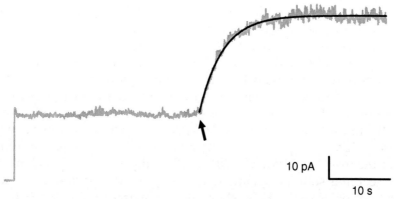

10 pA

10 s

Fig. 4. Chemical modification of R334C-CFTR increases macroscopic conductance in an excised, outside-out macropatch. Pipette solution at the tip contained (in mM): 150 NMDG-Cl, 1.1 MgCl$_2$, 2 Tris-EGTA, 10 TES; pH 7.4. Pipettes were back-filled with the same solution plus 1 mM MgATP and 100 U/mL PKA. Channels were phosphorylated over the course of 30 to 90 minutes. Bath/extracellular solution contained (in mM): 150 NMDG-Cl, 10 TES, pH 7.4, or the same solution plus 50 µM MTSET$^+$. At the beginning of the record, the membrane potential was stepped to $V_M = +80$ mV, leading to steady current carried by unmodified R334C-CFTR channels. At the time indicated by the arrow, the bath solution was changed rapidly to one containing MTSET$^+$. The increase in current reflects the deposition of positive charge at the engineered cysteine. The gray line shows the R334C-CFTR channel current itself, while the black line indicates a fit of the data to a first-order exponential function with $\tau = 3.1$ seconds in this experiment.

solution line containing extracellular solution without MTSET$^+$. Under control by pCLAMP software, the position was then abruptly switched to expose the patch to the same solution containing 10 to 50 µM MTSET$^+$ (Fig. 4). R334C-CFTR macroscopic current increased rapidly, reflecting deposition of charge by covalent modification. The steady-state current was increased by 2.3-fold, consistent with the change in conductance observed in single-channel patches. The kinetics of macroscopic modification were described best by a single exponential function, consistent with the presence of a single population of modifiable cysteines in R334C-CFTR. The rate of modification was sensitive to [MTSET$^+$], as expected for a first-order covalent reaction. Hence, the covalent modification of R334C-CFTR channels in excised, and outside-out macropatches recapitulate in every way the same process observed in single-channel patches.

The value of the outside-out macropatch configuration, in these experiments, derives from its ability to alter the conditions under which the covalent modification takes place. The CFTR channel open probability can be controlled in several ways. One approach is to present to the wild-type channel a solution containing a low [ATP], leading to low channel activity. Another approach is to use non- or poorly hydrolyzable analogues of ATP, such as adenosine 5'-(β,γ-imido)triphosphate (AMP-PNP), which lock the channel into a high open probability state (Verani et al., 2003). We found that inclusion of 2.75 mM AMP-PNP along with 1 mM ATP in the intracellular solution greatly decreased the rate constant for modification of R334C-CFTR channels by extracellular MTSET$^+$ in outside-out macropatches, indicating profound state-dependence of the reactivity of a cysteine engineered at this site (Zhang et al., 2005b). We are now using macropatch recordings to determine the state-dependent chemical reactivity of cysteines engineered at other positions in the outer vestibule of the pore, in efforts to understand how the pore changes conformation between open and closed states; this would be virtually impossible without the macropatch technique.

Macropatch recordings also work well for the study of voltage-gated channels under conditions that provide higher resolution than is possible with TEVC. Figure 5 shows a comparison of currents from the ClC-2 voltage-gated chloride channel in TEVC and outside-out macropatch configurations using identical voltage protocols, although the voltage pulses in macropatch experiments were only half the duration of those in TEVC experiments. ClC-2 currents show strong inward rectification in both configurations. This becomes problematic in TEVC experiments because, as described above, intracellular [Cl$^-$] is low in the oocyte. Of course, this problem does not exist in the macropatch configuration where one can control the composition of the solution on both sides of the membrane. Because the permeating ion provides the gating charge for ClC-type channels, judicious control of [Cl$^-$] makes it possible to control channel gating in such a way that maximal open probabilities can be obtained (Zuñiga et al., 2004). This is evident in the records shown in Fig. 5, where currents at each voltage more closely approached plateau levels in macropatch experiments even though the pulses were of shorter duration.

The ClC-2 channels are also activated by cell swelling in the whole-cell configuration. Therefore, solutions used for TEVC

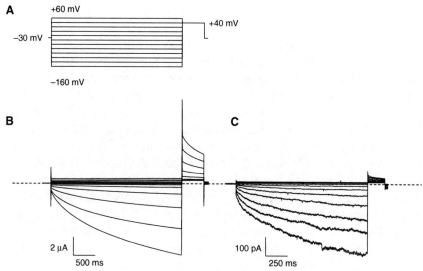

Fig. 5. Comparison of ClC-2 currents recorded by two-electrode voltage clamp (TEVC) and outside-out macropatch. (**A**) Voltage protocol used in both configurations, although the pulses were of different duration. (**B**) Currents measured from an oocyte in TEVC configuration. (**C**) Currents measured from an outside-out macropatch. The dashed line indicates the zero-current level.

experiments must contain mannitol to control cell volume and barium to inhibit calcium entry (Thompson et al., 2005); these maneuvers are not needed for macropatch experiments. Macropatch recordings also exhibit smaller capacitive transients in response to voltage jumps. Otherwise, ClC-2 channels appear to behave similarly in the macropatch and TEVC configurations, including sensitivity to inhibition by a peptide component of scorpion venom (Thompson et al., 2005). One added advantage of using the macropatch configuration as a cell-free configuration became clear during our study of ClC-2 channels in the presence of scorpion venom. We confirmed that the component of venom that inhibits ClC-2 is a peptide by incubating venom with trypsin followed by boiling to inactivate the protease. When trypsinized, boiled venom was applied to oocytes in the TEVC configuration, large background currents were activated, possibly due to release of metals or other components from the venom; this occurred in cells expressing ClC-2 and in uninjected cells as well. However, activation of background conductances did not occur when trypsinized, boiled venom was applied to the solution bathing outside-out macropatches.

Given the large membrane surface area within a macropatch, or even more so within a giant patch, the large number of proteins isolated within the pipette makes it possible to measure the activity not only of ion channels but also of electrogenic ion transporters (Hilgemann, 1995, 1996). We have studied the activity of the Calx1.2 $Na^+/Ca^{2+}$ transporter from *Drosophila melanogaster* (Omelchenko et al., 1998) using giant inside-out macropatches. The pipette solution contained 8 mM $Ca^{2+}$. When the bath (cytoplasmic) solution was rapidly changed from one containing 100 mM $Li^+$ to one containing 100 mM $Na^+$, outward currents were activated rapidly (Fig. 6) and then underwent $Na^+$-dependent self-inhibition (Hilgemann et al., 1992; Omelchenko et al., 1998). The $Na^+/Ca^{2+}$ exchanger provides an important mechanism for regulation of cytoplasmic $[Ca^{2+}]$

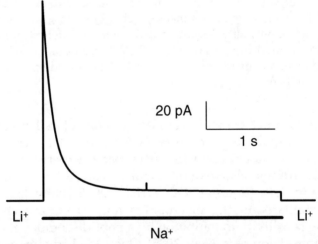

Fig. 6. Measurement of $Na^+$-$Ca^{2+}$ exchange current in an inside-out giant patch. Patch was pulled from an oocyte expressing Calx1.2 from *D. melanogaster*. Pipette solution contained (in mM): 100 NMDG-MES, 30 HEPES, 30 TEA-OH, 16 sulfamic acid, 8 $CaCO_3$, 6 KOH, 0.25 ouabain, 0.1 niflumic acid, 0.1 flufenamic acid; pH 7.0 (adjusted with MES). Bath/cytoplasmic solution contained: 100 Li-aspartate, 20 3-(N-Morpholino)propane-sulfonic acid (MOPS), 20 TEA-OH, 20 CsOH, 10 EGTA, 1.5 $Mg(OH)_2$; pH 7.0 (adjusted with MES). As indicated by the bar below the record, the bath/cytoplasmic solution was rapidly changed to one that contained Na-aspartate in place of Li-aspartate, resulting in outward exchanger current.

in cardiac myocytes; membrane macropatches or giant patches pulled from cells expressing these proteins provide a valuable means for testing drugs that target their activity.

## 6. Conclusion

Macropatch recording serves as a valuable approach for investigating the activity of ion channels and transporters with high resolution in both the temporal and amplitude domains. The use of a cell-free preparation such as this offers substantial advantages over whole-cell preparations, including facilitating the study of transport proteins and ion channel variants with very low unitary current amplitudes. In many cases, macropatch recordings allow the study of macroscopic consequences of microscopic kinetic events, including the analysis of chemical reactions such as covalent modification, under tightly controlled conditions. However, it is important to recognize that determination of the fine mechanistic details of the behavior of any ion channel requires approaches other than macropatch studies, such as traditional single-channel recording.

## Acknowledgments

The work reported here from the authors' lab was supported by the following grants to Nael A. McCarty: National Institutes of Health grants DK066409 and DK56483, American Heart Association Established Investigator Award 0140174N, and National Science Foundation grant MCB-0077575.

## References

Colquhoun, D. and Hawkes, A. G. (1995) The principles of the stochastic interpretation of ion-channel mechanisms, in *Single-Channel Recording,* 2nd ed. (Sakmann, B. and Neher, E., eds.), Plenum Press, New York, pp. 397–482.

Cui, G., Song, B., and McCarty, N. A. (2004) Differential block of CFTR pore by three members of the sulphonylurea family. *Biophys. J.* **86**, 586a.

Fuller, M. D., Zhang, Z.-R., Cui, G., Kubanek, J., and McCarty, N. A. (2004) Inhibition of CFTR channels by a peptide toxin of scorpion venom. *Am. J. Physiol.* **287**, C1328–C1341.

Fuller, M. D., Zhang, Z.-R., Cui, G., and McCarty, N. A. (2005) The block of CFTR by scorpion venom is state-dependent. *Biophys. J.* **89**, 3960–3975.

Gadsby, D. C., Vergani, P., and Csanády, L. (2006) The ABC protein turned chloride channel whose failure causes cystic fibrosis. *Nature* **440**, 477–483.

Hilgemann, D. W. (1995) The giant membrane patch, in *Single-Channel Recording*, 2nd ed. (Sakmann, B. and Neher, E., eds.), Plenum Press, New York, pp. 307–327.

Hilgemann, D. W. (1996) The cardiac Na-Ca exchanger in giant membrane patches. *Ann. N. Y. Acad. Sci.* **779**, 136–158.

Hilgemann, D. W., Matsuoka, S., Nagel, G. A., and Collins, A. (1992) Steady-state and dynamic properties of cardiac sodium-calcium exchange. Sodium-dependent inactivation. *J. Gen. Physiol.* **100**, 905–932.

Ikuma, M. and Welsh, M. J. (2000) Regulation of CFTR $Cl^-$ channel gating by ATP binding and hydrolysis. *Proc. Natl. Acad. Sci. USA* **97**, 8675–8680.

Liman, E. R., Tytgat, J., and Hess, P. (1992) Subunit stoichiometry of a mammalian $K^+$ channel determined by construction of multimeric cDNAs. *Neuron* **9**, 861–871.

Linsdell, P. and Hanrahan, J. W. (1996) Disulphonic stilbene block of cystic fibrosis transmembrane conductance regulator $Cl^-$ channels expressed in a mammalian cell line, and its regulation by a critical pore residue. *J. Physiol. (Cambr.)* **496**, 687–693.

Linsdell, P. and Hanrahan, J. W. (1998) Adenosine triphosphate-dependent asymmetry of anion permeation in the cystic fibrosis transmembrane conductance regulator chloride channel. *J. Gen. Physiol.* **111**, 601–614.

Machaka, K., Qu, Z., Kuruma, A., Hartzell, H. C., and McCarty, N. A. (2002) The endogenous $Ca^{2+}$-activated $Cl^-$ channel in *Xenopus* oocytes: a physiologically and biophysically rich model system, in *Chloride Channels of Excitable and Non-excitable Cells* (Fuller, C. M. and Benos, D. J., eds.), Academic Press, San Diego.

McCarty, N. A. (2000) Permeation through the CFTR chloride channel. *J. Exp. Biol.* **203**, 1947–1962.

McCarty, N. A., McDonough, S., Cohen, B. N., Riordan, J. R., Davidson, N., and Lester, H. A. (1993) Voltage-dependent block of the cystic fibrosis transmembrane conductance regulator $Cl^-$ channel by two closely related arylaminobenzoates. *J. Gen. Physiol.* **102**, 1–23.

McCarty, N. A. and Zhang, Z.-R. (2001) Identification of a region of strong discrimination in the pore of CFTR. *Am. J. Physiol.* **281**, L852–L867.

McDonough, S., Davidson, N., Lester, H. A., and McCarty, N. A. (1994) Novel pore-lining residues in CFTR that govern permeation and open-channel block. *Neuron* **13**, 623–634.

Omelchenko, A., Dyck, C., Hnatowich, M., et al. (1998) Functional differences in ionic regulation between alternatively spliced isoforms of the $Na^+$-$Ca^{2+}$ exchanger from *Drosophila melanogaster*. *J. Gen. Physiol.* **111**, 691–702.

Quick, M. W., Naeve, J., Davidson, N., and Lester, H. A. (1992) Incubation with horse serum increases viability and decreases background neurotransmitter uptake in *Xenopus* oocytes. *BioTechniques* **13**, 358–362.

Riordan, J. R., Rommens, J. M., Kerem, B.-S., et al. (1989) Identification of the cystic fibrosis gene: cloning and characterization of complementary DNA. *Science* **245**, 1066–1072.

Smith, S. S., Liu, X., Zhang, Z.-R. et al. (2001) CFTR: Covalent and noncovalent modification suggests a role for fixed charges in anion conduction. *J. Gen. Physiol.* **118**, 407–431.

Thompson, C. H., Fields, D. M., Olivetti, P. R., Fuller, M. D., Zhang, Z.-R., and McCarty, N. A. (2005) Inhibition of ClC-2 Cl⁻ channels by a peptide component of scorpion venom. *J. Membr. Biol.* **208**, 65–76.

Vergani, P., Lockless, S. W., Nairn, A. C., and Gadsby, D. C. (2005) CFTR channel opening by ATP-driven tight dimerization of its nucleotide-binding domains. *Nature* **433**, 876–880.

Vergani, P., Nairn, A. C., and Gadsby, D. C. (2003) On the mechanism of MgATP-dependent gating of CFTR Cl⁻ channels. *J. Gen. Physiol.* **120**, 17–36.

Welsh, M. J., Anderson, M. P., Rich, D. P., et al. (1992) Cystic fibrosis transmembrane conductance regulator: a chloride channel with novel regulation. *Neuron* **8**, 821–829.

Zhang, Z.-R., Cui, G., Liu, X., Song, B., Dawson, D. C., and McCarty, N. A. (2005a) Determination of the functional unit of the cystic fibrosis transmembrane conductance regulator chloride channel: one polypeptide forms one pore. *J. Biol. Chem.* **280**, 458–468.

Zhang, Z.-R., Cui, G., Zeltwanger, S., and McCarty, N. A. (2004a) Time-dependent interactions of glibenclamide with CFTR: Kinetically complex block of macroscopic currents. *J. Membr. Biol.* **201**, 139–155.

Zhang, Z.-R., Song, B., and McCarty, N. A. (2005b) State-dependent chemical reactivity of an engineered cysteine reveals conformational changes in the outer vestibule of the cystic fibrosis transmembrane conductance regulator. *J. Biol. Chem.* **280**, 41997–42003.

Zhang, Z.-R., Zeltwanger, S., and McCarty, N. A. (2000) Direct comparison of NPPB and DPC as probes of CFTR expressed in *Xenopus* oocytes. *J. Membr. Biol.* **175**, 35–52.

Zhang, Z.-R., Zeltwanger, S., and McCarty, N. A. (2004b) Steady-state interaction of glibenclamide with CFTR: evidence for multiple sites. *J. Membr. Biol.* **199**, 15–28.

Zuñiga, L., Niemeyer, M. I., Varela, D., Catalán, M., Cid, L. P., and Sepúlveda, F. V. (2004) The voltage-dependent ClC-2 chloride channel has a dual gating mechanism. *J. Physiol.* **555**, 671–682.

# 13

# Structure-Function Analyses of Single Cells by Combining Patch-Clamp Techniques with Reverse Transcription–Polymerase Chain Reaction

*Gerald Seifert and Christian Steinhäuser*

## 1. Introduction

One of the main challenges for a better understanding of signaling mechanisms in the normal and diseased central nervous system (CNS) is to unravel the molecular basics of function on the cellular and systemic levels. Several important cellular processes, such as proliferation, differentiation, and cell death, are directly or indirectly influenced or even controlled by the activity of ion channels and receptors. The patch-clamp technique has been proven to be a powerful method for studying functional peculiarities of such channels, which considerably improved our knowledge of mechanisms underlying neuronal excitation and inhibition.

Besides distinguishing ion channels according to their biophysical and pharmacological profile, molecular cloning technologies have been developed over the past two decades, enabling the identification of a wealth of structurally distinct subunits. These subunits comprise a number of ion channel families, and heterologous expression experiments in combination with electrophysiology have shown that the assembly of different subunits can cause distinct channel functioning. However, these studies have also demonstrated that in most cases, it is not possible to identify subunit combinations or even single subunits from functional studies alone.

From: *Neuromethods, Vol. 38: Patch-Clamp Analysis: Advanced Techniques, Second Edition*
Edited by: W. Walz @ Humana Press Inc., Totowa, NJ

Native cells usually possess very complex expression patterns, and so far our knowledge about their subunit assemblies and receptor stoichiometry is very limited.

The introduction of the single-cell reverse transcription-polymerase chain reaction (RT-PCR) (Lambolez et al., 1992) and its combination with the patch-clamp method and fast application techniques now make it possible to identify gene transcripts and to correlate them with functional data in the same individual cell. Unraveling alterations of ion channels and receptors on the functional and molecular level may open new perspectives for the development of target-specific drugs to improve the treatment of neurological diseases.

## 2. Optimizing the RT-PCR Procedure Using Messenger RNA from Brain Tissue

### 2.1. General Laboratory Practice

To avoid degradation of cellular RNAs and to maintain a ribonuclease (RNase)-free environment in the lab, all pipettes, tubes, glass, and plasticware have to be autoclaved with gas (ethylene oxide) or steam sterilization. The surface of the worktable tops and all used pipettes and tweezers must be frequently cleaned with RNase AWAY™ (Molecular BioProducts, San Diego, CA) to remove RNases and DNA contamination. Gloves should always be worn during the experiments. Only aerosol-resistant filter tips are recommended for the preparation of RT and PCR. RNAs should be processed on a laminar flow box to avoid any contamination of the samples by lab personnel.

Contamination with cDNAs, which are easily amplified by PCR and produce false-positive results, is of particular risk for transcript analysis. Therefore, RNA processing and PCR should be performed in separate rooms. Similarly all pipettes, equipment, and reagents should be separated to avoid cDNA transfer to reagents used for cell harvesting and first strand synthesis. For the same reason, amplified cDNAs must always be handled with care. These precautions absolutely have to be accompanied by appropriate negative control experiments to identify and prevent potential contamination (Dragon, 1993).

When performing semiquantitative real-time RT-PCR, all fluorescent probes must be accurately handled to prevent contamination of tubes, reagents, and samples to avoid false-positive fluorescence. For each PCR run, background fluorescence must be carefully checked with appropriate controls, for example, by omitting the template (RNA, cDNA) while leaving all other reagents unchanged, or by performing the reaction without DNA polymerase. Under these conditions, consecutive PCR cycles must necessarily fail to produce any increase in fluorescence intensity. In addition, fluorescence dyes are added to buffers of numerous commercial RT-PCR kits to overcome the background fluorescence problem (passive references).

## 2.2. Design of PCR Primers

To investigate the expression profile in single cells, the subunits of the corresponding gene family as well as closely related genes should be aligned to identify homologous and diverse nucleotide sequences. Primer design has particularly to consider all the putative insertions or alternatively spliced exons, editing, and mutation sites that might be of relevance for the subsequent analysis. A two-round PCR is recommended to increase yield and specificity using nested primers in the second amplification round. For qualitative approaches, it is appropriate to place the "outer" primers for the first PCR in conserved regions while at least one of the second-round, "nested" primers should be located in a subunit-specific region of the nucleotide sequence. Synthesis of long products is less efficient and hence the fragments to be amplified should not exceed 800 base pairs (bp). To avoid amplification of genomic DNA, ideal primers should be located on different exons separated by introns. The use of computer programs is recommended for the design of effective primers. These commercially available computer programs help to avoid primer pair dimers, primer self dimers, hairpins, or misannealing of primers. Primers can also be defined empirically, but the reaction conditions for optimal gene detection must always be optimized. The recommended guanine-cytosine content (GC-content) of the primers should exceed 50%.

## 2.3. mRNA Isolation and Reverse Transcription

To test and optimize the RT-PCR procedure, we prepare total RNA from freshly isolated rodent brain using guanidinium isothiocyanate and phenol (Trizol™, Invitrogen, Carlsbad, CA) (Chomczynski and Sacchi, 1987). The RNA pellet is solved in water and stored at −80°C. To estimate the RNA concentration, the optical absorbency of the solution is determined at 260 nm optical density ($OD_{260}$). To remove residual genomic DNA, we incubate 4 µg total RNA in PCR buffer, add dithiothreitol (DTT; 10 mM), $MgCl_2$ (2.5 mM), 40 U RNase inhibitor (RNasin™, Promega, Madison, WI), and 40 U RNase-free deoxyribonuclease (DNase)I (Roche, Mannheim, Germany; reaction volume 20 µL). After incubation in a bloc thermostat (37°C, 30 minutes), the reaction is stopped with ethylenediaminetetraacetic acid (EDTA) (2.5 mM), RNA extraction is repeated with phenol/chloroform/isoamyl alcohol (25:24:1; volume/volume), RNA is precipitated with isopropanol, and the RNA pellet is dissolved in water. The final concentration of the yielded RNA can be determined with absorption spectrometry ($OD_{260}$) and the solution should be stored at –80°C (see Ausubel et al., 1996).

Alternatively, mRNA can be isolated immediately after DNase treatment. Therefore, 50 µL of deoxythymidine oligonucleotide [oligo $(dT)_{25}$] linked Dynabeads™ (Dynal, Oslo, Norway) are used to segregate and purify messenger RNA (mRNA) from total RNA, the latter containing only about 3% of cytoplasmic mRNA. This method is based on the binding of polyadenylated mRNA tails to oligo $(dT)_{25}$, covalently linked to monodisperse polymer spheres that are coated with magnetic iron oxide. Then mRNA can be separated from total RNA by means of a permanent magnet. The mRNA bound to the beads has to be washed and suspended in water, frozen, and stored at –80°C. Removal of mRNA from the Dynabeads is realized by heating to 90°C (2 minutes). The beads are separated from the target-containing supernatant when the reaction tube is placed in a magnet stand, although the presence of the beads does not affect the RT-PCR. Subsequently, a fourth of the isolated mRNA amount (corresponding to 1 µg of the original total RNA) is transcribed into cDNA using the following mix: RT buffer, 10 mM DTT, 4 × 125 µM deoxyribonucleotide triphosphates (dNTPs), 50 µM random hexanucleotide primers (Roche), 40 U RNase inhibitor, 200 U Superscript II-RT (Invitrogen), reaction volume 20 µL, incubation for 1 hour at 37°C.

To adapt the procedure to single-cell conditions, 1 ng of DNA-free, total RNA is transcribed using an alternative RT protocol as described in Chapter 3, Section 3.3.

### 2.4. Two-Round PCR

For further PCR optimization, a two-round RT-PCR should be performed with 1/10th of the complementary DNA (cDNA) amount resulting from the RT. In the first PCR round, we usually perform a multiplex PCR for parallel detection of different genes, including a housekeeping gene as an internal positive control. This internal control is of particular relevance in the case of single-cell RT-PCR. However, different primer pairs sometimes interact with each other, which may reduce the yield of the PCR. Therefore, it is always recommended that the results obtained with primers for the target gene alone be compared with those obtained together with the housekeeping gene primers. Such type of analysis is shown in Fig. 1. In this example, increasing amounts of primers for glyceraldehyde-3-phosphate dehydrogenase (GAPDH) suppressed the amplification of cDNA encoding the $K^+$ inward rectifier (Kir) channel, Kir4.1, while the housekeeping gene $\beta$-actin did not. Obviously, in this case, GAPDH should not be used as an internal control because it can produce false-negative results.

After the first PCR round using multiplex primers, the PCR product has to be cleaned with commercial cleaning kits (e.g., PCR Purification Kit, Qiagen, Hilden, Germany). Therefore, we dissolve

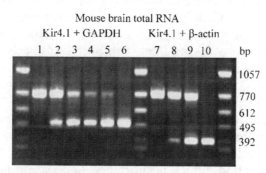

Fig. 1. Sensitivity and specificity of multiplex reverse transcription–polymerase chain reaction (RT-PCR). Mouse brain total RNA (DNA-free, 1 µg) was reverse transcribed into cDNA as described in Chapter 2, Section 2.3. This cDNA (2 µL) served as a template for PCR (35 cycles). PCR was run with Kir4.1 primers (concentration always 200 nM; lanes 1–5, 7–9) together with increasing concentrations of glyceraldehyde-3-phosphate dehydrogenase (GAPDH) (0, 50, 100, 150, and 200 nM; lanes 1–5) or $\beta$-actin primers (0, 100, and 200 nM; lanes 7–9). Lanes 6 and 10 represent respective PCR products using either GAPDH or $\beta$-actin primers alone (200 nM each). Agarose gel electrophoresis (1.5%, stained with ethidium bromide) confirmed the expected sizes of the amplified cDNAs (Kir4.1, 767 bp; GAPDH, 504 bp; $\beta$-actin, 387 bp). The molecular length marker was $\Phi$X-174–RF DNA *Hinc*II digest.

the PCR product in 40 μL of water. An aliquot of this solution (2 μL) is used as a target for the subsequent, parallel PCR runs. This second PCR usually runs with a reduced number of cycles (20 to 35, depending on the amount of existing gene copies).

Polymerase chain reaction is followed by ethidium bromide agarose gel electrophoresis (1.5%) to identify the lengths of the cDNAs. Ideally, only one band of the expected length should appear in each lane for the respective subunit. If an additional, weaker band of a putative by-product appears in the gel, the second PCR round has to be repeated. First, annealing temperature and $Mg^{2+}$ concentration should be empirically modified to find optimal conditions that result in specific PCR products. If this does not help, alternative primer pairs must be tested.

The first- and second-round PCR are each performed in a reaction volume of 50 μL, containing PCR buffer, 2.5 mM $MgCl_2$, 4 × 50 μM dNTPs, and 200 nM of the respective primer.

Our standard PCR protocol starts with initial denaturation at 95°C (4 minutes), and then the temperature is held at 85°C until *Taq* DNA polymerase (2.5 to 5 U; Invitrogen) is added to all tubes in the block ("hot start" PCR). For the second PCR round, a hot start *Taq* DNA polymerase was used that is inactive at room temperature due to binding of a thermolabile inhibiting antibody (2.5 U platinum *Taq* polymerase; Invitrogen). The latter *Taq* polymerase is only activated at 95°C through denaturation of the bound antibody. Five PCR cycles are performed, each of them with three temperature steps: denaturation at 94°C (25 seconds), primer annealing at 43°C to 65°C (2 minutes), and primer extension at 72°C (30 seconds). The next 20 to 40 cycles use the same temperatures, except the duration of primer annealing, which is shortened to 45 seconds. Final elongation runs at 72°C (7 minutes). The annealing temperature matches the melting temperature of the primers.

## 2.5. Restriction Analysis

Independent methods, such as nucleotide sequencing, sequence-specific probe hybridization (Southern blotting), and restriction analysis, have to be performed to verify the identity of the PCR products. Restriction analysis offers a fast and convenient possibility to confirm cDNA specificity. Here, cDNA double strands are cleaved by bacterial restriction endonucleases recognizing a defined nucleotide sequence. An advantage of this method is the almost

100% effectiveness of digestion. As a prerequisite for this type of analysis, the second PCR round must be repeated to produce a sufficient number of cDNA copies, and the PCR product has to be cleaned from buffers and salts. Restriction endonucleases need specific reaction conditions to reach maximal activity. We perform digestion in a small volume (15 µL), and add buffer and 10 to 20 U restriction enzyme to the PCR product, which is dissolved in water. Incubation takes at least 3 hours. The cDNA fragments are analyzed in agarose gels with high band resolution designed to detect even tiny length differences between double-stranded cDNAs (Agarose-7000 gel matrix; 2%; Invitrogen).

Suitable restriction endonucleases can be selected with computer programs. The enzyme should recognize a specific nucleotide sequence occurring in only one of the subunits to be analyzed, and should not cut related sequences. In addition, for easy detection the enzyme of choice should cut the cDNA only once or twice, to produce no more than three fragments.

# 3. Peculiarities of Single-Cell RT-PCR Techniques

## 3.1. Harvesting of Cellular RNA from Brain Slices

Patch-clamp analysis of voltage and ligand-gated ion channels in single neural cells and the subsequent identification of gene transcripts in the same cell require some attention to avoid degradation of RNA during recording. To reduce contamination with RNases and nucleotides other than from the tissue to be analyzed, solutions necessary for electrophysiological recording should be prepared in a room separated from that where PCRs are performed, and a sterile laboratory practice must be followed. Gloves should always be worn for the preparation of solutions and when performing electrophysiological recordings, particularly when touching the recording pipette. The surface of the pipette puller and the recording setup have to be disinfected with propanol or ethanol, and the Ag/AgCl electrode must be cleaned prior to recording. The recording chamber should be thoroughly washed daily with distilled water.

Filtrated millipore water should be used to prepare bath solutions perfusing the brain slice in the recording chamber. Reagents specifically prepared for use in RT-PCR should be utilized as far as possible, and the solutions have to be stored in sterilized glassware. Pipette solutions are prepared from aliquots of pyrogene-free water.

Separate salt aliquots are added and the pH can only be adjusted after cleaning the pH-electrode with water and ethanol. Pipette solutions are immediately aliquoted on a sterile bench and frozen. We observed a reduced success rate of transcript analysis when the preceding electrophysiological analysis took more than 15 minutes, probably due to washout and dilution of cytoplasmic RNA in the pipette or to its deletion by ubiquitously existing RNases. The latter can be counteracted by adding RNase inhibitor to the pipette solution (0.25 U/µL pipette solution; stock of 40 U/µL RNasin™; Promega) which should be daily prepared freshly. This precaution enabled successful mRNA identification even after prolonged electrophysiological analysis.

Our recording pipettes are fabricated from borosilicate capillaries, and we fill them with 6 µL of pipette solution using long polyethylene microloaders (Eppendorf, Hamburg, Germany), which restricts filling to the tip of the pipette. This tip can be considered sterile because it was heated during the pulling step. We refrain from adding buffers or dNTPs to the pipette solution to avoid changes of the ionic composition of pipette solution. The resistance of the recording pipettes is 4 to 6 MΩ, which allows both stable whole-cell recordings and subsequent aspiration of cytoplasm into the patch pipette. The cell content is harvested by applying negative pressure to the pipette interior while retaining the gigaohm seal resistance between the pipette tip and the membrane. It is advantageous to follow the flow of the cytoplasm into the pipette, preferably via a monitor connected to a CCD camera (e.g., C5405; Hamamatsu Photonics, Hamamatsu, Japan) (Jonas et al., 1994). This helps to avoid the aspiration of neighboring tissue. After sucking about half the cell content, we often lift the cell on top of the pipette above the slice and complete harvesting in the flow of the bath solution under visual control. Only single cells without any adhered other cells or tissue debris are selected for RT-PCR analysis (Fig. 2).

After aspiration of the cytoplasm, the pipette is transferred into a thin-walled reaction tube filled with 3 µL of diethylpyrocarbonate (DEPC)-treated water. The tip of the pipette is broken at the bottom, and only part of the pipette solution (about 3 µL) is expelled into the water with positive pressure. Thus, the cytoplasm is released into the water and the reaction tube is frozen in liquid nitrogen and stored at −80°C until the RT reaction is carried out (von Eggeling and Ballhausen, 1995).

Fig. 2. Harvesting of cytoplasm from glial cells. After recording in situ (brain slice, 150 μm thick, CA1 stratum radiatum) (**A**), the cell on top of the patch pipette was lifted out of the tissue (**B**) and subsequently was aspirated into the pipette (**C**). Cell harvesting was observed with infrared differential interference contrast (IR-DIC) optics and videomicroscopy.

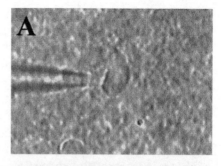

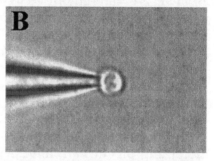

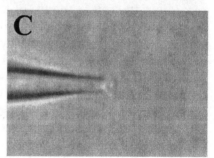

## 3.2. Harvesting of Acutely Isolated Cells

Brain slice recordings offer the great advantage of facilitating the analysis of membrane properties with the cellular environment left largely intact. However, voltage-clamp control is often limited under these conditions, particularly in cells with extensive branching. Also, the analysis of receptor currents with fast time constants of activation or inactivation is compromised by the relatively slow speed of agonist application due to diffusion barriers in situ. One possibility to overcome such limitations is the use of freshly isolated cells. Basically, RNA harvesting from isolated cells requires the same precautions as described above for the slice preparation. We use a combination of enzymatic and mechanical treatments to acutely isolate viable cells from the tissue. Brain slices are prepared (thickness 300–450 μm) and incubated in a protease (e.g., papain, trypsin, or pronase)-containing solution. The duration of enzymatic incubation depends on various parameters (e.g., age of the animal, cell type, enzyme concentration, temperature) and usually varies between 10 and 60 minutes (Seifert et al., 1997b, 2000). After thorough wash and recovery from protease incubation, the region of interest is carefully dissected with a scalpel. The cells are then disaggregated under microscopic control using fire-polished Pasteur pipettes (tip diameter 100–200 μm) or tungsten needles. Process-bearing cells without indications of osmotic swelling are selected and proved suitable for patch-clamp analysis and subsequent transcript investigation with RT-PCR. Careful comparison with recordings obtained in situ from the same cell type confirmed that characteristic electrophysiological parameters remain unchanged after the isolation procedure (Steinhäuser et al., 1994). Recording pipettes are filled with 6 μL of pipette solution. After recording, the cell at the tip of the pipette has to be transferred to a separate dish to be washed with fresh bath solution, and then is sucked into the recording pipette. The cell contents and about 3 μL of the pipette solution are expelled into a reaction tube containing 3 μL of DEPC-treated water. Alternatively, the whole cell can be harvested with a second pipette (diameter 10–20 μm) under microscopic control. The cell is easily removed from the tip of the recording pipette and sucked into the harvesting pipette, whereby one has to be cautious and avoid aspiration of too much bath solution. The tip of the harvesting pipette is broken, and about 3 μL of its content is expelled into a water-filled reaction tube (Seifert et al., 1997a). Harvesting with a second pipette can also be performed in situ if the cell is drawn out of the tissue after electrophysiological analysis (Seifert et al., 1999). The reaction tubes are frozen in liquid nitrogen and stored at −80°C. We always perform negative controls in parallel, by using cell-free bath solution for RT-PCR.

### 3.3. Preventing Contamination with Genomic DNA, Reverse Transcription of RNA

The nucleus is often aspirated into the pipette in our experiments and the harvested material therefore contains mRNA as well as genomic DNA. However, PCR approaches may use primers that do not span intron regions. In these cases, to prevent amplification of genomic DNA (Li et al., 1988), which would result in PCR products of the same length as the corresponding mRNA, RT-PCR has to be preceded by DNase treatment. To decide whether DNA digestion is necessary, for example, if the exon–intron structure of the gene to be analyzed is unknown, one should first determine the DNA size by amplifying genomic DNA (200 ng). In case the amplicon spans introns, this yields a larger PCR product than that of the corresponding cDNA.

DNase mastermix is added to the frozen cell contents in the reaction tube. This mastermix contains all reagents necessary for the RT reaction, except reverse transcriptase: first strand buffer, dNTPs ($4 \times 250 \mu M$), RNasin™ (20 U; Promega), random hexamer primers (50 μM; Roche), and RNase-free DNAseI (10 U; Roche) (reaction volume 10 μL) (Dilworth and McCarrey, 1992; Gurantz et al., 1996). The reaction mix has to be covered with freshly aliquoted paraffin oil, incubated at 37°C (30 minutes), and afterward DNaseI is heat-inactivated at 95°C (5 minutes). We observed a dramatic drop in the PCR success rate when trace amounts of active DNaseI remained during the RT reaction. Covering of the DNaseI reaction volume seems to be necessary to prevent evaporation. Subsequently, reverse transcriptase (0.5 μL) and RNase inhibitor (20 U) are added to the reaction volume, and the RT reaction is started at 37°C.

As a negative control for RT and verification of complete DNA degradation, the reverse transcriptase should be omitted. Prior to its application to single cells, the effectiveness of DNA digestion should be tested under the same conditions (buffers, reagents, volume) by adding increasing amounts of genomic DNA (10, 50, 100, 250, 500 ng; Promega) to water, treating with DNaseI, and amplifying the mixture in a two-round PCR. These experiments confirmed a reliable and complete digestion of up to 250 ng genomic DNA (Fig. 3), which by far exceeds the DNA content of a single cell (about 10 pg).

Efficient reverse transcription of RNA into cDNA is a prerequisite for the success of the subsequent PCR. Single-cell RT is performed in a small volume (about 10 μL) (Brenner et al., 1989; Lambolez et al., 1992). We use random hexamer nucleotide primers that bind to all possible matching sites and can transcribe even pieces of mRNA. Thus, the probability of receiving the first cDNA strand is greater as compared to corresponding experiments with

Mouse genomic DNA

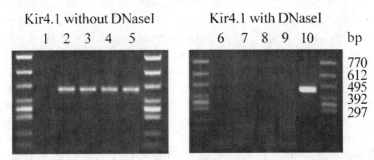

Fig. 3. Mouse genomic DNA (0, 50, 100, 250, and 500 ng; lanes 1–5, and 6–10, respectively) was used as a template for a two round-nested PCR with outer and inner Kir4.1 specific primers. A two-round PCR was performed (Schröder et al., 2002). Amplification yielded Kir4.1 cDNAs of the expected length (461 bp; lanes 2–5). No reaction products were observed when treating up to 250 ng genomic DNA with 5 U DNaseI (Roche) prior to the two round PCR (lanes 7–9). The length marker was ΦX-174–RF DNA HincII digest.

oligo $(dT)_{25}$ primers. For semiquantitative single-cell transcript analysis, the use of gene-specific antisense primers can be recommended. We use reverse transcriptase from Moloney murine leukemia virus (MMLV) carrying a RNase H deletion (RNase H⁻; Superscript II™; Invitrogen) or an enzyme from a new retroviral source, especially designed for the transcription of small RNA amounts (Sensiscript™; Qiagen Hilden, Germany).

### 3.4. Multiplex and Nested PCR

Single-cell RT-PCR provides the opportunity to detect several genes, including cell-type–specific markers or even a whole gene family in individual, electrophysiologically characterized cells, and thus enables investigation of structure/function relationships. Such an approach can be realized with a multiplex PCR using more than one primer pair for the first PCR round (Ruano et al., 1995). Careful optimization of primers and PCR conditions is strictly required.

For a simultaneous detection of several subunits of a gene family, we use nonspecific primer pairs and locate them in a conserved region of the gene that is common to all those subunits. Up- and downstream primers are designed to allow up to two mismatches while the last 6 bp of the primer's 3' end should fully match. In the case of more than two mismatches, degenerated primers should be used. Partition of cDNA after the RT should be avoided in the case of single cells because this often reduces the success rate of subsequent PCR amplification. Instead, we utilize the whole cDNA as a template in the first PCR round. The annealing temperature should always be set below the melting temperature of the primers (Lambolez et al., 1992). The separation of the initially co-amplified gene fragments is implemented in a second PCR round using nested primers. The first PCR round consists of about 30 to 45 cycles, with primer annealing being prolonged to 2 minutes for the first five rounds and lasting 45 seconds in the following cycles. A hot start PCR should be performed by adding *Taq* polymerase directly to the denatured reaction mix at 85°C, rather than by using hot start polymerase that requires preheating at 95°C (10 to 15 minutes). In single-cell approaches the PCR should be performed in reaction volumes of about 50 μL. After RT, mastermix containing PCR buffer, 2.5 mM $MgCl_2$, and 10 pmol of each primer is added to the cDNA. The dNTPs are not added (present at a concentration of $4 \times 50 \mu M$ due to dilution of the RT mix). *Taq* polymerase (4 U, Life Technologies) is added at 85°C. We use a PCR cycler (e.g., MJ Research, Waltham, MA) that permits very rapid temperature ramps, reducing the overall reaction time and limiting polymerase inactivation.

The second PCR runs with an aliquot (2 μL) of the cleaned PCR product resulting from the first round and uses either one (heminested) or two (nested) new primers. Both specificity and sensitivity of the PCR are improved with nested primers, which have to be set into the preamplified cDNA obtained from the first round. In the second round, sense or antisense primer must be gene-specific to prevent amplification of related subunits. To meet this postulation the primers of the first round should span a heterologous region, which is the target of the specific primers used in the second PCR.

Optimal conditions for the second PCR have to be found empirically whereby the variation of $Mg^{2+}$ concentration and annealing temperature is of particular relevance. Annealing temperatures should roughly match the melting temperatures of the respective primers. The number of cycles should be reduced as compared to the first round and range between 20 and 35. Again, the PCR should be performed in a reaction volume of 50 μL, adding mastermix containing PCR buffer, 2.5 mM $MgCl_2$, 10 pmol of each primer, $4 \times 50 \mu M$ dNTPs, and hot start platinum *Taq* polymerase (2.5 U, Invitrogen) to the preamplified product. The PCR product then can be analyzed in a 1.5% to 2% ethidium bromide labeled agarose gel.

As a negative control, we run the PCR with water instead of cDNA. Further controls concern the cell harvesting/RT-step (replacement of cell content by recording solution) and omission of reverse transcriptase in the RT-reaction (cf. Section 3.3).

## 3.5. Real-Time RT-PCR for Semiquantitative Single-Cell Transcript Analysis

### 3.5.1. Introduction to mRNA Quantification

An ABI PRISM™ 7700 Sequence Detection System (PE Applied Biosystems, Foster City, CA) can be used for mRNA quantification using a real-time RT-PCR amplification method. This system permits a quantitative detection of PCR product accumulation during each cycle by using the fluorogenic 5′-nuclease assay (Holland et al., 1991). Thereby, the real-time RT-PCR approach takes advantage of the 5′-nuclease activity of the DNA polymerase for cleavage of a dual-labeled fluorogenic hybridization probe (TaqMan™ probe). The TaqMan™ probe hybridizes in a sequence between the forward and reverse primers on one DNA strand. This probe consists of an oligonucleotide of about 25- to 35-bp length labeled on the 5′ end with a fluorescent reporter dye. The 3′ end of the TaqMan™ probe is labeled with a second fluorescent dye, which absorbs the emitted light of the reporter dye by fluorescent energy transfer (fluorescence quencher) when both dyes are attached to the oligonucleotide probe, as long as the probe is not digested. The DNA polymerase cleaves the TaqMan™ probe and separates reporter dye and fluorescence quencher, resulting in an interruption of fluorescent energy transfer and light emission by the reporter dye (Livak et al., 1995). Emission increase at dye-specific wavelengths during each PCR cycle is consecutively monitored. After normalization of emission intensity to an internal reference, the threshold cycle ($C_T$ value) is determined in the exponential phase of the PCR. The $C_T$ value is used for quantification of the input copy number of target mRNA normalized to a reference gene (Fink et al., 1998).

## 3.5.2. Cell Harvesting and mRNA Isolation

Application of this quantitative method to single cells requires the same precautions in respect to contamination with RNases and preamplified cDNAs as mentioned above. Harvesting of cytoplasm from single cells in situ should follow the instructions given in Chapter 3, Section 3.1, with an alteration in the last step. Since this method lacks an initial cDNA amplification round, and the cell content usually has to be split into several portions to enable parallel amplification of target and reference genes (see below), the success rate of single-cell analysis is lower for certain low-copy genes as compared with two-round, qualitative PCR approaches. We circumvent this problem by pooling cells with similar electrophysiological properties. To limit the liquid volume for the subsequent reactions, recording pipettes should be filled with 6 µL of internal recording solution and the cell contents should be harvested into an empty reaction tube. After breaking the tip of the pipette, about 2 µL of the pipette's contents are expelled under positive pressure, collected at the bottom by quick centrifugation, and immediately frozen in liquid nitrogen. In this way, four to six cells should be consecutively collected in the same tube (final volume about 10 µL). Oligo (dT)$_{25}$ linked Dynabeads™ (Dynal) can be used to separate mRNA from genomic DNA. Lysis buffer (20 µL) and Dynabeads suspended in lysis buffer (10 µL) are added to the reaction tube. After wash (50 µL), the beads with the adhered mRNA are suspended in water (20 µL), frozen, and stored at –80°C until amplification.

## 3.5.3. Primer and Probe Design

TaqMan™ probes and primers should be designed using commercial software (e.g., Primer Express, PE Applied Biosystems); they must be specific to the mRNAs to be detected. The usage of the TaqMan fluorescent probes additionally ensures specificity of the amplified products since the probe must be sequence-specific. Accordingly, increasing fluorescent intensity during consecutive PCR cycles is only observed if the probe is complementary to the target gene, but not during the formation of putative nonspecific amplicons. The designs of primers and probes are not independent of each other. As a general rule, the melting temperature of the probe should be 5°C to 10°C higher than that of the primers. The

5' end of the probe should not contain any "G" and should immediately follow the 3' end of the respective primer.

### 3.5.4. Optimization of Quantitative Transcript Analysis in Single Cells

The RT-PCR can be performed as a single-step, single-enzyme
procedure using recombinant DNA polymerase from the bacterium
*Thermus thermophilus* (rTth) (PE Applied Biosystems). In a single-
buffer system, this enzyme possesses efficient reverse transcriptase
activity at 60°C and acts as a DNA polymerase at temperatures
>60°C. To reach high amplification efficiency, primers should be
designed so that the size of the resulting amplicons do not exceed
150 bp. Generally, longer cDNA fragments are amplified with lower
and more variable efficiency, which may hinder semiquantitative
analysis relative to a reference gene.

Optimization of the PCR conditions should start with varying
the concentration of primers, probes, and the $Mn^{2+}$ content using
DNA-free mRNA. Next, since the different target and housekeeping genes to be compared are simultaneously run in separate tubes,
it is very important to ensure that the respective cDNAs are amplified with the same efficiency that is illustrated in the examples of
Sections 3.5.5, 4.2.5, and 4.3.2.

Semiquantitative RT-PCR protocols also use two-step reactions.
Here, the RT reaction prior to PCR amplification is accomplished
using a separate reverse transcriptase. For quantification, the transcripts from both the target gene and the reference gene must be
transcribed with equal efficiency. One has to consider that different
reverse transcriptases (e.g., from Moloney murine leukemia virus
or avian myeloblastosis virus) may display dissimilar efficiency in
first-strand synthesis, depending on the secondary structure of
mRNA and incubation temperature (Brooks et al., 1995; Bustin,
2000; Freeman et al., 1996; Wong et al., 1998). Moreover, different
RT primers can influence the ratio of target vs. reference cDNA.
A precise determination of mRNA copy numbers can be achieved
by performing the RT reaction with long, gene-specific antisense
primers rather than unspecific random hexamers and oligo $(dT)_{10-20}$
primers (Zhang and Byrne, 1999). Precipitation of cDNA resulting
from the RT reaction may improve specificity of the following PCR
(Liss, 2002).

The single-step/one-enzyme RT-PCR procedure using rTth polymerase with intrinsic RT activity benefits from the advantage of RT

priming with gene-specific primers allowing accurate mRNA quantification. However, the single enzyme method was reported to be less sensitive as compared with the two-enzyme RT-PCR approach, probably due to reduction of nucleotide incorporation in the presence of $Mn^{2+}$ ions (Bustin, 2000; Fromant et al., 1995).

## 3.5.5. Quantification of Real-Time RT-PCR

The real-time PCR technique is the method of choice to quantify low amounts of DNA or mRNA. An absolute estimation of gene copies can be accomplished by using DNA or RNA standards of given copy numbers. Standard DNA (recombinant, genomic, PCR product, synthesized oligonucleotides) or RNA (recombinant, messenger, total) must be accurately synthesized, purified, and their concentration precisely determined by spectrophotometry. Subsequently, a calibration curve must be generated to ensure a linear relation between threshold cycles, $C_T$, and the logarithm of copy numbers. $C_T$ is defined as the number of cycles necessary to reach a given level of fluorescence intensity beyond background, where PCR amplification is still in the exponential phase. The $C_T$ value of the target gene is then related to the $C_T$ value of the calibration curve, allowing determination of copy numbers (for review see Bustin, 2000; Pfaffl, 2004).

Frequently, a semiquantitative PCR method is used, comparing $C_T$ values of a target gene with those of a ubiquitous reference or housekeeping gene. This approach is sufficient to identify alterations in gene expression occurring during development or under pathophysiological conditions. Commonly used housekeeping genes include β-actin and glyceraldehyde-3-phosphate dehydrogenase (GAPDH). However, since the housekeeping genes can also be affected in the disease, cell-type–specific reference genes [e.g., synaptophysin for neurons and glial fibrillary acidic protein (GFAP), S100ß for glial cells] sometimes are preferable (Chen et al., 2001).

For relative quantification of different genes, the efficiency of PCR amplification has to be determined according to the following equation:

$$Y = X * E^n \tag{1}$$

or, in the logarithmic form,

$$\log Y = \log X + n * \log E \tag{1a}$$

with $Y$ being the amount of the PCR product, $X$ the input copy number, $E$ the efficiency of amplification, and $n$ the cycle number. $E$ varies between 1 and 2, and is determined through linear regression of the exponential phase of the PCR plot. Fig. 4A gives an example of the amplification kinetics of Kir4.1 and β-actin transcripts obtained from human hippocampus. According to equation 1, the fluorescence intensity, $R_n$, which is proportional to $Y$, can be determined at each cycle. The slope of the amplification curve is given by the ratio of two pairs of variates, $R_{n1} = X * E^{C_{T1}}$ and $R_{n2} = X * E^{C_{T2}}$, according to

$$R_{n2}/R_{n1} = E^{(C_{T2} - C_{T1})} \tag{2}$$

After taking the logarithm, it follows:

$$\log E = \log (R_{n2}/R_{n1})/(C_{T2} - C_{T1}) \tag{3}$$

and, taking the antilog,

$$E = (R_{n2}/R_{n1})^{1/(C_{T2} - C_{T1})} \tag{3a}$$

In Fig. 4A, the efficiencies for Kir4.1 and β-actin amounted to 1.79 and 1.77, respectively. Another model uses a Boltzmann equation to fit the increase of fluorescence intensity (Liu and Saint, 2002):

$$R_n/R_{max} = 1/(1 + \exp(-(n - n_{0.5})/k)) \tag{4}$$

where $R_n$ is the background-corrected fluorescence intensity at cycle $n$, $R_{max}$ the maximal fluorescence intensity (background corrected), $n_{0.5}$ the cycle number at half-maximal fluorescence intensity, and $k$ the slope of the curve. If $E$ is determined from $R_n$ values of consecutive PCR cycles, it follows from equations 3 and 4:

$$E_n = R_n/R_{n-1} \tag{5}$$

and

$$E_n = (1 + \exp(-(n - 1 - n_{0.5})/k))/(1 + \exp(-(n - n_{0.5})/k)) \tag{5a}$$

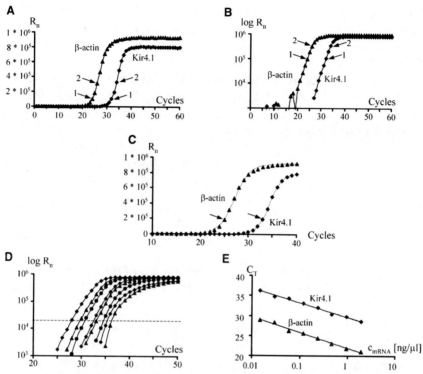

Fig. 4. Determination of real-time PCR efficiency. (**A,B**) DNA-free mRNA from human hippocampus was reverse transcribed and Kir4.1 and β-actin transcripts were amplified through one-enzyme real-time RT-PCR according to the instruction given in Section 4.3.1. The increase in fluorescence intensity was plotted against PCR cycles using linear (**A**) or logarithmic scaling (**B**) of $R_n$. Labeled data points were used to determine $E$ applying equation 3. Data points labeled in **A** and **B** are identical. (**C**) The fluorescence curves in **A** were fitted by a Boltzmann equation (equation 4; solid lines), yielding $n_{0.5} = 27.11$, $k = 1.63$ for β-actin, and $n_{0.5} = 34.44$, $k = 1.29$ for Kir4.1. (**D**) The logarithm of the increase in fluorescence intensity was plotted against PCR cycles using dilution series of human hippocampal mRNA and primers and probes for Kir4.1. The mRNA concentration ranged between 1.9 ng/μL and 14.8 pg/μL (obtained by 1:1 dilutions). (**E**) The threshold cycles for both transcripts (Kir4.1 obtained from the dashed line in **D**) at the respective mRNA concentration. Solid lines represent the best linear fit, with $S = -3.653$ for Kir4.1 and $S = -3.8793$ for β-actin (correlation coefficient 0.99 for both genes).

Here, $E_n$ depends on the parameters $k$ and $n_{0.5}$. Applying this method to the example shown in Fig. 4, the following efficiencies were obtained: $E_{33} = 1.87$ for Kir4.1 and $E_{25} = 1.67$ for β-actin (Fig. 4C).

Another method for evaluation of $E$ is based on dilution series of cDNA or RNA and plotting $C_T$ values in the exponential phase of the PCR against the logarithm of the concentration $c$ (Fig. 4D). According to equation (1) it follows:

$$n = \log R_n / \log E - \log c / \log E \qquad (6)$$

and hence

$$n = -1/\log E * \log c + \log R_n / \log E \qquad (6a)$$

In equation 6, $-1/\log E$ represents the slope $S$ of the linear function, and $\log R_n / \log E$, the intercept with the y-axis (Fig. 4E). It follows that

$$\log E = -1/S \qquad (7)$$

and hence

$$E = 10^{-1/S} \qquad (7a)$$

Applying this approach to the example displayed in Fig. 4, the efficiencies were 1.88 for Kir4.1 and 1.81 for β-actin (Fig. 4E).

It becomes obvious from this example that the three different methods yield similar efficiency values. The third method is used most commonly (see Section 4.2.4). However, it should be mentioned that this approach may result in an overestimation of real PCR efficiency, with $E > 2$ (Pfaffl, 2001); for review see Orlando et al., 1998; Wong and Medrano, 2005.

The ratio of the $E$ values of target and reference genes can be calculated according to the following equation:

$$\text{Ratio} = E_{\text{target}}{}^{\Delta C_T(\text{control–sample})} / E_{\text{reference}}{}^{\Delta C_T(\text{control–sample})} \qquad (8)$$

where "target" and "reference" indicate distinct genes, and "control" and "sample" refer to different conditions, for example, developmental or pathophysiological alterations (Pfaffl, 2001). This model considers different amplification efficiencies of target and reference genes.

Similar results are obtained with the $\Delta\Delta C_T$ method that compares differences in threshold cycles between samples and control tissue:

$$\text{Ratio} = ((E_{\text{target}})^{C_T\text{control}} / (E_{\text{reference}})^{C_T\text{control}}) /$$
$$((E_{\text{target}})^{C_T\text{sample}} / (E_{\text{reference}})^{C_T\text{sample}}) \qquad (9)$$

If amplification efficiencies of target and reference gene are equal or maximal ($E = 2$), a more simple relation results from equation 9 (see Livak and Schmittgen, 2001):

$$\text{Ratio} = E^{\Delta C_T\text{control}} / E^{\Delta C_T\text{sample}} = E^{-\Delta\Delta C_T} \qquad (10)$$

or, in the case of $E = 2$,

$$\text{Ratio} = 2^{-\Delta\Delta C_T} \qquad (10a)$$

The $\Delta\Delta C_T$ method yields only a first estimation of gene regulation, because it assumes identical efficiencies of target and reference gene amplification.

## 4. Examples: Changes of AMPA Receptor and Kir Current Properties in Glial Cells of Patients with Temporal Lobe Epilepsy

Glutamate is the principal excitatory neurotransmitter in the brain. Besides its physiological role it can cause overexcitation and cell destruction in neurological disorders, for example, epilepsy (Rogawski and Donevan, 1999). Evidence is available suggesting a role for ionotropic glutamate receptors, particularly the α-amino-3-hydroxy-5-methyl-4-isoxazole propionate (AMPA) subtype, in the generation and spread of seizure activity. In human epilepsy, hippocampal neurons of patients with Ammon's horn sclerosis (AHS) underwent changes in AMPA receptor subunit expression and increases in receptor density that may entail enhanced excitability and seizure susceptibility (De Lanerolle et al., 1998; Eid et al., 2002). AMPA receptors are also expressed by glial cells (Verkhratsky and

Steinhäuser, 2000) where they are able to sense glutamate released from neurons (Bergles et al., 2000; Jabs et al., 2005). In astrocytes of the hippocampus, AMPA receptor activation induced a depolarization not only via the cationic receptor current but in addition due to block of Kir channels following the $Na^+$ influx through the receptor pore (Schröder et al., 2002). Hence, in epilepsy and other pathological conditions characterized by elevated extracellular glutamate levels (Glass and Dragunow, 1995), this mechanism might enhance neuronal excitability by imposing an impaired $K^+$ buffer capacity on the astroglial network.

Kir channels constitute a large family with several subfamilies. These channels are present in many different tissues and cell types (Reimann and Ashcroft, 1999). Since they are selectively permeable for $K^+$ and have a high open probability at rest, they largely determine the resting membrane potential and stabilize it near the $K^+$ equilibrium potential. Kir channels are abundantly expressed in astrocytes and were implicated to play a major role for regulating $K^+$ homeostasis (Newman, 1986). Recent data suggested that alterations in astroglial Kir channel function might be involved in seizure generation in human epilepsy (Bordey and Sontheimer, 1998; Hinterkeuser et al., 2000; Schröder et al., 2000).

In the examples presented here we used single-cell RT-PCR subsequent to functional analysis to determine which subunit(s) make up human AMPA receptors (Section 4.2) and Kir channels (Section 4.3), and to identify those subunits that were potentially affected in human epilepsy. This was accomplished by comparing two forms of temporal lobe epilepsies: AHS and lesion-associated epilepsy. Focusing on the hippocampus, we considered tissue from lesion-associated epilepsy "control-like" because it presents no or only minor morphological abnormalities as compared to autopsy controls.

## 4.1. Astrocytes in Human Hippocampus Express Glutamate Receptors of the AMPA Subtype that Are Altered in Epilepsy Patients with Ammon's Horn Sclerosis

To quantitatively analyze ionotropic glutamate receptors in astrocytes of human hippocampus, glial cells were acutely isolated from the CA1 region of specimens neurosurgically removed from epilepsy patients, and investigated with a fast concentration-clamp method (Seifert and Steinhäuser, 1995). Rapid application of kainate

(1 mM) always induced nondesensitizing currents (Fig. 5A, upper panel). We used cyclothiazide (CTZ), a substance that selectively affects the flip versions of AMPA receptors (Partin et al., 1994), to uncover putative differences in receptor splicing in epilepsy patients with AHS or lesion-associated epilepsy. After a control application of kainate (1 mM) the cells were exposed to CTZ (100 μM, 30 seconds) and then kainate was rapidly applied again (Fig. 5A, lower panels), which produced a current increase in all cells analyzed. In astrocytes of AHS specimens, the CTZ-mediated potentiation reached 686% ± 69% (V = –70 mV, $n = 7$) while a significantly smaller enhancement was observed in the lesion group (510% ± 96%; $n = 18$; Fig. 5A,B). Fast application of glutamate to acutely isolated astrocytes induced rapidly desensitizing inward currents, with the amplitudes being the same in AHS and lesion cases. The current decay could be fitted by a single exponential, revealing significantly faster desensitization kinetics for cells of the lesion group ($\tau = 6.2 \pm 0.9$ ms; $n = 31$) as compared with AHS astrocytes (7.6 ± 1.4 ms; $n = 29$; Fig. 5C). These differences in current desensitization and modulation by CTZ suggested that in AHS, astroglial AMPA receptors underwent changes in alternative splicing (Seifert et al., 2002, 2004).

## 4.2. Molecular Characterization of AMPA Receptor Subunits and Splice Variants in Human Astroglial Cells

### 4.2.1. Methodical Considerations: Cell Harvesting, Single-Cell RT-PCR

After recording, the cell content was harvested by carefully sucking the cytoplasm into the recording pipette. The pipette contents were transferred into a tube filled with 3 μL of DEPC-treated water by breaking the tip of the pipette and expelling about 3 μL solution under positive pressure. The tube was frozen in liquid nitrogen and stored at –80°C. RT-PCR followed the strategy first described by Lambolez et al. (1992). RT was performed in a final volume of about 10 μL, adding Buffer RT (Qiagen, Hilden, Germany), deoxyribonucleotide triphosphates (dNTPs, final concentration 4 × 250 μM; Applied Biosystems, Darmstadt, Germany), random hexanucleotide primer (50 μM; Roche, Mannheim, Germany), 20 U RNasin (Promega, Madison, WI), and 0.5 μL Sensiscript reverse transcriptase (Qiagen). Single-strand cDNA synthesis was performed at 37°C for 1 hour.

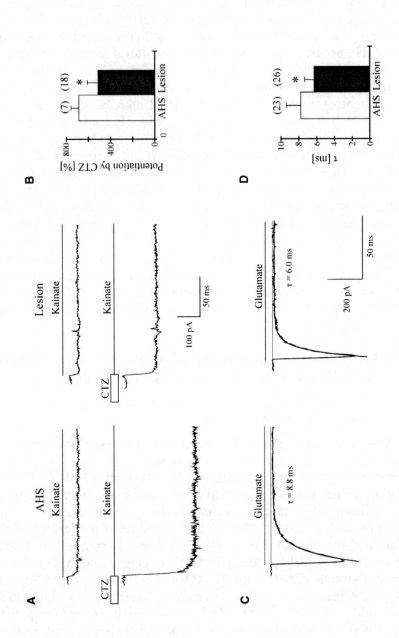

A two-round PCR was performed, using the product obtained after the first round as a template for the second round. The first PCR was run after adding PCR buffer, $MgCl_2$ (2.5 mM), and corresponding primers (200 nM each) to the RT product (final volume 50 μL). After denaturation (94°C, 4 minutes), 3.5 U Taq polymerase (Invitrogen) was added at 85°C. Forty-five cycles were performed (denaturation at 94°C, 30 seconds; annealing at 49°C, 2 minutes for the first five cycles, and 45 seconds for the remaining cycles; extension at 72°C, 25 seconds; final elongation at 72°C, 7 minutes). The primer pair used for the first PCR round amplified all AMPA receptor subunits, GluR1-GluR4. The sense primer was 5'-TGTGC ATTGTYTTTGCCTACATTGG, having one mismatch with GluR3 and GluR4, respectively. The antisense primer was 5'-CTCAGRG CRCTBGTCTTKTCCTT, having one mismatch with GluR2 and GluR4, respectively (gene bank accession numbers M64752, L20814, U10301, and U16129 for GluR1-4, respectively).

The PCR product was purified using the Ultra Clean DNA purification kit (Mobio, Solana Beach, CA), and dissolved in water (40 μL). An aliquot (3 μL) of the product was used as a template for the second PCR (35 cycles; annealing at 51°C (GluR1, 2, 4) and 54°C (GluR3); first five cycles: 2 minutes, remaining cycles: 45 seconds using nested, subunit specific sense primers and a common antisense primer. The nested sense primers were 5'-GGACGGGA CCAGACAACCAG for GluR1, 5'-TGAAGATGGAAGAGAAAC ACAAAG for GluR2, 5'-ACCCACAAAGTCCTCCTG for GluR3,

---

Fig. 5. Astrocyte glutamate receptor properties and their modulation in human epilepsy. (**A**) Receptor currents were evoked by rapid application of kainate (1 mM, V = −70 mV) to hippocampal astrocytes freshly isolated from neurosurgical specimens of AHS (left) and lesion patients (right). After preincubation in cyclothiazide (CTZ) (100 μM), the receptor currents were potentiated after applying kainate again. (**B**) The summary revealed a significantly (\*) higher CTZ potentiation in Ammon's horn sclerosis (AHS) vs. lesion-associated epilepsy. (**C**) Fast application of glutamate (1 mM) elicited inward currents in astrocytes from AHS and lesion specimens. The glutamate responses desensitized rapidly and almost completely, with current decay following a single exponential. Time constants of current desensitization as indicated. (**D**) Receptor desensitization was significantly slower in AHS. Holding potential was −70 mV. Cell numbers are given in parentheses. (Modified from Seifert et al., 2002, 2004, with permission.)

and 5′-GAAGGACCCAGCGACCAGCC for GluR4. The nested antisense primer was the same for all subunits: 5′-TCGTACCAC-CATTTGTTTTTCA (one mismatch with GluR1). Products were identified by agarose gel electrophoresis (1.5%; stained with ethidium bromide) using a molecular weight marker (ΦX174 *Hinc*II digest; Eurogentec, Seraing, Belgium). The expected amplified fragment lengths were 632 bp, 639 bp, 630 bp, and 626 bp for GluR1-4 (Seifert et al., 2004).

### 4.2.2. Methodical Considerations: Comparison of GluR Splice Variants in Both Types of Epilepsy by Restriction Analysis

For restriction analysis, the second PCR round was again performed, with a reduced cycle number (25 cycles). The product was purified and dissolved in 25 µL of water. The splicing status of the GluR cDNAs was identified with the splice variant-specific restriction endonucleases *Alu*I (cut GluR1 flip, yielding a 609-bp and a 23-bp fragment), *Hpa*I (GluR2 flop, 571 and 68 bp), *Hpa*I (GluR3 flop, 562 and 68 bp), and *Bsa*HI (GluR4 flip, 590 and 36 bp), all purchased from New England Biolabs (Frankfurt, Germany); 7 µL of the respective purified PCR product were incubated in 10 U restriction enzyme (total volume 15 µL; 6 hours, 37°C). The cDNA fragments were separated with electrophoresis and agarose, allowing resolution of 10 bp differences (2% Agarose-7000 gel matrix, 50 bp length marker; Invitrogen, Karlsruhe, Germany). The gels were stained with ethidium bromide and evaluated with a digital imaging system (AlphaImager, San Leandro, CA). Exposure times were adapted to the respective staining intensities, and the optical densities of the fragments were calculated after background correction. The fractions of flip vs. flop (flip/flop ratios) were obtained for each restriction enzyme by comparing the optical densities of corresponding digested and undigested bands (normalized to 100%). To enhance reliability, the second PCR and restriction analysis were always performed twice.

For a positive control, total RNA was prepared from human hippocampal specimen using Trizol™ (Invitrogen). Then a two-round RT-PCR was performed with 2 ng total RNA and primers as described above. Subsequent gel analysis did not detect unspecific products. PCR products obtained from human total RNA and from cloned GluRs were digested with the restriction enzymes above, yielding fragments of the expected lengths. In the case of cloned GluRs, digestion was complete. Omission of reverse transcriptase or aspiration of bath solution served as negative controls.

### 4.2.3. Identification of AMPA Receptor Subunits and Splice Variant Expression in Human Astrocytes

After recording, the cytoplasm of the respective astrocyte was harvested for transcript analysis. Two-round single-cell RT-PCR revealed a relative frequency of 59%, 97%, 41%, and 72% for the astroglial GluR1-4 subunits, respectively (Fig. 6A). The GluR2 subunit, controlling the $Ca^{2+}$ permeability of AMPA receptors, was found in almost all cells (28 of 29 cells) and 20 cells (69%) coexpressed GluR2 together with GluR4. In individual cells, the subunit combination GluR1, 2, 4 was most abundant (Fig. 6B). This expression pattern did not differ between astrocytes from AHS or lesion cases.

To investigate the flip/flop status of the respective subunit and identify the subunit(s) that potentially underwent changes in alternative splicing, the PCR products were subjected to restriction analysis. Overall, the GluR subunits were preferentially expressed in the flip form, which agreed with the robust current potentiation by CTZ (Fig. 6C). Strikingly, astrocytes from AHS patients expressed a significantly enhanced level of GluR1 flip compared to lesion cases, while flip/flop splicing of GluR2–4 did not differ in the two forms of epilepsy (Fig. 6D) (Seifert et al., 2004).

### 4.2.4. Methodical Considerations: Quantification of Flip/Flop Splice Variant Ratios Using Real-Time PCR

An ABI PRISM™ 7700 Sequence Detection System (PE Applied Biosystems, Foster City, CA) was used for cDNA quantification. GluR flip and flop transcripts were investigated with TaqMan™ PCR Core Reagent Kit (PE Applied Biosystems). Primer sequences and TaqMan probes were purchased from Eurogentec. The reaction volume was 12.5 µL. The cDNA of the first single-cell PCR round was used as a template for the following reactions (aliquots of 0.5 µL). In each PCR run, GluR flip and flop of single cells were recorded in parallel with mixtures of cloned GluR flip/flop cDNAs for calibration (96 well plates). The optimized reaction mixture contained TaqMan buffer, $3 \times 200\,\mu M$ dNTPs (dATP, dCTP, dGTP),

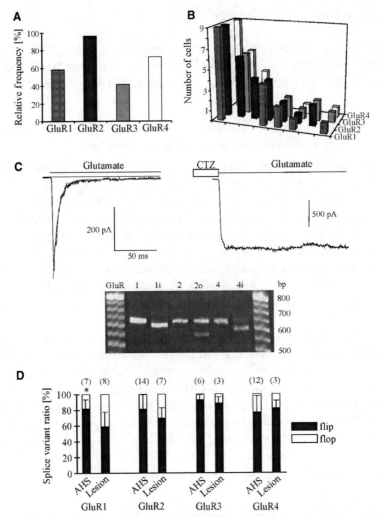

Fig. 6. α-amino-3-hydroxy-5-methyl-4-isoxazole propionate (AMPA) receptor subunit expression and splicing by human astrocytes. (**A**) After electrophysiological characterization, the cytoplasm of the respective cell was harvested and analyzed with single-cell RT-PCR. The histogram shows the relative frequency of GluR subunit expression in human astrocytes ($n = 29$). (**B**) The subunit combinations expressed by individual cells. Coexpression of GluR1, -2, -4 was found most frequently. (**C**) A cell from an AHS patient was exposed to glutamate (1 mM; left panel) and, after wash and preincubation of CTZ, the agonist was applied again (right). CTZ abolished receptor desensitization and significantly enhanced current amplitude. In this cell, subsequent RT-PCR and restriction analysis revealed a preferential expression of the flip variants of GluR1, -2, and -4. (**D**) Summary of restriction analysis revealed prevailing expression of flip versions. GluR1 flip was more abundant in AHS (*), cell numbers in parentheses. (Modified from Seifert et al. 2004, with permission.)

400 µM dUTP, 2.5 mM MgCl$_2$, 0.3 U AmpliTaq Gold Taq DNA polymerase, 0.125 U uracil-N-glycosylase (UNG), 100 nM fluorogenic TaqMan probe, and 300 nM of each PCR primer. The used GluR splice form-specific oligonucleotides were for GluR flip: sense, 5'-TCAGTGAGSMAGGCGTCTTAGA; antisense, 5'-GTCTT GTCCTTACTTCCRGAGTC; GluR flop: sense, 5'-TTAACCTSGC AGTATTAAAACTGA; antisense, 5'-TGGARTCACCTCCCCCGC TG; probes: GluR flip sense, FAM-5'-AAGCTGAAAARCAAATGG TGGTAC GATAAAGG-TAMRA; GluR flop sense, FAM-5'-TGGA CAAATTGAAAAACAAATGG TGGTACGA-TAMRA. After incubation (50°C, 2 minutes), UNG was denatured at 95°C (10 minutes), followed by 60 PCR cycles (denaturation at 95°C, 15 seconds; primer annealing and extension at 59°C, 60 seconds). The fluorescence intensity was read out during each annealing/extension step of the respective PCR cycle. Fluorescence emission at dye-specific wavelengths was consecutively monitored during each PCR cycle. After normalization of emission intensity to an internal reference, $\Delta$Rn, the threshold cycle ($C_T$) was determined in the exponential phase of the PCR.

Specificity and efficiency of the reaction conditions were optimized using cloned human AMPA receptor cDNA. The primers and Taqman probes were designed for amplification of the respective splice variant, either flip or flop, and no PCR products were amplified when the cloned human GluR cDNA of the opposite splice form was used as PCR target. The PCR conditions were optimized to amplify both splice forms of GluR1–4 with similar efficiency, using serial dilutions of GluR cDNA (from $4 \times 10^{-17}$ mol to $3.125 \times 10^{-19}$ mol). Fig. 7A (left) illustrates that increasing amounts of the respective cloned human cDNA resulted in a continuous decrease of the $C_T$ values. The resulting logarithmic functions for GluR1 flip and flop displayed the same slope. The calculated amplification efficiencies according to equations 6 and 7 were 2.14 and 2.20 for the flip form of GluR1 and GluR2, respectively, and 2.11 and 2.16 for the flop form of GluR1 and GluR2, respectively. Next, defined flip/flop ratios were used (total cDNA concentration was kept constant at $2 \times 10^{-17}$ mol) and the flip and flop primers and probes were tested to obtain the respective $C_T$ values. The differences in $C_T$ values between flip and flop PCR runs were plotted against the flip/flop ratio of the target cDNA, yielding calibration curves for estimation of this ratio in cDNAs obtained from epilepsy patients (Fig. 7B).

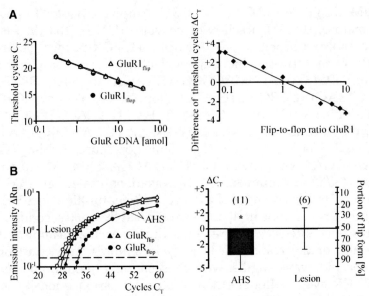

Fig. 7. Determination of AMPA receptor flip/flop ratios in human astrocytes using semiquantitative real-time PCR. (**A**) Cloned human GluR1 flip and flop cDNAs were used to calibrate real-time PCR. Threshold cycles were determined at various dilutions, and the almost identical slopes of the curves indicated the same amplification efficiency for both splice variants (left). Standard $\Delta C_T$ curves were obtained for defined GluR1 flip/flop ratios at constant cDNA concentration ($2 \times 10^{-17}$ mol). $\Delta C_T$ was determined by subtracting $C_{T,GluRflop}$ from $C_{T,GluRflip}$ at corresponding flip/flop ratios (right). (**B**) Amplification plots of two individual astrocytes from AHS (filled symbols) and lesion specimens (open symbols; left). Note the higher cycle number necessary to amplify flop transcripts (circles) in the AHS cell. The threshold for detection was set at $\Delta R_n = 0.17$ (indicated by the dashed line). The bar graph (right) sums up, demonstrating larger $\Delta C_T$ values (*) and an enhanced portion of flip splice variants in AHS vs. lesion-associated epilepsy. (Modified from Seifert et al. 2004, with permission.)

## 4.2.5. Relative Increase of AMPA Receptor Flip Splice Variants in Ammon's Horn Sclerosis

Single-cell real-time PCR was performed to get a quantitative measure of the differences in astroglial receptor splicing in AHS vs. lesion-associated epilepsy. The optimized PCR protocol was applied to cDNA obtained from individual, functionally characterized hippocampal astrocytes. Fig. 7C depicts the amplification curves of GluR flip and flop variants of cDNA obtained from two astrocytes, one from a lesion case and one from an AHS patient. The detection threshold was set at $\Delta R_n = 0.17$ (indicated by the dashed line). While the $C_T$ values for GluR$_{flip}$ were almost the same in the two cells, the initial cDNA level encoding GluR$_{flop}$ was lower in AHS, since a higher $C_T$ number was necessary to reach the threshold (Fig. 7B, left). Comparison of the difference $\Delta C_T = C_{T,GluRflip} - C_{T,GluRflop}$ revealed a significant difference for cells from AHS ($\Delta C_T = -3.3 \pm 2.2$, $n = 11$, 11 patients) vs. lesion patients ($\Delta C_T = -0.1 \pm 2.6$, $n = 6$, 4 patients). After calibration of $\Delta C_T$ values using the flip/flop ratios of cloned human GluR cDNA, flip/flop splicing in individual cells could be evaluated (Fig. 7B, right). Comparing the above mean $\Delta C_T$ values with the calibration curve corroborated a high expression level of flip variants in AHS (about 80%) and a more balanced flip/flop ratio in astrocytes from lesion-associated epilepsy (53% flip) (Seifert et al., 2004).

## 4.3. Properties of Kir Channel Subunits in Astrocytes of Human Hippocampus

### 4.3.1. Methodical Considerations: Real-Time Single-Cell RT-PCR

Analysis basically followed the strategies described in Chapter 3, Section 3.5. TaqMan™ EZ RT-PCR Kit (PE Applied Biosystems) was used to investigate human Kir4.1 and β-actin transcripts with real-time RT-PCR. The reaction volume was 12.5 μL. DNA-free mRNA was isolated with Dynabeads™ (Dynal) and used as a template for the following reactions (aliquots of 4.9 μL). RT-PCRs for Kir4.1 and β-actin were run in parallel for each of the samples to be analyzed. The optimized reaction mixture contained TaqMan EZ buffer, $3 \times 300$ μM dNTP (dATP, dCTP, dGTP), 600 μM dUTP, 3 mM Mn(OAc)$_2$, 1.25 U rTth DNA polymerase, 0.125 U uracil-N-glycosylase (UNG), 100 nM fluorogenic TaqMan probe, and 50 nM of each primer. Human β-actin as well as Kir4.1 primers and TaqMan probes were designed: Kir4.1: sense primer, 5'-GCGCAAAA GCCTCCTCATT; antisense primer, 5'-CCTTCCTTGGTTTGGT GGG; probe FAM-5'-TGCCAGGTGACAGGAAAACTGCTTCAG-TAMRA, gene bank accession number U52155; β-actin: sense primer, 5'-CAAGTACTCCGTGTGGATCGG; antisense primer, 5'-GCTGATCCACATCTGCTGGAA; probe FAM-5'-TCCATCCTG

GCCTCGCTGTCCA-TAMRA, gene bank accession number NM 001101. During the first step (50°C, 2 minutes), UNG hydrolyzes dU-containing DNA to prevent carryover contamination during the following PCR. Subsequent first-strand synthesis was performed at 60°C (20 minutes). The antisense primers also served as primers for the reverse transcriptase. UNG was denatured at 95°C (5 minutes), followed by 60 PCR cycles (denaturing at 94°C, 15 seconds; primer annealing and extension at 59°C, 60 seconds). The fluorescence intensity was read out during each annealing/extension step. RT-PCR was performed in optical tubes (PE Applied Biosystems) covered with optical caps for optimal fluorescence excitation and detection of emitted light.

### 4.3.2. Correlation of Kir Current Density and Kir4.1 Transcript Expression in Human Astrocytes

We have previously reported that astrocytes in the hippocampus of patients with AHS possess a reduced Kir current density as compared to cells from nonsclerotic epilepsy patients (Hinterkeuser et al., 2000). Moreover, subsequent single-cell RT-PCR identified Kir4.1 transcripts in about 44% of the cells analyzed (Schröder et al., 2000). Here we performed real-time RT-PCR to figure out whether the reduction of Kir currents in the sclerotic tissue was due to diminished expression of Kir4.1 channels.

Since analysis of Kir4.1 and β-actin was simultaneously run in separate tubes, first we checked whether the corresponding cDNAs were amplified with the same efficiency (see Section 3.5.5 and Fig. 4 for calculation of efficiency). Real-time RT-PCR was applied to two samples of pooled, electrophysiologically characterized human hippocampal astrocytes. Kir current densities were determined at $-130\,mV$ by dividing steady-state amplitudes by the respective membrane capacitance as described elsewhere (e.g., Hinterkeuser et al., 2000). Cell pool 1 contained cytoplasm of five cells from an AHS patient, with a mean current density of $29.2 \pm 13.7\,pA/pF$. Cell pool 2 was collected from another epilepsy patient with lesion-associated epilepsy. These cells possessed a higher Kir current density at $-130\,mV$ ($31.8 \pm 5.5\,pA/pF$; $n = 4$). After TaqMan analysis, $\Delta R_n$, the fluorescence intensity normalized to a passive fluorescent reference, and was plotted against PCR cycles (Fig. 8). Threshold for detection was set at $\Delta R_n = 0.02$ (indicated by the dashed line). While the $C_T$ values for β-actin were almost the same in the two

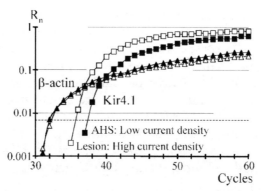

Fig. 8. Semiquantitative transcript analysis of Kir4.1 in single astrocytes of epilepsy patients. Amplification plot of two samples of pooled astrocytes from patients with AHS (filled symbols) and lesion-associated epilepsy (open symbols). While β-actin expression (triangles) was identical in both cell pools, Kir4.1 transcript (squares) amount of the AHS sample was lower ($\Delta C_T = 2.7$) than that of the lesion patient ($\Delta C_T = 1.6$). The threshold for detection was set at $\Delta R_n = 0.007$ (indicated by the dashed line).

cell pools, mRNA levels for Kir4.1 were clearly different, with the difference $\Delta C_T = C_{T,Kir4.1} - C_{T,\beta\beta\text{-actin}}$ of the sample from the AHS patient being significantly higher than that of the control cells from lesion-associated epilepsy (2.7 vs. 1.6; the $C_T$ values for Kir4.1 were 32.9 and 33.8 and for β-actin 31.1 and 31.3, for AHS and lesion, respectively). These $\Delta C_T$ values were then used to compare the relative amount of Kir4.1 transcripts in the respective cell pool ($\Delta\Delta C_T$-method; cf. Fink et al., 1998). Using equations 8 and 9 and determining efficiencies through dilutions series revealed a reduction of Kir4.1 mRNA in AHS specimens to 50.3% compared to lesion-associated epilepsy. Thus, cells with a larger Kir current density also possessed a higher Kir4.1 mRNA level, suggesting that down-regulation of this gene might underlie insufficient functioning of hippocampal astrocytes in epilepsy patients with AHS.

## 5. Conclusion

The combination of the patch-clamp method with single-cell RT-PCR proved to be a powerful approach to identify the molecules underlying biophysical properties of individual cells. Moreover, this technique simultaneously allows detection of transcripts encoding proteins other than transmembrane channels, for example, cell-type–specific markers. However, since the presence of transcripts does not necessarily entail the presence of the corresponding proteins and its incorporation into the plasma membrane, such investigations should always be flanked by other, independent methods, such as immunocytochemistry or electron microscopy. The detailed analysis of structure-function relations in single native cells is now within reach, and this information will be important to explore mechanisms that govern changes in cell–cell signaling during normal development and in pathogenesis.

## Acknowledgments

The authors' work was supported by the Deutsche Forschungsgemeinschaft (SFB/TR3, SE 774/3). We thank Dr. S. Hinterkeuser, who provided the photographs shown in Fig. 2.

## References

Ausubel, F. M., Brent, R., Kingston, R. E., et al. (eds.). (1996) *Current Protocols in Molecular Biology*, John Wiley & Sons, New York.

Bergles, D. E., Roberts, J. D., Somogyi, P., and Jahr, C. E. (2000) Glutamatergic synapses on oligodendrocyte precursor cells in the hippocampus. *Nature* **405**, 187–191.

Bordey, A. and Sontheimer, H. (1998) Properties of human glial cells associated with epileptic seizure foci. *Epilepsy Res.* **32**, 286–303.

Brenner, C. A., Tam, A. W., Nelson, P. A., et al. (1989) Message amplification phenotyping (MAPPing): a technique to simultaneously measure multiple mRNAs from small numbers of cells. *BioTechniques* **7**, 1096–1103.

Brooks, E. M., Sheflin, L. G., and Spaulding, S. W. (1995) Secondary structure in the 3' UTR of EGF and the choice of reverse transcriptases affect the detection of message diversity by RT-PCR. *BioTechniques* **19**, 806–815.

Bustin, S. A. (2000) Absolute quantification of mRNA using real-time reverse transcription polymerase chain reaction assays. *J. Mol. Endocrinol.* **25**, 169–193.

Chen, J., Sochivko, D., Beck, H., Marechal, D., Wiestler, O. D., and Becker, A. J. (2001) Activity-induced expression of common reference genes in individual CNS neurons. *Lab. Invest.* **81**, 913–916.

Chomczynski, P. and Sacchi, N. (1987) Single-step method of RNA isolation by acid guanidinium thiocyanate-phenol-chloroform extraction. *Analyt. Biochem.* **162**, 156–159.

De Lanerolle, N. C., Eid, T., Von Campe, G., Kovacs, I., Spencer, D. D., and Brines, M. (1998) Glutamate receptor subunits GluR1 and GluR2/3 distribution shows reorganization in the human epileptogenic hippocampus. *Eur. J. Neurosci.* **10**, 1687–1703.

Dilworth, D. D. and McCarrey, J. R. (1992) Single-step elimination of contaminating DNA prior to reverse transcriptase PCR. *PCR Methods Appl.* **1**, 279–282.

Dragon, E. A. (1993) Handling reagents in the PCR laboratory. *PCR Methods Appl.* **3**, S8–S9.

Eid, T., Kovacs, I., Spencer, D. D., and De Lanerolle, N. C. (2002) Novel expression of AMPA-receptor subunit GluR1 on mossy cells and CA3 pyramidal neurons in the human epileptogenic hippocampus. *Eur. J. Neurosci.* **15**, 517–527.

Fink, L., Seeger, W., Ermert, L., et al. (1998) Real-time quantitative RT-PCR after laser-assisted cell picking. *Nature Med.* **4**, 1329–1333.

Freeman, W. M., Vrana, S. L., and Vrana, K. E. (1996) Use of elevated reverse transcription reaction temperatures in RT-PCR. *BioTechniques* **20**, 782–783.

Fromant, M., Blanquet, S., and Plateau, P. (1995) Direct random mutagenesis of gene-sized DNA fragments using polymerase chain reaction. *Anal. Biochem.* **224**, 347–353.

Glass, M. and Dragunow, M. (1995) Neurochemical and morphological changes associated with human epilepsy. *Brain Res. Rev.* **21**, 29–41.

Gurantz, D., Ribera, A. B., and Spitzer, N. C. (1996) Temporal regulation of *Shaker*- and *Shab*-like potassium channel gene expression in single embryonic spinal neurons during $K^+$ current development. *J. Neurosci.* **16**, 3287–3295.

Hinterkeuser, S., Schröder, W., Hager, G., et al. (2000) Astrocytes in the hippocampus of patients with temporal lobe epilepsy display changes in potassium conductances. *Eur. J. Neurosci.* **12**, 2087–2096.

Holland, P. M., Abramson, R. D., Watson, R., and Gelfand, D. H. (1991) Detection of specific polymerase chain reaction product by utilizing the 5′–3′ exonuclease activity of Thermus aquaticus DNA polymerase. *Proc. Natl. Acad. Sci. USA* **88**, 7276–7280.

Jabs, R., Pivneva, T., Huttmann, K., et al. (2005) Synaptic transmission onto hippocampal glial cells with hGFAP promoter activity. *J. Cell Sci.* **118**, 3791–3803.

Jonas, P., Racca, C., Sakmann, B., Seeburg, P. H., and Monyer, H. (1994) Differences in $Ca^{2+}$ permeability of AMPA-type glutamate receptor channels in neocortical neurons caused by differential GluR-B subunit expression. *Neuron* **12**, 1281–1289.

Lambolez, B., Audinat, E., Bochet, P., Crépel, F., and Rossier, J. (1992) AMPA receptor subunits expressed by single Purkinje cells. *Neuron* **9**, 247–258.

Li, H. H., Gyllensten, U. B., Cui, X. F., Saiki, R. K., Erlich, H. A., and Arnheim, N. (1988) Amplification and analysis of DNA sequences in single human sperm and diploid cells. *Nature* **335**, 414–417.

Liss, B. (2002) Improved quantitative real-time RT-PCR for expression profiling of individual cells. *Nucleic Acids Res.* **30**, e89.

Liu, W. and Saint, D. A. (2002) Validation of a quantitative method for real time PCR kinetics. *Biochem. Biophys. Res. Commun.* **294**, 347–353.

Livak, K. J., Flood, S. J., Marmaro, J., Giusti, W., and Deetz, K. (1995) Oligonucleotides with fluorescent dyes at opposite ends provide a quenched probe system useful for detecting PCR product and nucleic acid hybridization. *PCR Methods Appl.* **4**, 357–362.

Livak, K. J. and Schmittgen, T. D. (2001) Analysis of relative gene expression data using real-time quantitative PCR and the 2(-Delta Delta C(T)) Method. *Methods* **25**, 402–408.

Newman, E. A. (1986) High potassium conductance in astrocyte endfeet. *Science* **233**, 453–454.

Orlando, C., Pinzani, P., and Pazzagli, M. (1998) Developments in quantitative PCR. *Clin. Chem. Lab Med.* **36**, 255–269.

Partin, K. M., Patneau, D. K., and Mayer, M. L. (1994) Cyclothiazide differentially modulates desensitization of a-amino-3-hydroxy-5-methyl-4-isoxazolepropionic acid receptor splice variants. *Mol. Pharmacol.* **46**, 129–136.

Pfaffl, M. W. (2001) A new mathematical model for relative quantification in real-time RT-PCR. *Nucleic Acids Res.* **29**, e45.

Pfaffl, M. W. (2004) Quantification strategies in real-time PCR, in *A-Z of quantitative PCR* (Bustin, S. A., ed.), International Universities Line, La Jolla, CA, pp. 1–23.

Reimann, F. and Ashcroft, F. M. (1999) Inwardly rectifying potassium channels. *Curr. Opin. Cell Biol.* **11**, 503–508.

Rogawski, M. A. and Donevan, S. D. (1999) AMPA receptors in epilepsy and as targets for antiepileptic drugs. *Adv. Neurol.* **79**, 947–963.

Ruano, D., Lambolez, B., Rossier, J., Paternain, A. V., and Lerma, J. (1995) Kainate receptor subunits expressed in single cultured hippocampal neurons: molecular and functional variants by RNA editing. *Neuron* **14**, 1009–1017.

Schröder, W., Hinterkeuser, S., Seifert, G., et al. (2000) Functional and molecular properties of human astrocytes in acute hippocampal slices obtained from patients with temporal lobe epilepsy. *Epilepsia* **41** S6, 181–184.

Schröder, W., Seifert, G., Hüttmann, K., Hinterkeuser, S., and Steinhäuser, C. (2002) AMPA receptor-mediated modulation of inward rectifier K⁺ channels in astrocytes of mouse hippocampus. *Mol. Cell Neurosci.* **19**, 447–458.

Seifert, G., Huttmann, K., Schramm, J., and Steinhauser, C. (2004) Enhanced relative expression of glutamate receptor 1 flip AMPA receptor subunits in hippocampal astrocytes of epilepsy patients with Ammon's horn sclerosis. *J. Neurosci.* **24**, 1996–2003.

Seifert, G., Kuprijanova, E., Zhou, M., and Steinhäuser, C. (1999) Developmental changes in the expression of *Shaker-* and *Shab*-related K⁺ channels in neurons of the rat trigeminal ganglion. *Mol. Brain Res.* **74**, 55–68.

Seifert, G., Rehn, L., Weber, M., and Steinhäuser, C. (1997a) AMPA receptor subunits expressed by single astrocytes in the juvenile mouse hippocampus. *Mol. Brain Res.* **47**, 286–294.

Seifert, G., Schröder, W., Hinterkeuser, S., Schumacher, T., Schramm, J., and Steinhäuser, C. (2002) Changes in flip/flop splicing of astroglial AMPA receptors in human temporal lobe epilepsy. *Epilepsia* **43** S5, 162–167.

Seifert, G. and Steinhäuser, C. (1995) Glial cells in the mouse hippocampus express AMPA receptors with an intermediate Ca²⁺ permeability. *Eur. J. Neurosci.* **7**, 1872–1881.

Seifert, G., Zhou, M., Dietrich, D., et al. (2000) Developmental regulation of AMPA-receptor properties in CA1 pyramidal neurons of rat hippocampus. *Neuropharmacology* **39**, 931–942.

Seifert, G., Zhou, M., and Steinhäuser, C. (1997b) Analysis of AMPA receptor properties during postnatal development of mouse hippocampal astrocytes. *J. Neurophysiol.* **78**, 2916–2923.

Steinhäuser, C., Kressin, K., Kuprijanova, E., Weber, M., and Seifert, G. (1994) Properties of voltage-activated sodium and potassium currents in mouse hippocampal glial cells in situ and after acute isolation from tissue slices. *Pflugers Arch.* **428**, 610–620.

Verkhratsky, A. and Steinhäuser, C. (2000) Ion channels in glial cells. *Brain Res. Rev.* **32**, 380–412.

von Eggeling, F. and Ballhausen, W. (1995) Freezing of isolated cells provides free mRNA for RT-PCR amplification. *BioTechniques* **18**, 408–410.

Wong, L., Pearson, H., Fletcher, A., Marquis, C. P., and Mahler, S. (1998) Comparison of the efficiency of Moloney murine leukaemia virus (M-MuLV) reverse transcriptase, RNase H⁻M-MuLV reverse transcriptase and avian myeloblastoma leukaemia virus (AMV) reverse transcriptase for the amplification of human immunoglobulin genes. *Biotechnol. Tech.* **12**, 485–489.

Wong, M. L. and Medrano, J. F. (2005) Real-time PCR for mRNA quantitation. *BioTechniques* **39**, 75–85.

Zhang, J. and Byrne, C. D. (1999) Differential priming of RNA templates during cDNA synthesis markedly affects both accuracy and reproducibility of quantitative competitive reverse-transcriptase PCR. *Biochem. J.* **337**, 231–241.

# 14

# Planar Patch Clamping

## Jan C. Behrends and Niels Fertig

## 1. Introduction

The technique of patch clamping can be seen in retrospect as a combination of two separate lines of development that both originated in the 1960s and 1970s. The classical biophysics of the nerve impulse had by then been established in the squid giant axon using a combination of (1) voltage clamping with axial wire electrodes and (2) internal perfusion or dialysis. This combination had given experimenters control of both the electrical and the chemical gradients governing membrane ion flux. The problem of the day was to extend this type of analysis to smaller, noncylindrical, cellular structures (such as neuronal somata) that would not allow insertion of metal wires, let alone tolerate any of the procedures used for internal perfusion or dialysis of squid axons. While intracellular glass microelectrodes (Ling and Gerard, 1949) afforded intracellular electrical access to most cellular somata, two independent electrodes for current passing and voltage recording, respectively, were initially necessary, until time-sharing systems made single-microelectrode voltage clamping possible (Wilson and Goldner, 1975). Even then, however, two severe problems remained: (1) spatially nonuniform voltage control (the so-called space-clamp problem), and (2) the lack of control over intracellular ionic composition. Each of these provided the driving force for one of the two branches of development mentioned above.

The need for perfect space clamp led to the development of patch clamping in its original sense, that is, extracellular voltage clamping of an electrically isolated patch of membrane by means of a smooth-tipped glass pipette filled with electrolyte (Huxley and Taylor, 1958; Pratt and Eisenberger, 1919) that was connected to a voltage-clamp

From: *Neuromethods, Vol. 38: Patch-Clamp Analysis: Advanced Techniques, Second Edition*
Edited by: W. Walz @ Humana Press Inc., Totowa, NJ

circuit pressed against the cell membrane (Frank and Tauc, 1963; Neher and Lux, 1969; Strickholm, 1962). The accomplishments of this original mission of the patch-clamp pipette (e.g., Eckert and Lux, 1976; Heyer and Lux, 1976), however, were soon eclipsed by the well-known scientific triumph of the first direct recordings of single-ion-channel currents through very small areas of a cellular membrane, which was enabled by a miniaturized and simplified version of the same device (Neher and Sakmann, 1976).

Solutions to the problem of internal perfusion, on the other hand, were pioneered in the Bogolometz Institute of Physiology in Kiev (Kostyuk et al., 1975; Krishtal and Pidoplichko, 1975). The first version of this internal perfusion method used 250- to 300-µm-thick sheets of polyethylene forming partitions between two electrolyte-filled compartments and containing single pores produced by thermal moulding with a hot metal microneedle (Fig. 1). These relatively large openings (10–50 µm) could be used to trap large, enzymatically isolated neurons from snail ganglia so that their membrane sealed off the opening. The membrane inside the aperture could then be destroyed by a short pressure transient in order to establish access to the whole cell. The seal (or, more appropriately, shunt) resistances thus achieved were rather low but sufficient for recording because the snail neurons had very low input resistances themselves. Internal perfusion, on the other hand, was quite efficient. However, this first appearance of the apertures in planar substrates in cellular electrophysiology remained a short episode. The Kiev group rapidly abandoned the planar geometry in favor of a pipette-like structure made from plastic tubing (Kostyuk et al., 1984), while in another laboratory, that of Arthur M. "Buzz" Brown in Galveston, TX, the internal perfusion principle was adopted for use with smaller vertebrate and mammalian cells by using carefully fire-polished, perfusable glass suction pipettes (Lee et al., 1980).

The final synthesis of both strands of development happened when the gigaohm seal and whole-cell patch clamp recording were discovered in Göttingen: applying suction instead of just pressing the pipette against the membrane did an important part of the trick for the gigaseal, and rupturing the membrane patch with further suction pulses to obtain whole cell access provided the long-sought full control of the ionic driving forces, even with very small cells

---

[1] In fact, in Hamill et al., 1981, the reader is introduced to the whole-cell configuration of the patch-clamp technique: "The technique to be described can be viewed as a microversion of the internal dialysis techniques originally developed for molluscan giant neurons."

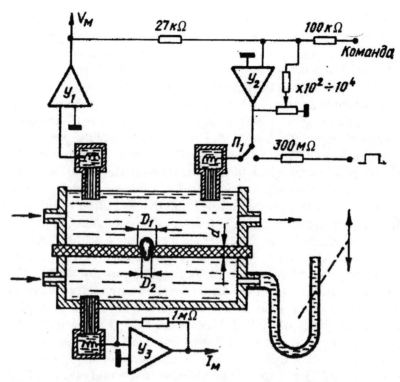

Fig. 1. The historical forerunner of planar patch clamping. A polyethylene membrane of thickness ($d$) 250 to 350 µm separates two electrolyte-filled compartments. A central, conical aperture (D1, 200–300 µm; D2, 25–80 µm) is depicted containing a single neuronal soma of a diameter of 40 to 200 µm. (Krishtal and Pidoplichko, 1975, with permission.)

(Hamill et al., 1981).[1] Whole-cell patch clamping, in particular, developed from a niche method into a mass movement and the pipette became the natural tool of the electrophysiologist. The alternative structure, the aperture in a planar partition, however, lay unused for almost two decades. In retrospect, the main reason for this is that there was no good method to make suitably small openings in planar substrates. Clearly, diameters need to be in the low micrometer range in order to be useful for small cells, and the elegance and ease of micropipette fabrication by pulling heated glass tubes is such that one tends to eschew a search for alternative designs.

It is therefore logical that the "pore in a planar partition" resurfaced more or less simultaneously in a number of laboratories when solid-state microstructuring technologies became more generally known and available to scientists outside industrial microchip fabrication. While the true primary motive for the attempt to produce a planar version of gigaseal patch clamping probably—in our case, certainly—was the presentiment that it could be done and that it would be both entertaining and informative to try, several advantages of the planar design over the pipette were envisaged, of which a potential for automation, parallelization, and high throughput of cellular electrophysiology was almost immediately realized by all groups involved. The renaissance of the planar partition in the late 1990s coincided with a heightened focus on ion channels as pharmacological targets mediating desirable as well as undesirable effects (e.g., acquired LQT syndrome),[2] which prompted the Food and Drug Administration (FDA) and other regulators to stipulate, for example, human Ether-A-Go-Go Related Gene (hERG) testing for all candidate drugs. Thus, for instance, a pharmaceutical company funded what were probably among the earliest attempts at reviving the planar concept at the Institute for Natural and Medical Sciences Institute (NMI) of Tübingen University in Germany. This work, made public only much later, started out from plastic (polyimide) films into which holes were drilled using focused ion beam milling. The results of these first trials, however, were not particularly encouraging with regard to formation of gigaohm seals (Stett et al., 2003). The same has obviously to be said of several other first attempts, including our own using apertures formed by reactive ion etching in free-standing silicon-nitride/silicon-oxide layers (Fertig et al., 2000). It seems remarkable, therefore, that, less than a decade later, planar patch clamping is an established technology with approximately five different commercially available devices producing useful data in a growing number of industrial and academic laboratories worldwide.

## 2. Designs and Concepts

### 2.1. Planar Patch Clamp Chips Based on Polymer Substrates

The laboratory of Fred Sigworth, one of the original Göttingen pioneers of patch clamping, brought forward the idea of using the

---

[2] This situation at the end of the 1990s also produced technologies to automate pipette-based patch clamping such as the Apatchi by Sophion SA (a spin-off from Denmark's Neurosearch) and the Interface patch clamp invented by the late David Owen of CeNeS in England, a great pioneer in the field (see review by Mathes, 2006).

heat-curable polymer polydimethylsiloxane (PDMS), better known to electrophysiologists under its trade name of Sylgard™ (Dow Corning, Midland, MI) to fabricate planar patch clamp partitions. Inspired by the work of Whitesides and others (1998), they teamed up with a neighboring solid-state physics lab to apply PDMS-micromolding technology to the problem. Micromolding either from a micromachined silicon master or from a quartz rod pulled into the shape of a patch pipette and subsequent serial sectioning of the molded polymer was used (Klemic et al., 2002). In this way, apertures with diameters down to about 4 µm could be produced. After oxidation in a plasma oven, the PDMS assumes glass-like surface properties and becomes sufficiently hydrophilic to allow filling of the aperture with salt solutions. These apertures were used to produce gigaohm seals on *Xenopus* oocytes with a success rate of 13% (Klemic et al., 2002). The technique using pipette-shaped rods was expanded for volume production of apertures by the company Axon Instruments Inc. (Molecular Devices, Sunnyvale, CA) in order to produce the planar patch clamp substrate for their automated patch-clamp project, and whole-cell recordings from rat neuroblastoma cells were obtained using such devices (Osipchuk et al., 2001).

While commercial development of the PDMS technology at Axon was later discontinued in favor of a glass substrate (see below), the Sigworth group found yet another ingenious way of making small apertures in flexible patch partitions made from this cheap and easy to use material; they used a stream of pressurized nitrogen made to flow through a 2-µm opening in a steel plate supporting a thin layer of PDMS during its curing process (Klemic et al., 2005). With these apertures, which have also been obtained in array format (Li et al., 2006), whole-cell and cell-attached measurements from rat basophilic leukemia (RBL) cells have been performed. It is tempting to follow the author's suggestion that using this approach, planar patch-clamp apertures could be made in the lab in the image of the traditional "electrophysiologist who makes things himself" using relatively inexpensive machinery, much like classical patch pipettes are made with a puller standing next to the patch rig. Because the effects of plasma treatment wear off relatively quickly and shelf life is limited even under water, local production of the devices is clearly mandatory.

Recently, Luke Lee and colleagues at Berkeley published the first promising experiments with a device where the patch aperture is not a through-hole in the substrate but is formed where a micrometer-sized horizontally oriented channel joins a larger compartment, thus forming what the authors call a "lateral cell trapping junction" (Ionescu-Zanetti et al., 2005; Seo et al., 2004). The device is made by combination of lithographic microstructuring of the photosensitive polymer SU8 and PDMS micromolding from a silicon master. This lateral-patch principle makes it possible to accommodate multiple, electrically independent apertures within a few hundreds of micrometers from each other. Even if only 5% of the seals obtained in the most recent study were gigaohm seals, this technique certainly holds great promise for the future, because it opens the way for a very close integration of microfluidics and automated patch clamping.

Polyimide films, which had performed unpromisingly at the NMI, were nevertheless used in a first commercial product, the Ionworks platform originally developed by Kirk Schroeder and Brad Neagle at Essen Instruments in Ann Arbor, MI (Kiss et al., 2003; Schroeder et al., 2003). Here, arrays of micrometer-sized apertures produced in polyimide films by laser irradiation were arranged in a 384-well format. Seal resistances obtained are in the range of 100 M$\Omega$, which agrees well with the NMI experience. Nevertheless, in the true spirit of our Kievan forebears, even with high leak conductances, useful whole-cell recordings from mammalian cell lines can be obtained using leak current compensation circuitry, and several further studies using the Ionworks device have been published. Recently, a version of the substrate containing 64 apertures per well has been introduced. Here, the recorded membrane current is summed over multiple cells and the relative importance of the large shunt conductance reduced (Finkel et al., 2006). The device is today considered a useful screening tool, even if the recordings obtained are not of the highest quality.

## 2.2. Planar Patch-Clamp Designs Based on Silicon Microstructuring

Silicon micromachining was a well-established technology and had relatively recently become available in academic settings when the planar partition was rediscovered for electrophysiology. It is natural, therefore, that when the search began for a suitable technology to produce an insulating partition with a micrometer-sized orifice, silicon micromachining suggested itself.

Two generally different approaches emerged. In the first, the aperture is formed by electron beam or ultraviolet (UV) lithography and reactive ion etching in a thin (100–200 nm) suspended silicon-nitride membrane, which has been obtained by a one-sided KOH-etch of the underlying silicon wafer. The thin layer facilitates the production of very small apertures even well into the submicrometer range. We attempted to use such structures to form giga-seals on cells by suction and, while we were able to trap cells in these apertures (Fertig et al., 2000), obtained very little shunt resistance increase. However, Christian Schmidt, Michael Mayer, and Horst Vogel in Lausanne, using essentially the same structure plus an additional layer of $SiO_2$ and polylysin treatment or silanization, were able to form suspended bilayers from giant unilamellar vesicles on these apertures using electrophoretic positioning (Schmidt et al., 2000). This procedure gave very tight electrical seals (from 1 to 200 G$\Omega$), enabling the recording of ionic current through, for example, channels formed by the peptaibol alamethicin. Later on, this design was used by a start-up company called Cytion SA in Lausanne to form gigaohm seals on cells. The company's procedure to form seals did not rely on suction, but rather on a proprietary coating of the chip surface and promotion of strong surface adhesion of the cell membrane. Whole-cell access could be obtained by adding pore formers to the lower compartment to obtain a perforated patch configuration (Horn and Marty, 1988). Cytion developed a single-channel planar patch-clamp automat that is reported to have worked quite well. Cytion was acquired by Molecular Devices Inc. (Sunnyvale, CA) in 2001, but shut down shortly after.

There is now a general consensus in the field that forming high-resistance seals requires sufficient contact area between the membrane and the wall of a pore. Apertures in very thin (<1–2 μm) diaphragms such as a passivation layer on a silicon wafer, therefore, are not suitable for this traditional seal formation strategy, while, as shown by Schmidt et al. (2000), promoting strong membrane adhesion to the surface of the chip can yield good results.

However, the problem might in principle be overcome by thickening the diaphragm with deposition of suitable materials, for example, $SiO_2$. This strategy is currently being pursued by a group in Grenoble, France, and has led to a 16% success rate in forming seals >1 G$\Omega$ (Picollet-D'hahan et al., 2004; Sordel et al., 2006).

The second approach based on silicon micromachining starts out with a thin layer of "naked" silicon, into which an aperture is etched by deep reactive ion (DRI) etching of silicon, the so-called Bosch process, which enables high aspect ratios. In a second step, the silicon is covered with an insulating layer of $SiO_2$ to provide electrical insulation and reduce the size of the aperture. In the first version of these devices, the insulating layer was built using plasma-enhanced chemical vapor deposition (Pantoja et al., 2001, 2004). These devices were very suitable for recording from suspended lipid bilayers (Pantoja et al., 2001), while the probability of forming gigaohm scals with cells was low, probably due to a considerable roughness of the pore's inner walls. Recently, Matthews and Judy (2006) described a more involved process for making the insulating layer inside and around DRI-etched pores, which consists of thermal wet oxidation (heating to 1100°C in the presence of saturated water vapor) to form a first $SiO_2$ layer that carries all the original roughness of the parent silicon surface, then removing this layer with hydrofluoric acid to expose the underlying smooth crystalline silicon surface, low-pressure chemical vapor deposition of amorphous silicon and a second thermal wet oxidation step to produce smooth, round openings. Seal resistances exceeding 1 G$\Omega$ have been obtained using this aperture.

The great advantage of silicon as a starting substrate is the fact that one can draw from the well-stocked toolbox of an established and mature industry for structuring it and modifying its surface properties. On the other hand, no amount of machining can remove the fact that silicon itself is not an insulator but a semiconductor. Therefore, any silicon-based strategy for making a patch partition has to grapple with the problem of high density of free charge carriers that leads to large capacitances, slow voltage transients, and high noise, as well as that of the photoelectric effect by which light induces even more charge carriers and can give rise to capacitive current flow across the partition.

Nevertheless, careful minimization of fluid contact area (Matthews and Judy, 2006) and packaging of the chips can be used to reduce these problems significantly. Proof of this is the silicon-based planar patch-clamp chip used in the QPatch, a commercial 16-channel planar patch-clamp automat (Kutchinsky et al., 2003; Mathes, 2006) that was successfully introduced into the market by Sophion SA of Ballerup in Denmark, in 2004. This chip device is also produced starting from silicon substrates. Another company, Cytocentrics AG, has been spun off from the NMI to commercially develop a highly ambitious concept called "cytocentering" (Stett et al., 2003), where, in order to imitate the sequence of the separate steps of contacting and sealing that are characteristic of classic pipette-based patch clamping, the cell is first positioned and held in place by suction through a larger, outer aperture before being electrically contacted by separately controlled negative pressure through a second aperture inside the first. The complex structure required for this approach has been realized in a thick (15-μm) $SiO_2$ layer on top of a silicon substrate. While the commercial product is not yet available, 10 of 11 chips used were reported to have formed a gigaseal in one test run (Van Stiphout et al., 2005).

## 2.3. Planar Patch-Clamp Chips Based on Glass Microstructures

Our initial unsuccessful trials with silicon-nitride membranes had also acquainted us with the disadvantages of the silicon as bulk material (see above), which led us toward the one material that is a good insulator and known to seal with membranes: glass. This idea, however, engendered a somewhat lengthy search for ways of making a micrometer-sized aperture into a glass substrate layer. After visiting several dead ends, a cooperation with the Gesellschaft für Schwerionenforschung (Society for Heavy Ion Research) in Darmstadt, resulted in a viable method: after locally thinning a glass or quartz wafer by a one-sided hydrofluidic acid (HF) etch, a single heavy ion can be shot through and leave a so-called ion track, a local structural disturbance that is etched much more quickly by HF than the undisturbed material. Thereby, a further, carefully timed one-sided etching step results in a conical pore with a clean, smooth-rimmed opening (Fertig et al., 2001). With these structures, we were able to perform the first whole-cell (Fertig et al., 2002a) and, a little later, single-channel recordings (Fertig et al., 2002b) on a planar chip device.

Today, glass-based planar patch-clamp chips are commercially used by Nanion Technologies GmbH, our own spinoff, in their semiautomated single-recording device (the Port-a-Patch) as well as in their fully automatic, multirecording patch-clamp robot (the Patchliner) and by Aviva Biosciences Inc., of San Diego, CA, which produce the chips for the Axon Instrument (now part of Molecular Devices Corp.) PatchXpress, another multirecording robot. The Aviva Sealchip™ is chemically treated using a proprietary process and is delivered and stored in a liquid of equally proprietary composition, which has to be removed by washing before use, while Nanion's borosilicate Nano-Patch-Clamp (NPC) chips are shipped dry and are used as delivered. They do not degrade even over long periods of time (>1 year) as long as they are protected against moisture and contamination. We typically vacuum package the chips and store them dry at room temperature.

## 3. Experimental Considerations

In conventional patch clamping, the patch pipette is manually maneuvered under optical control via a microscope, and a specific cell is chosen to place the pipette tip onto the surface of the cell membrane. Then a tight seal is established by gentle suction and an omega-shaped protrusion of membrane is drawn into the patch pipette.

The procedure is very similar but still somewhat different in the case of planar patch-clamp chips. Here, a suspension of cells is placed on top of the chip. Then a single cell is positioned onto the aperture in the chip by application of suction, typically by execution of an automated, software-controlled suction protocol. In contrast to the classic patch-clamp technique, it is the cell that is moved to the aperture and not the pipette that is moved to the cell. As in conventional patch clamping, a seal is obtained by application of suction and the membrane can then be ruptured for whole-cell access with suction or voltage pulses. The result is an electrical connection to the inside of the cell allowing for current recording.

### 3.1. Cell Culture

As with planar patch clamping, a single cell is randomly chosen from a suspension by application of suction (essentially a blind patch approach!), and good cell quality and viability are mandatory for obtaining optimal results. Hence, the requirements for cell culture and preparation in planar patch clamp tend to be somewhat greater than for the conventional technique. The cells provided in suspension must also be well isolated, as cell clusters are detrimental for the success rate of automated gigaseal formation.

Most experiments to date have been made using chinese hamster ovary (CHO), human embryonic kidney (HEK), Jurkat, and rat basophillic leukemia (RBL) cells. The cells are either used as wild-type cells or stably transfected with various types of ion channels. The culture conditions used were the standard conditions described in the American Type Culture Collection (ATTC) catalogue for each cell line.

Optimal results were obtained by splitting the cells every second to third day, avoiding confluent cells. Cells such as HEK293 often build clusters, when they are left for longer than 3 days on the plate or in the flask. For splitting it is advisable to use a long trypsin treatment to obtain single cells. Cells should be uniformly distributed throughout the dish. For cell harvesting, a short treatment with a cell detacher such as trypsin or various substitutes (see below) is sufficient and results in healthier cells. While both HEK293 and CHO cells are adherent cells, the CHO are more strongly attached to the substrate than the HEK293 cells. With RBL cells, typically one half of the cells is attached to the plastic while the other half is growing in suspension, which makes them very amenable to cell suspension preparation for automated patch-clamp applications.

## 3.2. Cell Harvesting

The confluency of the cells should be in the range of 50% to 80%. For harvesting of the cells for an experiment, no significant difference in percentage of gigaseals could be found for using trypsin, Accutase, enzyme-free solution, or phosphate-buffered saline ethylenediaminetetraacetic acid (PBS-EDTA) for lifting the cells. In some cases, treatment with trypsin made the cells more fragile, and the cells sometimes needed a recovery time of approximately 30 minutes.

### 3.2.1. Harvesting Protocol

Using T75 flasks or plates with diameter = 96 mm and a surface of 60 cm$^2$):

- Wash two times with 10 mL PBS (without $Ca^{2+}$ and $Mg^{2+}$).
- Add 2 mL of detacher (PBS-EDTA 2 mM, trypsin/EDTA, Accutase, . . .).
- Incubate for 3 minutes at 37°C and in 5% $CO_2$ for detaching of the cells.
- Check the detachment of cells under a microscope. Move the plate or flask gently to detach all cells from the bottom (do not hit the flask to detach cells).
- Add 10 mL of HEK medium and FCS.
- Pipette the cells gently up and down with a 10-mL pipette.

- After pipetting five times, look at the cells under a microscope. If the cells are already single (~80–90%), no further pipetting is needed.
- If cells still form clusters, gently pipette cells another 10 times. Repeat this step until cells are single (~80–90%).
- Perform 2 minutes of centrifugation (1000 U/min, 100 g).
- Discard the supernatant.
- Resuspend the cells in 10 mL medium with fetal calf serum (FCS).
- Perform 2 minutes of centrifugation (1000 U/min, 100 g).
- Discard the supernatant.
- Resuspend the cells in ~200 µL of external recording solution (resulting in a cell density of approximately $1 \times 10^6 - 5 \times 10^7/$mL medium).

An optical control of the cells under the microscope should then reveal single, round cells with smooth membrane edges and no cell clusters.

### 3.3. Cell Application and Positioning on the Chip

In a very simple manner, the actual aperture containing the partition is mounted onto a twist cap, which is placed onto a holder that contains the reference electrode and a suction line (Fig. 2). Before placing the chip-containing cap onto the holder, a droplet of intracellular electrolyte solution is pipetted to the bottom part of the chip. Another droplet of extracellular electrolyte solution is pipetted onto the top part of the chip, where in both cases about 5 µL of a sterile filtered solution is sufficient. This approach allows for fast exchange of chips after each experiment and enables experiments with low volume consumption.

Fig. 2. The chip is filled with saline solutions and mounted in a holder that allows applying pressure/suction. The electrolyte solution on both sides of the chip is electrically contacted via Ag/AgCl electrodes.

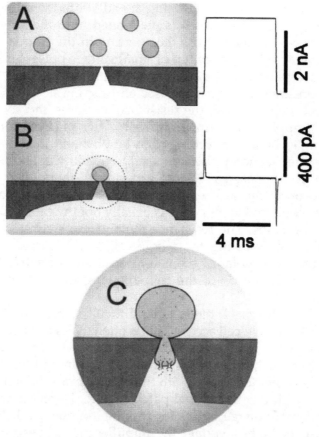

Fig. 3. Schematic of the chip and the procedure of cell contacting. (**A**) The measured current response to a voltage pulse is shown and again (**B**) after a cell is sealed onto the aperture by suction. (**C**) Close-up view of the mechanically and electrically tight contact of the cell membrane and the chip in cell attached mode. (Fertig et al., 2002, with permission. Copyright 2002, American Institute of Physics.)

For making electrical contact, an Ag/AgCl electrode inside the holder reaches into the electrolyte solution and another silver wire is placed in the saline on top of the chip, serving as the ground electrode. Due to the miniaturized arrangement, a small metallic container can be used as a Faraday cage for very effective shielding (Fig. 3).

Before the delivery of cell suspension, slightly positive pressure is applied to the chip to prevent contamination of the aperture.

Then, about 5 μL of cell suspension are added to the saline on top of the chip, and a single cell is positioned onto the aperture by suction. The resistance across the aperture in the chip is continuously monitored by application of small voltage pulses and is used as feedback parameter for suction application. After cell application, the positive pressure is released and suction slowly increased. Once a cell is positioned, the resistance increases and continuous suction is applied to form a gigaseal. After gigaseal formation, either single-channel recordings can be obtained in cell-attached mode or brief suction pulses are used to obtain whole cell access.

In a conventional patch-clamp experiment, there often is a significant drop of the seal resistance when breaking the membrane to obtain whole-cell access. This is usually not the case with planar patch-clamp chips. The resistance remains more constant during the process of accessing the whole cell or often even increases further. Due to the planar geometry a very short electrical path is provided to the inside of the cell, allowing for very low access resistances.

For some cell lines it tends to be difficult to rupture the membrane to gain whole-cell access. This is often the case when the cells were grown for more than 3 days on the plate. In this case, it appears to be helpful to use a zapping pulse in addition to some suction. This method works with the patch-clamp chips more reliably than with a conventional patch pipette. Different amplifiers allow different voltage pulses for zapping. In our hands, it works best to use the highest voltage the amplifier can deliver and a short duration.

Typical success rates in obtaining gigaseals and stable whole-cell recordings are in the range of 60% to 90%. Generally, the recordings are somewhat more stable and long-lasting as compared to conventional patch clamping. Any relative movements of the patch pipette with regard to the cell are detrimental to the seal and overall stability; hence, a solid vibration isolation is required in conventional patch clamping. In cases where a chip is used, there are no relative movements and the whole arrangement is mechanically more insensitive, requiring no vibration isolation or suspension.

Pharmacological experiments are carried out by simply adding compounds to the patch clamped cell on the chip either by a pipetting step or by means of a laminar flow chamber for perfusing the cell. Depending on the holder/cartridge design of the chip, microfluidic means can be used for compound applications. We have, for example, developed a microfluidic cartridge that contains a glass substrate with 16 patch-clamp apertures, each of which is individually addressable by microfluidic channels on the intra- as well as on the extracellular side. This design allows perfusion of cells and compounds by robotic pipetting means, making the whole approach very suitable for automation. In this case of multiple apertures on a single chip, individual, feedback-controlled, suction lines are required for positioning and sealing of the cells.

In addition to scaling up the number of recording channels, the throughput capability is increased by automated application of drugs by a pipetting robot. This is shown in Fig. 4, where a con-

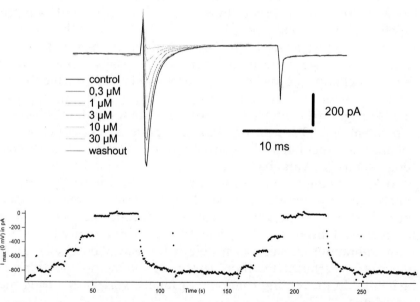

Fig. 4. Experimental stability. The data show the maximum current amplitude elicited with a voltage pulse to a holding potential of 0 mV as recorded in Nav1.5 expressing CHO cells. Five concentrations of tetrodotoxin (TTX) (0.3, 1, 3, 10, 30 μM) have been applied, then washed out and applied again to demonstrate the stability of whole-cell recordings. The different current amplitude plateaus correspond to the different TTX concentrations applied, whereby the current is completely blocked at 30 μM, the highest concentration. (*Source:* Small, 2006, with permission. Copyright 2006, Wiley-VCH Verlag GmbH & Co KG.)

centration series of a compound (tetrodotoxin, TTX) is applied to a whole-cell recording from a cell expressing a sodium channel to generate a dose–response curve and determine the concentration that inhibits 50% ($IC_{50}$) for the compound. In these recordings, a standard voltage pulse is repetitively applied to activate the sodium current such that the blocking action of TTX can be observed as downward steps in the overall sodium current amplitude of the cell. With increasing concentrations of TTX, the amplitude diminishes further and is finally completely blocked. A washout step with control solution shows the complete recovery of the original current amplitude. Following this, the entire protocol is repeated, showing the stability and reproducibility of the experiments.

A further interesting capability of the planar approach is the application of different solutions not only from the extracellular side but also from the intracellular side. For recording with a patch pipette, this is not really possible, due to the long and thin shaft of the pipette, which leads to a diffusion-limited and hence very slow solution exchange. Intracellular perfusion is easily achieved with the patch-clamp chip, as both sides of the cell are accessible due to the flat geometry of the chip. This possibility permits the application of drugs on the intracellular side for the investigation of ligand-dependent ion channels or signaling pathways. As many ion channels are regulated via internal binding sites for second messengers, this is a valuable tool.

Another example of a useful application for internal perfusion capabilities is perforated patches. Here, instead of breaking open the cell membrane for whole-cell access, a pore-forming compound (such as nystatin or amphotericin) is applied internally to render the membrane electrically permeable. In this way, whole-cell recordings can be performed without rupturing the membrane; this can be helpful in preventing rundown or kinetic variances during the recordings.

In Fig. 5, experiments on Jurkat cells expressing a potassium channel (Kv1.3) are shown. Current flow through these channels is known to be blocked by cesium ions, and in the experiment $Cs^+$-containing solution is applied internally, effectively blocking the current through the potassium channel as expected. Two recordings are performed in parallel, demonstrating the scalability of the chip-based approach by simultaneous measurements from different cells on a single chip.

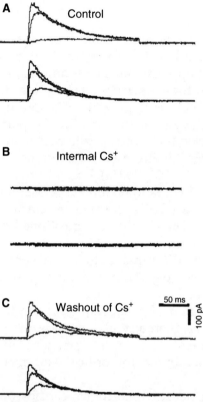

Fig. 5. Internal solution exchange. A nice feature of the planar patch clamp chips is the internal solution exchange during the experiment. The figure shows two simultaneously recorded Jurkat cells in the presence of a control internal solution (**A**), after the exchange of the internal solution with a Cs⁺-containing solution (**B**) and a subsequent washout step (**C**). The complete block of current by cesium and the recovery after washout with control solution well documents the internal perfusion possibility. (*Source:* Small, 2006, with permission. Copyright 2006, Wiley-VCH Verlag GmbH & Co KG.)

## 4. Outlook

The planar patch-clamp principle is at present used mainly for automated whole-cell patch-clamp recordings. It may, however, have other useful applications. For example, planar lipid bilayers can be formed on these apertures, facilitating the study of reconstituted ion channels and possibly transporters. Schmidt et al. (2000) formed bilayers on their $SiN_3/SiO_2$ partitions from giant unilamellar vesicles that were positioned using an electric field gradient, while we initially used the painting method to produce bilayers (Fertig et al., 2001, 2002a). Recently, we have used the Port-a-Patch and NPC-1 chips to obtain suspended bilayers from giant unilamellar vesicles (GUVs) formed by electroswelling (Sondermann et al., 2006). The GUVs were positioned on the aperture by suction exactly like the cells were. When they rupture, they form a suspended bilayer. With this method, bilayers can be formed on very small apertures (e.g., 1 µm in diameter), where painted bilayers do not form readily.

Besides automation of patch clamping and throughput, advantages of the planar geometry also include the potential for more sensitive recordings of ionic currents and an increased accessibility of the membrane for optical and mechanical detection techniques.

To make more sensitive current recordings, noise needs to be minimized. Given high seal resistances, one of the main sources of noise is any random voltage fluctuation in the voltage clamp amplifier, which forces an equally random current to flow across the total capacitance of the input circuit, including the recording electrode. By reducing the capacitance of the electrode, considerable reduction in noise should be possible. In planar patch clamping, the geometry of the electrode can be chosen more freely than with pulling pipettes and adapted to minimize capacitance. In particular, with pulling a pipette, the wall becomes inevitably thinner as the dimensions get smaller, increasing electrical capacitance. This is obviously not the case when apertures are etched or otherwise machined into bulk material. In fact, optimists have voiced the hope that planar patch-clamp chips may provide a way of approaching the "holy grail" of current measurement: the resolution of elementary charges (Klemic and Sigworth, 2005). The NPC glass chip, when contacted with recording solution on both sides in the way described above, contributes approximately 0.5 pF to the total input capacitance (Sondermann et al., 2006). Using a capacitive feedback

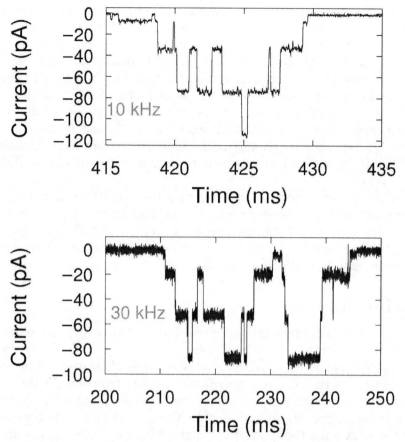

Fig. 6. Sample recordings of alamethicin-mediated currents from a suspended bilayer formed by sucking a giant unilamellar vesicle onto the aperture of a Nanion glass chip, where it bursts and forms a seal. 85 mM KCl; resistance of open aperture, 4 MΩ; seal resistance, 50 GΩ; capacitive feedback (Axopatch 200B). Upper trace: driving force, –140 mV; cutoff, 10 kHz; root mean square (rms) noise, 0.54 pA. Lower trace: driving force, –90 mV; cutoff, 30 kHz; rms noise, 1.6 pA.

amplifier and without optimizing any other parameters, we have recently performed single-channel recordings with <2 pA root mean square (rms) noise at 30 kHz cutoff frequency (Fig. 6).

For several years, simultaneous electrophysiological and fluorescence measurements have been used to obtain information about state-dependent conformational changes in ion channel proteins. These measurements were performed on large ensembles of channels expressed in cells (Bezanilla, 2005), where only average properties of, for example, open and closed states can be detected. It would be highly desirable to perform such combined measurements with single in channel resolution to obtain information on individual state—dependent transitions (Selvin, 2002). Pilot studies have been performed using the model channel gramicidin in painted lipid bilayers using a planar patch-clamp–like configuration (Borisenko et al., 2003; Harms et al., 2003) and it seems that similar experiments should be feasible on protein ion channels, either in native cell membranes or reconstituted in planar lipid bilayers.

Finally, the planar patch-clamp situation might be amenable to simultaneous use of electrophysiology and atomic force microscopy or some other scanning probe technique to study movement and shape changes in membrane proteins that accompany their functions (e.g., Gullo et al., 2005).

## Acknowledgments

We thank Dr. Markus Sondermann of the University of Freiburg for permission to use unpublished data (Fig. 6). Niels Fertig particularly thanks the whole team at Nanion for their dedication and excellent work. We are grateful for support from the Center for Nanoscience and the start-up program (FLÜGGE) of the University of Munich, the German Research Council (DFG) Collaborative Research Center (SFB) 468, the Federal Ministry for Education and Research, and the Bavarian State Ministry for the Economy. Current work in Jan C. Behrends's laboratory is funded by the DFG SFB 505 and the Medical Faculty of the University of Freiburg.

## References

Bezanilla, F. (2005) Voltage-gated ion channels. *IEEE Trans. NanoBioSci.* **4**, 34–48.

Borisenko, S., Lougheed, T., Hesse, J., et al. (2003) Simultaneous optical and electrical recording of single gramicidin channels. *Biophys. J.* **84**, 612–622.

Eckert, R. and Lux, H. D. (1976) A voltage-sensitive persistent calcium conductance in neuronal somata of Helix. *J. Physiol.* **254**, 129–151.

Fertig, N., Blick, R. H., and Behrends, J. C. (2002) Whole cell patch clamp recording performed on a planar glass chip. *Biophys. J.* **82**, 3056–3062.

Fertig, N., Klau, M., George, M., Blick, R. H., and Behrends, J. C. (2002) Probing single ion channel proteins on a planar microstructure. *Appl. Phys. Lett.* **81**, 4865–4867.

Fertig, N., Meyer, C., Blick, R. H., Trautmann, C., and Behrends, J. C. (2001) Microstructured glass chip for ion channel electrophysiology. *Phys. Rev. E* **64**, 040901 (Rap. Comm.).

Fertig, N., Tilke, A., Blick, R. H., Behrends, J. C., ten Bruggencate, G., and Kotthaus, J. P. (2000) Stable integration of isolated cell membrane patches in nanomachined aperture. *Appl. Phys. Lett.* **77**, 1218–1220.

Finkel A., Wittel A., Yang N., Handran S., Hughes J., and Costantin J. (2006) Population patch clamp improves data consistency and success rates in the measurement of ionic currents. *J. Biomol. Screen.* **11**(5), 488–496.

Frank, K. and Tauc, L. (1963) Voltage clamp studies on molluscan neuron membrane properties, in *The Cellular Functions of Membrane Transport* (Hoffmann, J. F., ed.), Prentice Hall, Englewood Cliffs, NJ, pp. 113–135.

Gullo, M. R., Akiyama, T., Frederix, P. L. T. M., et al. (2005) Towards a planar sample support for in situ experiments in structural biology. *Microelectron. Eng.* **78–79**, 571–574.

Hamill, O. P., Marty, A., Neher, E., Sakmann, B., and Sigworth, F. J. (1981) Improved patch-clamp techniques for high-resolution current recording from cells and cell-free membrane patches. *Pflügers Arch.* **391**, 85–100.

Harms, G. S., Orr, G., Montal, M., Thrall, B. D., Colson, S. D., and Lu, H. P. (2003) Probing conformational changes of gramicidin ion channels by single-molecule patch-clamp fluorescence microscopy. *Biophys. J.* **85**, 1826–1838.

Heyer C. B. and Lux, H. D. (1976) Properties of a facilitating calcium current in pace-maker neurones of the snail, *Helix pomatia*. *J. Physiol.* **262**, 319–348.

Horn, R. and Marty, A. (1988) Muscarinic activation of ionic currents measured by a new whole-cell recording method. *J. Gen. Physiol.* **92**, 145–159.

Huxley, A. F. and Taylor, R. E. (1958) Local activation of striated muscle fibres. *J. Physiol.* **144**, 426–441.

Ionescu-Zanetti, C., Shaw, R. M., Seo, J., Jan, Y. N., Jan, L. Y., and Lee, L. P. (2005) Mammalian electrophysiology on a microfluidic platform. *Proc. Natl. Acad. Sci. USA* **102**(26), 9112–9117.

Kiss, L., Bennett, P., Uebele, V. N., et al. (2003) High throughput ion-channel pharmacology: planar-array-based voltage clamp. *Assay Drug Dev. Technol.* **1**, 127–135.

Klemic, K. G., Klemic, J. F., Reed, M. A., and Sigworth, F. J. (2002) Micromolded PDMS planar electrode allows patch clamp electrical recordings from cells. *Biosens. Bioelectron.* **17**(6–7), 597–604.

Klemic, K. G., Klemic, J. F., and Sigworth, F. J. (2005) An air-molding technique for fabricating PDMS planar patch-clamp electrodes. *Pflugers Arch.* **449**(6), 564–572.

Klemic, K. G. T. and Sigworth, F. J. (2005) Microchip technology in ion-channel research. *IEEE Trans. NanoBioSci.* **4**, 121–127.

Kostyuk, P. G., Krishtal, O. A., and Pidoplichko, V. I. (1984) Perfusion of isolated neurons fixed in plastic film, in *Intracellular Perfusion of Excitable Cells* (Kostyuk, P. G. and Krishtal, O. A. ed.), Wiley, New York, pp. 35–51.

Kostyuk, P. G., Krishtal, O. A., and Pidoplichko, V. I. (1975) Effect of internal fluoride and phosphate on membrane currents during intracellular dialysis of nerve cells. *Nature* **257**, 691–693.

Krishtal, O. A., and Pidoplichko, V. I. Intracellular perfusion of giant snail neurons [Russian]. *Neirofisziologija* **7**, 327–329.

Kutchinsky, J., Friis, S., Asmild, M., et al. (2003) Characterization of potassium channel modulators with QPatch automated patch-clamp technology: system characteristics and performance. *Assay Drug Dev. Technol.* **1**, 685–693.

Lee, K. S., Akaike N., and Brown, A. M. (1980) The suction pipette method for internal perfusion and voltage clamp of small excitable cells. *J. Neurosci. Methods* **2**, 51–78.

Li, X., Klemic, K. G., Reed, M. A., and Sigworth, F. J. (2006) Microfluidic system for planar patch clamp electrode arrays. *Nano. Lett.* **6**, 815–819.

Ling, G. and Gerard, R. W. (1949) The normal membrane potential of frog sartorius fibres. *J. Cell. Comp. Physiol.* **34**, 383–396.

Mathes, C. (2006) Qpatch: the past, present, and future of automated patch clamping. *Expert Opin. Ther. Targets* **10**, 319–327.

Matthews, B. and Judy, J. W. (2006) Design and fabrication of a micromachined planar patch-clamp substrate with integrated microfluidics for single-cell measurements. *J. Microelectromechanical Syst.* **15**, 214–222.

Neher, E. and Lux, H. D. (1969) Voltage clamp on Helix pomatia neuronal membrane; current measurement over a limited area of the soma surface. *Pflügers Arch.* **336**, 87–100.

Neher, E., and Sakmann, B. (1976) Single-channel currents recorded from membrane of denervated frog muscle fibres. *Nature* **29**, 799–802.

Osipchuk, Y., Dromaretcky, A., Savtchenko, A., et al. (2001) Whole cell recordings from new planar patch clamp electrodes. *Soc. Neurosci. Abstr.* **27**, 606.

Pantoja, R., Nagarah, J. M., Starace, D. M., et al. (2004) Silicon chip-based patch-clamp electrodes integrated with PDMS microfluidics. *Biosens. Bioelectron.* **15**, 509–517.

Pantoja, R., Sigg, D., Blunck, R., Bezanilla, F., and Heath, J. R. (2001) Bilayer reconstitution of voltage-dependent ion channels using a microfabricated silicon chip. *Biophys. J.* **81**(4), 2389–2394.

Picollet-D'hahan, N., Sordel, Y., Garnier-Ravéaud, S., et al. (2004) A silicon-based multi-patch device for ion channel current sensing. *Sensor. Lett.* **2**, 2.

Pratt, F. H. and Eisenberger, J. P. (1919) The quantal phenomena in muscle: methods, with further evidence of the all-or-none principle for the skeletal fibre. *Am. J. Physiol.* **499**, 1–54.

Schmidt, C., Mayer, M., and Vogel, J. (2000) A chip-based biosensor for the functional analysis of single ion channels. *Angew. Chem. Int. Ed.* **39**, 3137–3140.

Schroeder, K., Neagle, B. Trezise, D. J., and Worley, J. (2003) IonWorksTMHT: a new high-throughput electrophysiology measurement platform. *J. Biomol. Screen.* **8**, 50–64.

Selvin, P. R. (2002) Lighting up single ion channels. *Biophys. J.* **84**, 1–2.

Seo, J., Ionesco-Zanetti, C., Diamond, J., Lal, R., and Lee, L. P. (2004) Integrated multiple patch-clamp array chip via lateral cell trapping junctions. *Appl. Phys. Lett.* **84**, 1973–1975.

Sondermann, M., George, M., Fertig, N., and Behrends, J. C. (2006) High resolution electrophysiology on a chip: transient dynamics of alamethicin channel formation. *Biochim. Biophys. Acta.* **1758**, 545–551.

Sordel, T., Garnier-Raveaud, S., Sauter, F., et al. (2006) Hourglass SiO(2) coating increases the performance of planar patch-clamp. *J. Biotechnol.* **125**(1), 142–154.

Stett, A., Bucher, V., Burkhardt, C., Weber, U., and Nisch, W. (2003) Patch-clamping of primary cardiac cells with micro-openings in polyimide films. *Med. Biol. Eng. Comput.* **41**(2), 233–240.

Strickholm, A. (1962) Excitation currents and impedance of a small electrically isolated area of the muscle cell surface. *J. Cell Comp. Physiol.* **60**, 149–167.

Van Stiphout, P., Knott, T., Danker, T., and Stett, A. (2005) 3D microfluidic chip for automated patch clamping, in *Mikrosystemtechnik Kongress* (G, M. M., V, D. E., V, D. I., eds., Berlin, pp. 435–438). http://www.cytocentrics.com/images/microfluidic-chip.pdf.

Whitesides, G. M., Xia, Y. N. (1998) Soft lithography *Ann. Rev. Sci.* **28**, 153–184.

Wilson, W. A. and Goldner, M. M. (1975) Voltage clamping with a single microelectrode. *J. Neurobiol.* **6**, 411–432.

# 15

# Automated Glass Pipette–Based Patch-Clamp Techniques

*Michael Fejtl, Uwe Czubayko, Alexander Hümmer, Tobias Krauter, and Albrecht Lepple-Wienhues*

## 1. Introduction

Patch clamping facilitates directly controlling the voltage of a cell and thus has been acknowledged to be the gold standard for assessing ion-channel physiology in academic and industrial laboratories. The method was developed 30 years ago by Neher and Sakmann (1976), and its various different configurations with glass electrodes have been successfully applied in many labs worldwide (Hamill et al., 1981). The standard approach of patch clamping is done completely manually and requires expensive equipment, such as microscopes, micromanipulators, antivibration tables, and Faraday cages. Further, manual operation is time-consuming, requires experienced personnel, and accommodates only a low amount of throughput per day.

Thus automation of the patch-clamp technique has been the focus of several researchers and biotech companies over the last several years. Interestingly, most of the available systems use a chip-based design rather than standard glass patch electrodes. It was assumed that the mass production of glass electrodes would be labor-intensive and expensive, although it is well established that the standard glass patch electrode is the best substrate for patch clamping. In that regard, several companies offer a variety of systems for the pharmaceutical industry built around the patch-on-a-chip design. Molecular Devices (IonWorksHT, IonWorks-Quattro, PatchXpress), Sophion (QPatch16), and Nanion Technologies (Port-a-Patch, NPC-16 PatchLiner) are three of the companies that have systems on the

From: *Neuromethods, Vol. 38: Patch-Clamp Analysis: Advanced Techniques, Second Edition*
Edited by: W. Walz @ Humana Press Inc., Totowa, NJ

market, and other potential systems are in development. Overall, these systems utilize silicon-, polyethylene-, polydimethylsiloxane (PDMS)-, quartz-, Teflon-, or polymer-based planar arrays or single chips, with micrometer-size holes accommodating loose or tight seal formation (for a review of the available patch-on-a-chip approaches see Shieh, 2004; Wang and Li, 2003; Wood et al., 2004; and the other chapters in this book).

This chapter describes the development of several approaches to automate patch clamping with standard glass electrodes. We demonstrate that mass production of standard glass patch electrodes is possible, rendering the patch-clamp electrode as the recording consumable of choice for automation. The goal is to provide full stand-alone walk-away unattended performance, unlike several of the available patch-on-a-chip operations. Four approaches to automate patch clamping with glass electrodes have been undertaken by commercial companies, and they are described here.

## 2. The AutoPatch®

CeNeS introduced the AutoPatch® system, which is composed of a so-called interface patch, whereby a regular patch pipette directed with the tip upward hits a "hanging drop" of a glass tubing filled with a cell suspension (Fig. 1). Once inserted into the drop, the patch pipette hits a cell and forms a gigaseal and subsequently a whole-cell configuration, which is similar to manual patching. This technology was later adopted by the company Xention and is now part of their technology platform.

## 3. The Apatchi-1™

Sophion developed the Apatchi-1™ system, a visual cuing system to identify a cell in a regular culture dish via a CCD camera and image recognition, and to subsequently direct a patch pipette onto the cell automatically. Up to 18 prefilled patch electrode holders can be placed in a rack and they are then taken one by one by a robotic arm for conducting the experiment. Thus several cells can be patched consecutively (Asmild et al., 2003).

## 4. The RoboPatch

The Discovery Neuroscience section of Wyeth Research have recently published two papers describing the invention of the RoboPatch, a home-made platform utilizing standard equipment

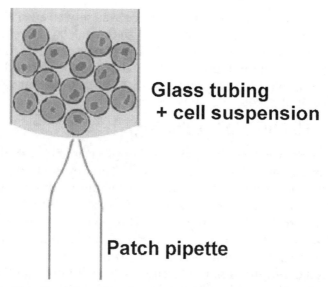

**Glass tubing + cell suspension**

**Patch pipette**

Fig. 1. The "hanging drop" interface blind-patching method developed by CeNeS. A glass tube is filled with a cell suspension and then a patch electrode is moved toward the liquid–air interface. Once inserted into the drop, the tip hits a cell and upon suction the gigaseal and subsequent whole-cell configuration is achieved.

for patch clamping. Here two applications are described. First is an inside-out whole-cell configuration utilizing a similar method originally put forward by Lepple-Wienhues et al. (2003) that dispenses cells inside a regular patch pipette. The difference is that rather than applying suction at the tip as described by Lepple-Wienhues et al, Discovery Neuroscience used pressure from the back opening of the patch pipette to push cells inside the pipette toward the tip to establish the gigaseal configuration (Vasilyev et al., 2005). The inside-out whole-cell patch-clamp configuration was then achieved by exposing the patch pipette briefly to air. This process ruptures the membrane and leaves the inside-out whole-cell configuration intact. Then the patch pipette is moved to various compound plates for studying the effects on either voltage-activated or on ligand-operated currents. The second approach also utilizes a blind method by inserting a regular patch electrode into a culture well that has been filled with a cell suspension. Upon finding a cell by observing an increase in pipette resistance, suction was applied and the cell formed a gigaseal and subsequently a whole-cell formation. Then a small-volume (20 µL) plastic interface was moved over the patch pipette as a sleeve when the pipette with the patch is lifted up away from the cell suspension into the air. The plastic sheath is filled with extracellular solution to preserve the patch and the electrode is then physically moved to compound wells to obtain current responses (Vasilyev et al., 2006).

## 5. The "Flip-the-Tip" Technology

This approach was invented by Lepple-Wienhues et al. (2003) and commercialized by flyion GmbH. The key invention of this technology is to invert the general principle of the patch-clamp technique, that is, to dispense a few hundred cells into the back of a standard borosilicate patch pipette. Then suction is applied at the tip to draw a single cell at the very end of the tip to form a classical gigaseal. Further suction pulses lead to breakthrough and to the whole-cell configuration. It is necessary to point out that the extracellular face of the cell is directed toward the inside of the pipette, and the intracellular compartment faces outside into the inner chamber of the recording tip socket, which is filled with intracellular solution (Fig. 2). Compounds or agonists are then applied via an

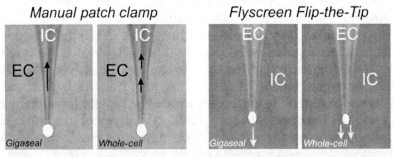

Fig. 2. The core technology of flyion's Flip-the-Tip approach is shown in comparison to the classic standard patch-clamp procedure. Cells are dispensed inside the patch electrode filled with extracellular (EC) solution. Then continuous suction of 10 to 20 mbar (indicated by the long arrow) is applied outside the tip to draw a single cell to the very end of the tip. Subsequently brief suction pulses (indicated by short arrows) are applied, and the cell enters the whole-cell configuration. The extracellular side faces inside the pipette and the intracellular face is directed outside to the intracellular compartment filled with intracellular (IC) solution.

internal perfusion fused silica needle directly onto the cell. The term *internal perfusion* is used here because compound solution is applied inside the pipette from the back. No internal solution is used but rather blockers/activators of voltage-gated ion channels or agonists/antagonists for ligand-operated channels are used instead, nonetheless, the term *internal perfusion* is used because it is an established process in the patch-clamp community. It is well known that conventional patch electrodes have a long shank, and hence the time course of solution exchange for internal perfusion is on the order of minutes and thus is not suitable for recording of fast ligand-activated channels. Moreover, using a conventional patch-clamp approach takes several minutes to establish a perforated patch configuration because a seal must be achieved first before the pore-forming agents then reach the cell membrane. Flyion has taken up this challenge and developed a new tool to boost the speed of internal perfusion into the millisecond range. Thus the recording of ligand-operated channels and the fast generation of perforated patches is now possible with the "Flip-the-Tip" technology using ChipTips. A detailed technical review and laboratory procedures of the Flip-the-Tip technology have been published elsewhere (Fejtl et al., 2005, 2006).

## *5.1. The Feedback Microforge*

The Feedback Microforge is based on a method called pressure polishing. Here, pressurized air is blown into the back of a patch electrode during the fire-polishing process, reshaping the tip (Goodman and Lockery, 2000). While Goodman and Lockery (2000) describe the principle of the operation, the Feedback Microforge is the only system available that enables complete automation of the process via visual feedback. A regular pulled-patch electrode is placed inside the Feedback Microforge, and upon pushing the start button, the pressure-polishing process is initiated. A built-in CCD camera and sophisticated imaging software recognizes the location of the tip within the heating filament automatically. A previously obtained image of the final shape of the pipette serves as the reference image of the system. Images of the tip during the flaming and polishing process are taken in real time at video rate and compared to the reference image. Up to eight multistep cycles are possible, depending on the shape the user wants to obtain. The following images show a regular patch pipette and a reshaped patch pipette,

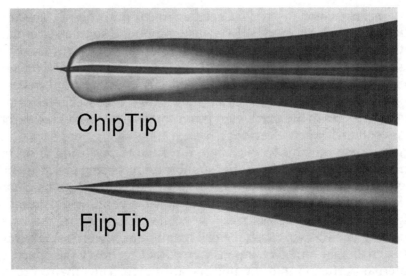

Fig. 3. Pressure-polishing reshapes a standard patch-clamp electrode referred to as a FlipTip flyion standard consumable into a ChipTip with a much wider shank, reducing the shank resistance and increasing the accessibility for an internal perfusion pipette. Thus an internal perfusion needle can reach the cell close to 150 μm, and it allows fast solution exchange in the millisecond range with a time constant of <50 ms.

which we have termed the ChipTip (Fig. 3). Virtually any shape of the shank can be obtained, and in this regard the patch glass electrode is unsurpassed in its flexibility, because the recording tip can be tuned to fit any cell size and purpose (Fig. 4.).

## 5.2. The ChipTip

The ChipTip production at flyion is performed robotically, allowing the production of several hundred recording tips per day. Thus making ChipTips on a large-scale mass operation is established. Patch electrodes are pulled to a pipette resistance of 1 to 1.5 MΩ and are 100% quality-controlled. This is achieved by measuring the amount of air (cm$^3$/min) pushed through every manufactured patch electrode. A previously established correlation curve between measured air volume and electrically measured pipette resistance serves as a reference for quality control of the pipettes. The patch electrodes are then glued into a molded plastic jacket, now termed the ChipTip, and 48 pieces are packed into a rack and stored in a box. The racks and the box are labeled with a bar code, a silicagel pad is placed in the box to take up humidity, and the box is shipped tight-sealed, resulting in a shelf life of several weeks (Fig. 5). Given the flexibility of producing various chip diameters and shank shapes of the ChipTips, several advantages are apparent: a fast solution exchange time constant of <50 ms, a low volume consumption of <1 μL for a complete solution exchange at the cell, a low stray capacitance of <10 pF, strong reduction in space-clamp problems, and almost a complete elimination of the tip shank resistance. This leads to an unprecedented performance of the ChipTips for

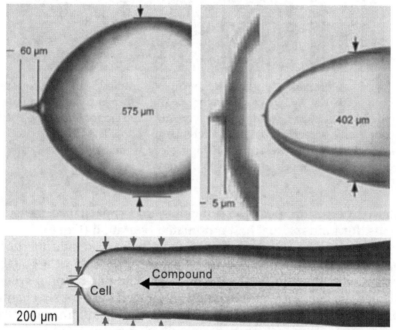

Fig. 4. Using the Feedback Microforge, a variety of tip openings and shapes of a patch pipette are possible. The upper panel depicts examples to show the effect of pressure-polishing to achieve different types of tip openings and shanks. The lower panel shows the modified patch electrode used for making the ChipTip. Errors indicate the measuring points of the Feedback Microforge. A cell is depicted at the tip of the pipette, and the direction of compound application is indicated with the long arrow pointing toward the cell.

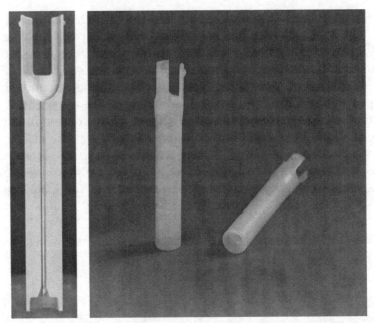

Fig. 5. A borosilicate glass electrode was pulled and subsequently treated with the Feedback Microforge. The latter is then glued into a molded plastic jacket, forming the recording ChipTip for either the Patch-Box or the Flyscreen automated patch-clamp robot. The ChipTip production, including the pressure-polishing step, is managed robotically to produce several hundred ChipTips per day.

internal perfusion, perforated patches, and capacitance measurements, for both manual and automated operation (Fig. 6).

Understanding the need of users in both the academic and pharmaceutical communities led us to the development of a scalable system with the Flip-the-Tip technology. The one-channel system is called the PatchBox, and the fully automated three- or six-channel system is then referred to as the Flyscreen®8500 automated patch-clamp robot. These systems are described in more detail in the following subsections.

### 5.3. The PatchBox

One of the main reasons to develop the PatchBox was to allow users to employ their own pulled-patch electrodes. The PatchBox gives the user more flexibility to record cells with different diameters, because electrodes can be adapted to a different tip opening and shape simply by changing the pulling and pressure-polishing parameters. The self-made electrodes are then screwed in a ChipTip-like electrode holder, which is similar to the molded plastic form used for the standard flyion consumable. The electrode is then inserted into the recording socket of the PatchBox. The box has a small footprint of $30 \times 30 \times 14$ cm, can be placed on any lab bench or table, does not require a Faraday cage, connects either to a supplied add-on vacuum unit or to an in-house vacuum line (max: 1 bar), and connects to any standard patch-clamp amplifier (Fig. 7). The operation principle is as follows: The inner chamber is primed with intracellular solution, and a patch electrode (ChipTip or self-made electrode) prefilled with extracellular solution is placed into the recording socket. Then extracellular solution is added, and upon validation of the correct pipette resistance by the patch-clamp software, a few hundred cells are dispensed into the tip socket. Up to this point, these steps are all performed manually. The subsequent gigaseal formation, whole-cell break in, and experimental voltage protocol are then all performed automatically. Addition of compounds is then performed either manually or semiautomated, via either a single or multiple standard perfusion system connected to the application needle for internal perfusion.

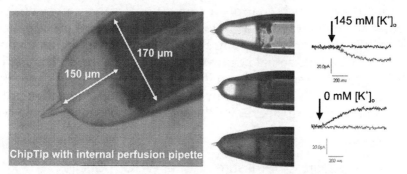

Fig. 6. A ChipTip is shown with an inserted application pipette for internal perfusion. The pipette can reach the cell close to 150 μm, and, assuming a cylinder model, a total volume of less than 1 μL needs to be exchanged. This is the lowest compound consumption for a patch-clamp recording consumable. Thus, given the time constant of less than 50 ms, ligand-operated channels can be recorded and fast generation of a perforated patch-clamp configuration is achieved. The middle panel depicts the solution exchange in a ChipTip, and the right panel shows a shift in the steady state outward hERG K-current caused by applying 145 mM K$^+$, evoked in a CHO cell at a holding potential of $-20$ mV.

Fig. 7. The PatchBox with the recording tip socket and an inserted ChipTip.

This is the most flexible system, ranging from manual operation to a semiautomated approach. Script-based protocols facilitate creating completely user-specific recording situations. In general, the PatchBox consists of the following control circuits (Fig. 8): an analog input (0–5 V), which controls a proportional valve to establish a preset negative pressure of 10 to 20 mbar for the seal process, and 60 to 100 mbar for establishing the whole-cell break-in; an analog output to read the actual negative pressure at the inner chamber of the recording tip socket; digital input 1, which opens and closes the valve for a specific time window to allow the preset negative pressure to reach the tip socket for the seal and subsequent whole-cell

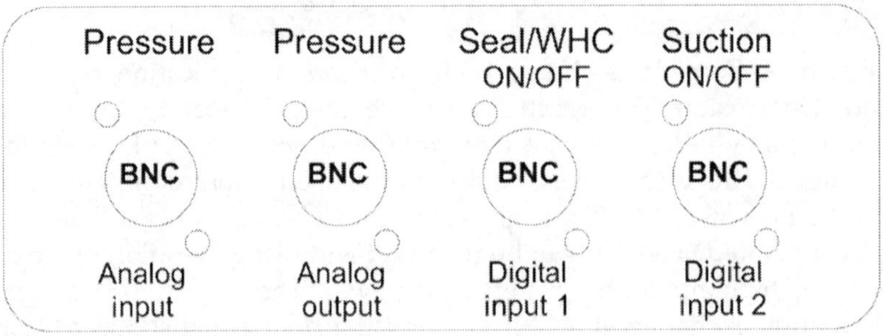

| Pressure | Pressure | Seal/WHC ON/OFF | Suction ON/OFF |
|---|---|---|---|
| ○ | ○ | ○ | ○ |
| BNC | BNC | BNC | BNC |
| ○ | ○ | ○ | ○ |
| Analog input | Analog output | Digital input 1 | Digital input 2 |

Fig. 8. The front panel of the PatchBox is shown with the main input and one output BNC connector. For a description of the latter, see text.

pressure; and digital input 2, which opens and closes a valve connected directly to the suction line for rinsing and cleaning of the recording socket (0.5 bar). All the inputs can be controlled via either a script-based protocol executed in the patch-clamp software program, or by manually operating digital/analog switches or potentiometers. Thus full flexibility is given to the user, taking into account the various available hardware and software configurations in research laboratories. Sample traces of ionic currents recorded using the Flip-the-Tip approach with the PatchBox are shown in Fig. 9.

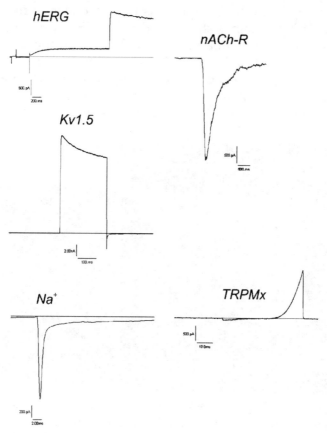

Fig. 9. Sample raw data traces are shown recorded with the PatchBox and the HEKA PatchMaster software: a Na$^+$ current expressed in CHL cells, a hERG potassium current expressed in a CHO cell, a Kv1.5 potassium current expressed in a LTK cell, an ACh-evoked current expressed in a Te671 cell line, and a TRPMx current expressed in a HEK cell.

Based on the approach with the PatchBox, a fully automated system with three or six recording channels was developed, mostly to be used in screening campaigns in the pharmaceutical industry or at contract research organizations.

## 5.4. The Flyscreen®8500 Automated Patch-Clamp Robot

The goal was to develop a system that would allow unattended operation, full automation of liquid handling and cell manipulation, and replacement of recording consumables without user intervention. Thus a standard liquid handling system was adopted and a fully digitally controlled amplifier (HEKA EPC10/3) was employed to fulfill these requirements. A complete three-channel Flyscreen®8500 system is shown in Fig. 10A. Referring to the same steps as described for the PatchBox, all manual operations have been replaced by liquid handling systems 0 and 1 and the gripper. These tools together make up the toolhead (Fig. 10B). In general, a complete experiment is conducted the following way: First, the inner chambers of the tip sockets are primed with intracellular (IC)

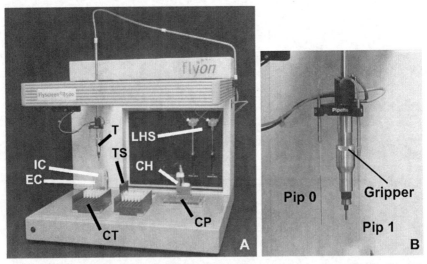

Fig. 10. (**A**) The Flyscreen®8500 automated patch-clamp robot is shown. LHS, liquid handling systems, connected to pipette 0 and pipette 1, respectively. T, toolhead; TS, a 3-channel recording tip socket module; CH, cell hotel, which triturates 2.5 mL of a cell suspension; CP, compound plate; CT, ChipTip recording rack; IC, cup for intracellular solution; EC, cup for extracellular solution. (**B**) The toolhead, depicting the gripper for handling the ChipTips, and pipettes 0 and 1, respectively.

solution. Second, the ChipTips are filled with extracellular (EC) solution and transported by the gripper from the ChipTip rack into the tip sockets. The software validates the correct pipette resistance and subsequently cells from the cell hotel are taken with pipette 0 and dispensed into the ChipTips. Then automated gigaseal formation and whole-cell break-in are performed the same way as described for the PatchBox. The system continues by checking a positive current response before continuing with the application of drugs. A more detailed description of the actual laboratory procedure, including cell culture handling, has been published elsewhere (Fejtl et al., 2006). Thus the focus of this chapter is on the theoretical concept of patch-clamp automation using the Flyscreen®8500 robot.

A strict parallel operation does not comply with the practical application of patch clamping, given that every single cell has its own time course of behavior during an experiment, including the time it takes to reach the gigaseal state, the time to reach the whole-cell configuration, and the time the cell needs to fulfill user-set criteria to be accepted for screening purposes. In essence, every experiment has its own individual time course. Hence we developed a time-scheduled asynchronous operation of the Flyscreen robot. That is, although a parallel operation takes place and the liquid handler serves every tip socket, software markers are placed before a specific operation to indicate the priority of the next scheduled task. Each operation to be performed is executed via so-called segment templates, small programs that are bundled together and make up a complete sequence file to run the entire experiment. An experienced electrophysiologist will recognize that the segment templates resemble operations similar to the ones used in manual patch clamping, that is, to adjust the stray capacitance of the pipette via the CFast segment template, or to adjust the capacitance of the cell in the whole-cell mode via the CSlow segment template.

To define an experimental run, the sequence editor is used to insert the necessary segment templates. A few of the available segment templates are shown in Fig. 11. The channel expression template is shown in more detail in Fig. 12. In practice, if the robot arm has to decide to either perform a compound application in tip socket 1 or to replace a ChipTip in tip socket 2, a higher priority for the task "compound measurement" ensures that this operation will be performed first. Moreover, flag conditions allow user-specific inputs to be specified in order to increase the performance of

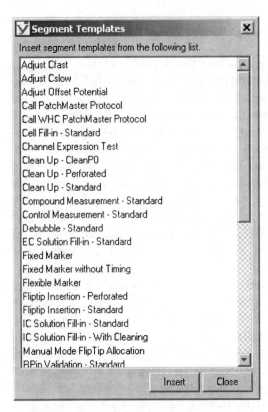

Fig. 11. An abbreviated list of the available segment templates in the Flyscreen software is shown. Most of them are understood intuitively by an electrophysiologist, such as Adjust CFast and Adjust Cslow, while others are related to robot operations such as Cell Fill-in Standard.

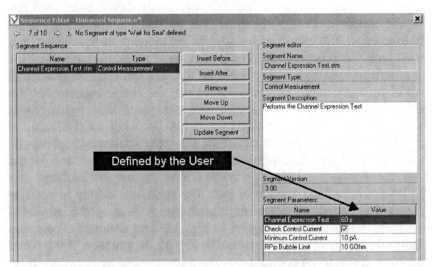

Fig. 12. The channel expression test segment template is shown in the sequence editor. Here, the user can edit parameters to set flag conditions for on-line validating of an ongoing experiment. The software then decides, based on the measured criteria (minimum control current), whether to continue with drug application or to start a new run with a new ChipTip.

the system. As an example, if the cell does not express a reasonable current amplitude, the software will recognize that and will not perform the scheduled compound application but rather start a new experiment with a new ChipTip. Several of these flag conditions can be set by the user, and the most important ones are the maximal accepted series resistance, the lowest acceptable seal resistance, the highest acceptable leak current, and a current run-up/run-down check implementation. Given this asynchronous but parallel operation, a high experimental success rate of about 60% to 80% (whole cell, and depending on cell type) can be achieved, resulting in a data throughput of several hundred data points per day.

## 6. Discussion

It is astounding that it took almost 30 years since the first introduction of patch clamping with glass electrodes for new methods and automated tools to be developed for the neuroscience community and the pharmaceutical industry. One major impetus certainly has been the ever-expanding high throughput screening protocols used in the drug development cycle. Assuming a compound library of 500,000 compounds and a hit rate of only 1% after the initial primary screening process, a total of 5,000 compounds are left to enter the secondary screening process. Moreover, the problem of drug withdrawal over the last 10 years has led many companies to rethink the drug development process, in particular to shift the stage of safety screening downstream before entering preclinical and clinical trials. Hence, patch clamping applies not only for secondary screening and lead optimization, but also for safe pharmacology. Manual patch clamping with glass electrodes is still used today in the pharmaceutical industry as the gold standard for functional characterization of ion channels, in particular when good laboratory practice (GLP)-compliant studies are required. Only recently has the shift begun to move away from the manual patch-clamp operation to automated systems. Since glass electrodes are known to be the best substrate for patch clamping, the Flip-the-Tip approach developed by flyion has been established in several drug companies and also universities. Given that the Feedback Microforge now allows reshaping a conventional patch electrode to almost any shape, a broad range of cells with different diameters can be recorded with the Flip-the-Tip technology. Moreover, since the mass production of 100% quality-controlled recording consumables has been established, further applications using automated patch clamping with real glass electrodes will certainly emerge over the years to come.

## References

Asmild, M., Oswald, N., Krzywkowski, K. M., et al. (2003) Upscaling and automation of electrophysiology: toward high throughput screening in ion channel drug discovery. *Receptors and Channels* **9** (1), 49–58.

Fejtl, M., Czubayko, U., Hümmer, A., Krauter, T. and Lepple-Wienhues, A. (2005) Automating true manual patch-clamping. *Genet. Eng. News* **25**(14), 37–43.

Fejtl, M., Czubayko, U., Hümmer, A., Krauter, T., and Lepple-Wienhues, A. (2006) Flip-the-Tip: automated patch clamping based on glass electrodes, in *Patch Clamp Methods and Protocols* (Molnar, P. and Hickman, J. J., eds,), Humana Press, Totowa, NJ.

Goodman, M. B., and Lockery, S. R. (2000) Pressure polishing: a method for reshaping patch pipettes during fire polishing. *J. Neurosci. Methods* **100**(1–2), 13–15.

Hamill, O. P., Marty, A., Neher, E., Sakmann, B., and Sigworth, F. J. (1981) Improved patch-clamp techniques for high-resolution current recording from cells and cell-free membrane patches. *Pflugers Arch.* **391**(2), 85–100.

Lepple-Wienhues, A., Ferlinz, K., Seeger, A., and Schafer, A. (2003) Flip the tip: an automated, high quality, cost-effective patch clamp screen. *Receptors and Channels* **9**(1), 13–17.

Neher, E. and Sakmann, B. (1976) Single-channel currents recorded from membrane of denervated frog muscle fibres. *Nature* **260**(5554), 799–802.

Shieh, C. C. (2004) Automated high-throughput patch-clamp techniques. *Drug Discov. Today* **9**, 551–552.

Wang, X. and Li, M. (2003) Automated electrophysiology: high throughput of art. *Assay Drug Dev. Technol.* **1**, 695–708.

Wood, C., Williams, C., and Waldron, G. J. (2004) Patch clamping by numbers. *Drug Discov. Today* **9**, 434–441.

Vasilyev, D. V., Merrill, T. L., and Bowlby, M. R. (2005) Development of a novel automated ion channel recording method using "inside-out" whole-cell membranes. *J. Biomol. Screen* **10**(8), 806–813.

Vasilyev, D. V., Merrill, T., Iwanow, A., Dunlop, J., and Bowlby, M. (2006) A novel method for patch-clamp automation. *Pflugers Arch.* **452**, 240–247.

# Index